Ecology of Dakota Landscapes

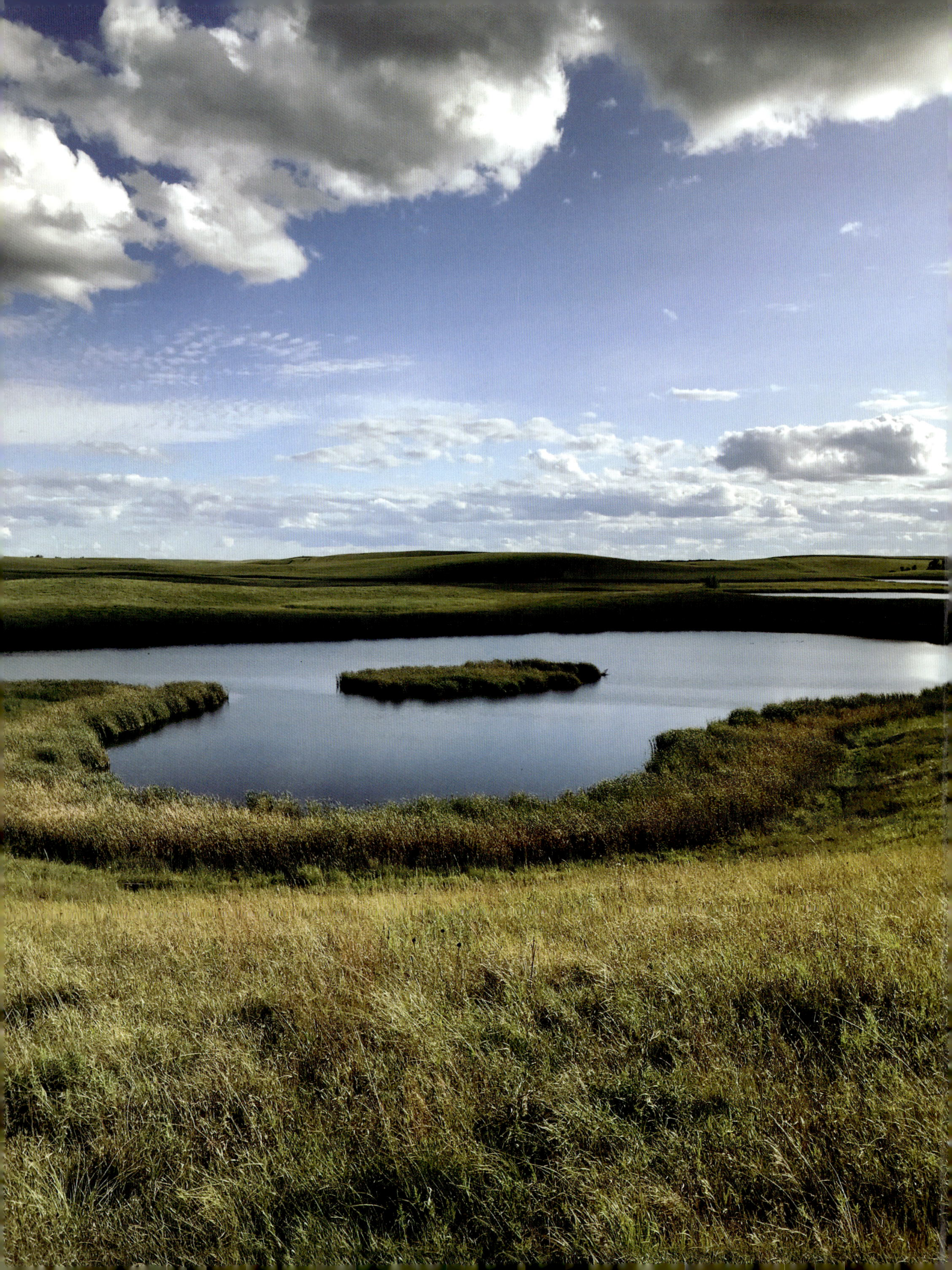

W. Carter Johnson

and Dennis H. Knight

Ecology of Dakota Landscapes

Past, Present, and Future

Yale UNIVERSITY PRESS
New Haven and London

UW | Biodiversity Institute

Laramie, Wyoming

Published with financial support from the University of Wyoming Biodiversity Institute (wyomingbiodiversity.org).

Yale University Press books may be purchased in quantity for educational, business, or promotional use. For information, please e-mail sales.press@yale.edu (U.S. office) or sales@yaleup.co.uk (U.K. office).

Designed by Nancy Ovedovitz and Mary Valencia.
Set in Stone and Beton type by Newgen North America.
Printed in China.

Library of Congress Control Number: 2021945805
ISBN 978-0-300-25381-8 (paperback: alk. paper)

A catalogue record for this book is available from the British Library.

This paper meets the requirements of ANSI/NISO Z39.48-1992 (Permanence of Paper).

10 9 8 7 6 5 4 3 2 1

Frontispiece: Natural wetland in native prairie on the Bien Ranch, Prairie Coteau.

For our early mentors,

Sven G. Froiland and Robert L. Burgess

Nature is an open book. . . . Each grass-covered hillside is a page on which is written the history of the past, conditions of the present, and predictions of the future. Let us look closely and understandingly, and act wisely, and in time bring our methods of land-use and conservation activities into close harmony with the dictates of nature.

—John Weaver, 1954

Contents

Preface

Our purpose in writing this book is to share our enthusiasm for the natural history of North Dakota and South Dakota—two states on the Northern Great Plains that inspired us to pursue careers in ecology. As youngsters we were skeptical when told that boulders in farm fields were dropped there by glaciers from Canada. We also were fascinated by the millions of colorful waterfowl that migrated through our home counties in the spring and fall, and we were amazed by the fossils of now-extinct reptiles and mammals found in the badlands of both states. Pronghorn on the western plains were as exotic as the antelope of Africa that we read about in magazines; we enjoyed hiking through pine forests while on family trips to the Black Hills. Later we welcomed the opportunity to help our students learn about the research that has been done in the region, and now we have enjoyed writing about what we think of as the highlights. Much of the book is pertinent to neighboring states and provinces.

The first chapter includes an overview of Dakota landscapes and a brief introduction to the science of ecology. This is followed by a chapter on how the natural resources of the region developed through geologic time, from inland seas to the present. Thinking about the future benefits from understanding the past. We summarize climate change over millions of years, the dramatic effects of glaciers, and the immigration of people—first from Asia and much later from Canada, Europe, and the rest of the world. The climate and how it is changing is summarized in chapter 3.

Subsequent chapters focus on the ecology of native grasslands, the farms that replaced them over large areas, and how droughts, fire, grazing, burrowing animals, and invasive plants have had widespread effects. This is followed by chapters on places where trees are found on the upland, such as in ravines and shelterbelts, on rocky escarpments, and in the highlands of Turtle Mountain, the Killdeer Mountains, and the Black Hills. Trees are now more widespread than prior to the arrival of European immigrants. Highly valued rivers, lakes, and wetlands are the subject of three chapters—one on the ecology of rivers in general, another on the iconic Missouri River and its reservoirs, and a third on the lakes, marshes, fens, and other wetlands of the Prairie Pothole Region. Picturesque buttes, badlands, and sandhills add diversity to the Great Plains and are the subject of chapter 12.

Exploring the Northern Great Plains over the years has enhanced our appreciation for the rich variety of grasslands, croplands, woodlands, wetlands, and badlands found in the region, all used and enjoyed in various ways by people from many walks of life. Opportunities to learn about the region's natural history abound, but there also are reasons to be concerned about the future. Throughout the book we have tried to provide a balanced overview of various land management problems and how they might be resolved, always considering the importance of sustaining agricultural economies while conserving soil and maintaining biological diversity. Our last chapter is about how undesired trends might be curbed and degraded land restored. Nearly every chapter ends with a section on the probable effects of climate change in the future.

Recognizing the diverse backgrounds of people who are likely to read our book, the text is essentially free of technical terms. Definitions of those that remain can be located using the index. English units of measure and the common names of plants and animals are used in the text. Numerous endnotes guide readers to our sources of information and more detail.

Acknowledgments

Completing this book depended on the assistance of artists, cartographers, colleagues, and land stewards. We are indebted to them all. Allory Deiss, Ken Driese, Bruce Millett, Christopher Nicholson, Gray Tappan, and Malia Volke produced maps, graphs, and drawings that by themselves have created a page-turner for readers interested in the Northern Great Plains. When our photos were inadequate, numerous photographers provided what we needed. Their names are in the captions accompanying their contributions. And for the privilege of using images of classic paintings by Karl Bodmer and George Catlin, we thank the Joslyn Art Museum in Omaha. Other paintings in the book were done by John James Audubon, Ron Backer, Nancy Carlsen, and Karen Carr. We have a greater appreciation for how fine art can enhance a science book.

Also inspirational were visits to numerous museums, large and small. We spent many hours at the North Dakota Heritage Center and State Museum in Bismarck, the Knife River Indian Villages National Historic Site near Stanton, the South Dakota Cultural Heritage Center in Pierre, The Agricultural Heritage Museum in Brookings, and the Museum of Geology at the South Dakota School of Mines in Rapid City. Conversations with farmers, ranchers, conservationists, and state and federal agency personnel influenced our perspective on the challenges facing Dakotans in the twenty-first century.

Our chapters are much better because of suggestions made by colleagues on early drafts. One or more of the chapters on history, climate, and grasslands were reviewed by Chamois Anderson, Alan Ashworth, Pete Bauman, John Bluemle, Don Boyd, Shawn DeKeyser, Eric Grimm, Bonnie Heidel, Norman Henderson, Jason Lilligraven, Alan Knapp, David Ode, Fern Swenson, Amy Symstad, and Kathryn Yurkonis. The agriculture and trees-on-the-farm chapters, one or both, were read by Dave Archer, Dwayne Beck, John Bluemle, Scott Kronberg, Jon Lundgren, Dacia Meneguzzo, and Cody Zilverberg. Comments by John Bluemle, Bonnie Heidel, Norman Henderson, Lezlee Johnson, and David Ode helped us refine our chapter on hardwood forests and woodlands. Kurt Allen, Mike Battaglia, John Bluemle, John Freeman, Bonnie Heidel, Jason Lilligraven, and Amy Symstad commented on some of what we wrote about badlands, buttes, sandhills, and the Black Hills. One or more of our chapters on rivers, wetlands, and lakes were reviewed by Jane Austin, Steven Chipps, Mark Dixon, Daniel Engstrom, Kurt Forman, Jonathan Friedman, Todd Mortenson, Michael L. Scott, Craig Spencer, Fern Swenson, and Arnold van der

Valk. Chamois Andersen, Pete Bauman, Steve Buskirk, Gary Clambey, Kristiana Hanson, and Daniel Licht helped us improve the last chapter. We thank Bonnie Heidel, Hollis Marriott, David Ode, Jeff Printz, and Stacy Swenson for botanical assistance and Phil White for his careful attention to editorial details.

Publishing the book required another talented team. At Yale University Press, Jean Thomson Black was encouraging and very helpful, Elizabeth Sylvia, Amanda Gerstenfeld, and Mary Pasti guided us through the submission and editorial process, and Nancy Ovedovitz and Mary Valencia designed the book. Charlie Clark and Katherine Faydash of Newgen North America coordinated copyediting and numerous other details. Brent Ewers, director of the Biodiversity Institute at the University of Wyoming, secured funding that enabled color illustrations in every chapter.

Over the years our research on the Northern Great Plains would not have been possible without grants from the Army Corps of Engineers, Environmental Protection Agency, South Dakota Department of Game, Fish, and Parks, The Nature Conservancy, U.S. Department of Energy, U.S. Fish and Wildlife Service, U.S. Geological Survey, and others. The Department of Natural Resource Management at South Dakota State University contributed to the book's production costs.

Finally, we gratefully acknowledge the support and encouragement of Janet Johnson and Judy Knight, our incredible companions for so many years.

Ecology of Dakota Landscapes

Chapter 1 The Land, Ecosystems, and Ecology

The states of North Dakota and South Dakota, on the northern Great Plains of North America, are located where the summers are warm and often dry, where the winters are cold, and where prairies at one time extended to the horizon. The first people arrived about 13,500 years ago and lived among bison, wolves, grizzly bears, elk, pronghorn, deer, prairie dogs, black-footed ferrets, badgers, prairie chickens, and numerous other animals.[1] Some of them may have seen now-extinct mammoths and the remnants of melting glaciers. Trees grew here and there on highlands, in ravines, and where water was more readily available or fires burned less frequently. European immigrants in the 1800s would think of grassland soils as "black gold" because their crops often exceeded expectations. Consequently, the prairie over large areas was cultivated. Groves of trees were planted to provide a source of wood and shelter from wind.

The Dakotas can be divided into two regions: glaciated terrain to the north and east of the Missouri River, and unglaciated terrain on the other side—West River country for South Dakotans (figs. 1.1 and 1.2).[2] Glaciated landscapes are relatively young, having been subject to erosion for only about 10,000 years. Consequently, the land is mostly undulating or flat. Exceptions are highlands known as the Prairie Coteau, Missouri Coteau, and Turtle Mountain, all part of the Prairie Pothole Region and much appreciated by hunters, anglers, and other outdoor enthusiasts for the thousands of lakes and marshes found there. Tallgrass prairie in the east and mixed-grass prairie in the west covered most of the region before the introduction of modern agriculture in the late 1800s (figs. 1.3 and 1.4, table 1.1). Some of the land was remarkably flat and easily farmed, such as the lowlands of the Red and James Rivers—the location of prehistoric glacial lakes that existed thousands of years ago. In the western Dakotas, beyond the reach of glacial ice, the land has been carved by erosion for 60 million years or more, long enough to create buttes, escarpments, badlands, and canyons (figs. 1.5–1.9). The Black Hills, an outlier of the Rocky Mountains, rise several thousand feet above the plains and are heavily forested with ponderosa pine.

The Dakotas are largely rural in nature, with the two states each having fewer than a million people. The largest cities are Sioux Falls (190,000), Fargo (126,000), Rapid City (80,000), and Bismarck (74,000). Corn and soybeans are the most important crops where growing seasons are relatively long and warm. Wheat is important elsewhere. Other crops include hay, canola, sunflower, sugar beets, potatoes, oats, flax, rye, and barley.

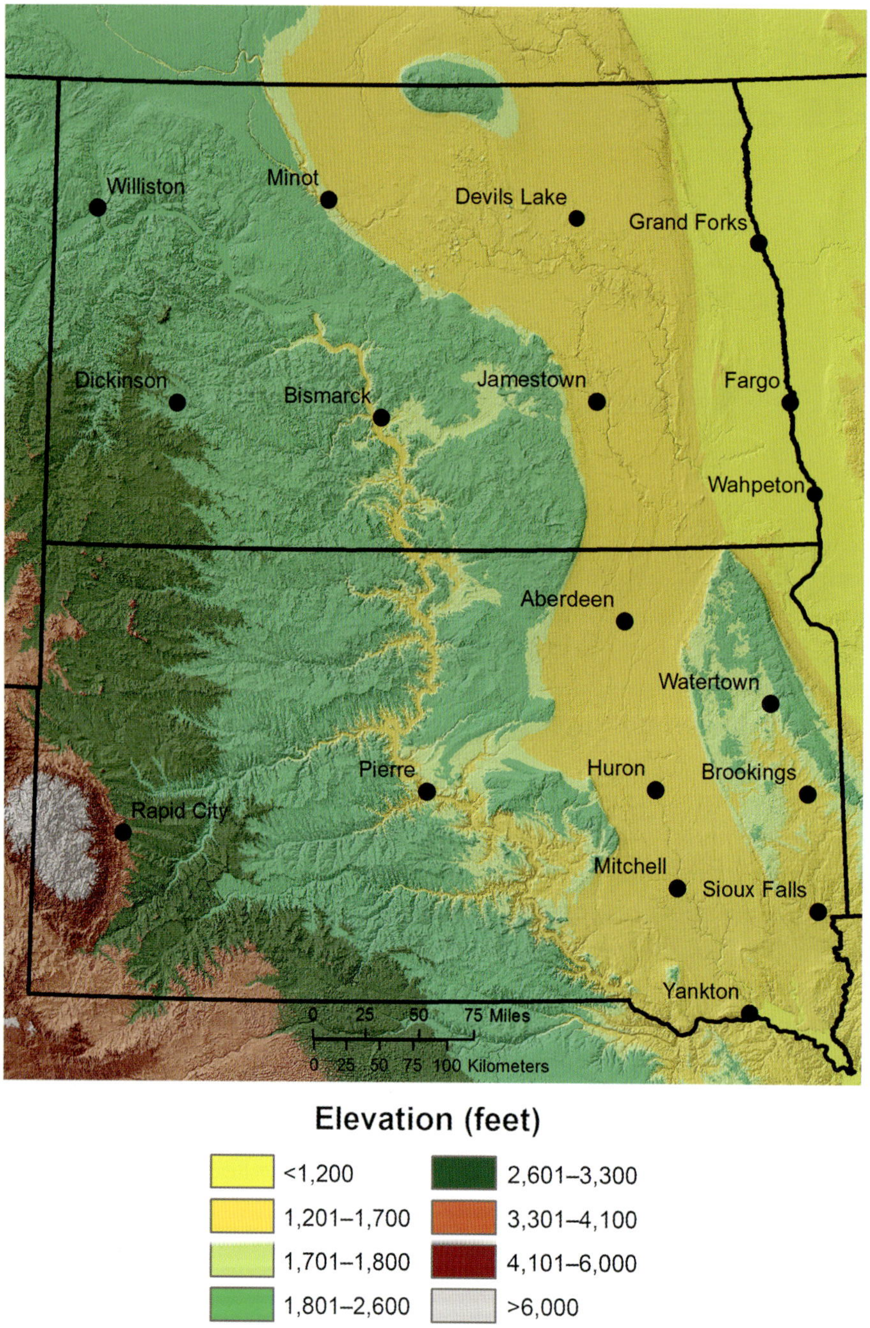

Fig. 1.1. Elevation above sea level slopes downward from west to east in North and South Dakota. The highest point is Black Elk Peak (formerly Harney Peak) in the Black Hills (7,242 feet above sea level). White Butte is the highest point in North Dakota (3,506 feet). The lowest points in each state are at or near where the Red River flows into Manitoba (778 feet) and at or near Big Stone Lake (966 feet), the origin of the Minnesota River in northeastern South Dakota. See fig. 1.2 for the names of topographic features. Cartography by Ken Driese.

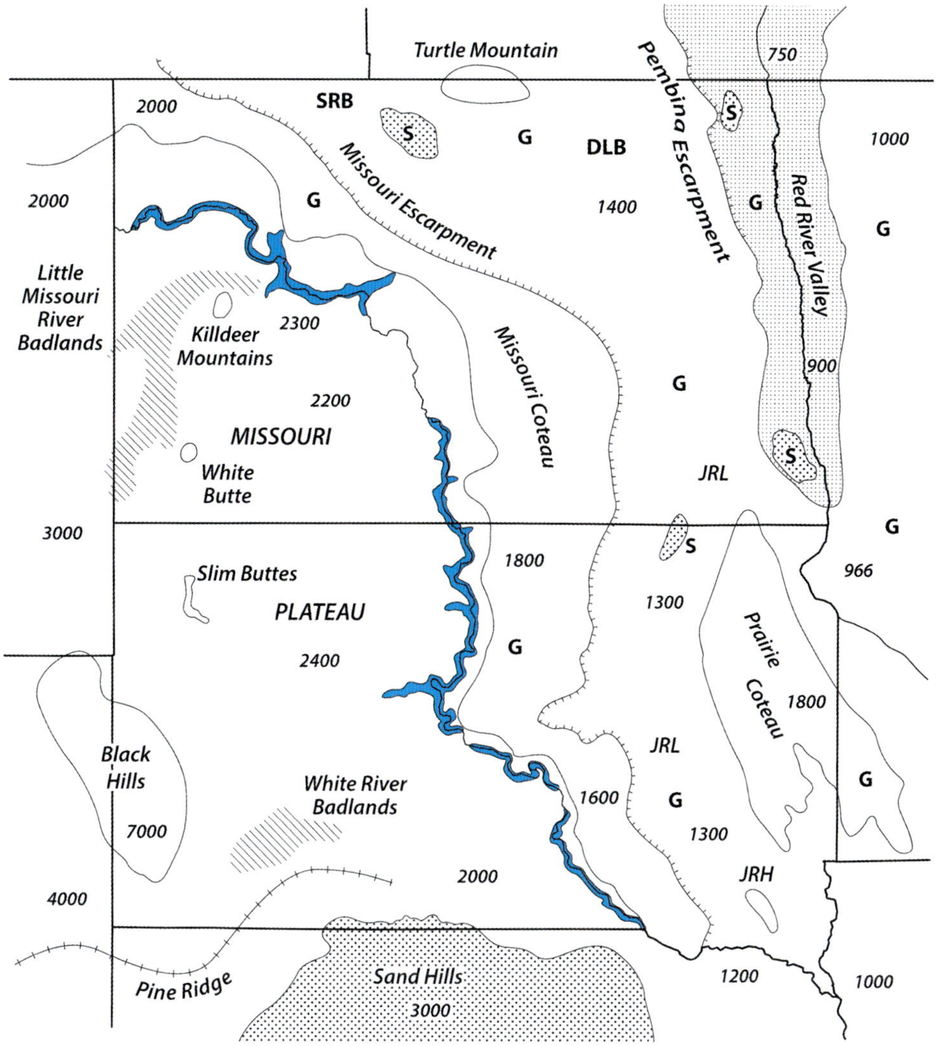

Fig. 1.2. Primary landforms in North Dakota and South Dakota, including the Missouri River reservoirs and the Red River Valley. Numbers indicate approximate elevation (feet above sea level). The Missouri Coteau and Prairie Coteau are glacial moraines that rise several hundred feet above the surrounding plains (see fig. 1.1). Turtle Mountain is of similar origin. Shown with symbols are glaciated terrain (G), sandhills (S), James River Lowlands (JRL), James River Highlands (JRH), Souris River Basin (SRB), and Devils Lake Basin (DLB). In North Dakota, the glaciated land between the Missouri Coteau and Red River Valley is sometimes referred to as *drift prairie*. Adapted from Bluemle (2000, 2016) and Gries (1996).

Some fields are irrigated, especially near the Missouri River and westward. Native grasslands still exist where irrigation is not feasible or the soils are not suitable for cultivation. To reduce soil erosion, many landowners have received federal funding to stabilize highly erodible fields with perennial grasses and forbs. These lands conserve soil and provide other benefits. In addition to agriculture, some areas have the roads and infrastructure required for oil and gas extraction, which is principally from the Bakken Formation in northwestern North Dakota.

The present-day landscapes of the Dakotas are much different from what Meriwether Lewis and

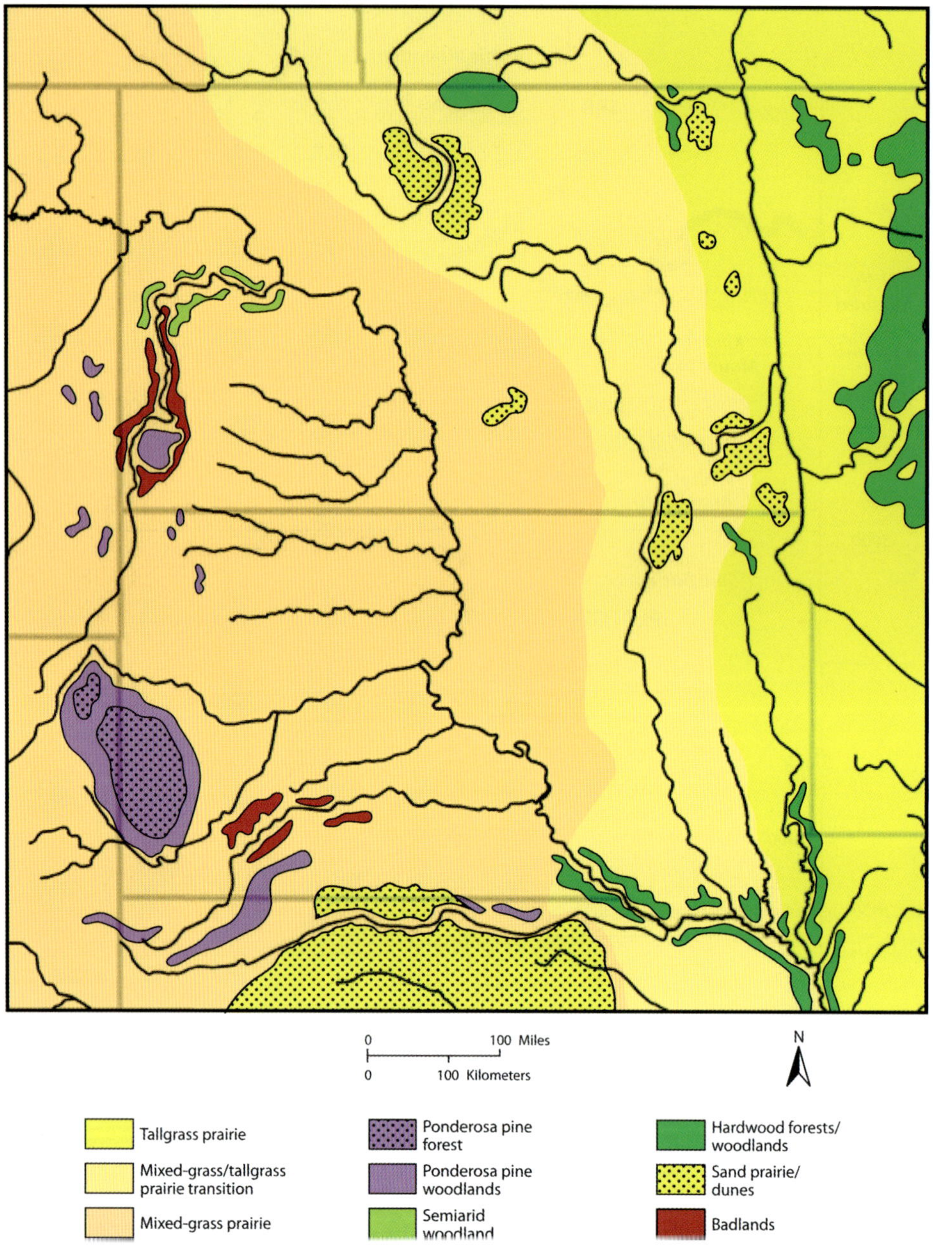

Fig. 1.3. Approximate land cover in North and South Dakota during the 1700s, when the first explorers of European origin arrived. Lakes, wetlands, and small tracts of woodland are not shown on this map but are less common in the western half of both states, where the climate is drier (see chapter 3). Present-day land cover is shown in fig. 1.4. Prairie grasslands (including tallgrass, mixed-grass, and sand prairie grass) were most widespread, covering about 96 percent of both states. Forests and woodlands dominated by deciduous hardwood trees in the east and evergreen ponderosa pine and Rocky Mountain juniper in the west covered about 2.5 and 1.5 percent of the two-state area, respectively. Only large patches of each vegetation type can be illustrated at this scale. Adapted from maps produced by Küchler (1964), Barker and Whitman (1988), Bryce et al. (1998), Chapman et al. (1998), Savage (2004), and Johnson and Larson (2016).

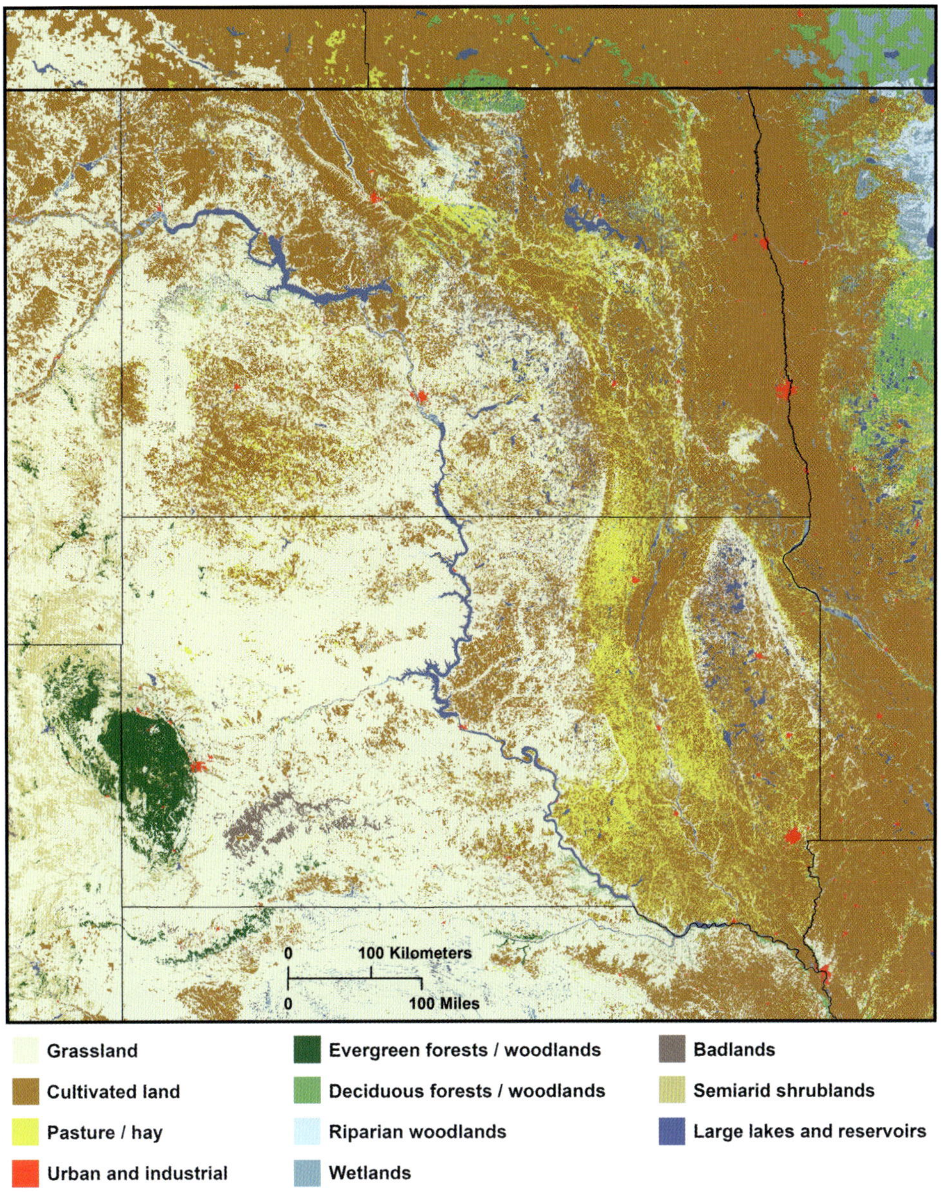

Fig. 1.4. Present-day land cover and land use based on Landsat satellite images for 2016. Most cultivated land is on glaciated terrain. Small lakes, wetlands, creeks, and woodlands are difficult to detect at this scale. See table 1.1 for acres of land in each cover type. Upland deciduous forests in western Minnesota have been described by Clambey (1980) and Tester et al. (2020). Adapted by Gray Tappan from Homer et al. (2020). U.S. Geological Survey, EROS Center, https://doi.org/10.1016/j.isprsjprs.2020.02.019.

William Clark saw in 1804. With their crew of thirty-three and one dog, they paddled, pushed, and pulled flatboats up the Missouri River. Cattle and cultivated fields have largely replaced the bison herds and prairie, and the wild, flood-prone Missouri has been tamed by the construction of six massive dams that store more water than any other reservoir system in North America (fig. 1.10). These reservoirs are appreciated for irrigation water, flood control, hydroelectric power, and the

Table 1.1. Area occupied by different kinds of ecosystems in South Dakota and North Dakota

	South Dakota		North Dakota		Both states	
Ecosystem	Acres	Percentage	Acres	Percentage	Acres	Percentage
Grassland	24,783,800	50.2	13,202,427	29.2	37,986,227	40.2
Cultivated land	12,958,913	26.2	21,347,290	47.2	34,306,203	36.3
Pasture and hay	5,020,103	10.2	3,771,823	8.3	8,791,926	9.3
Large lakes and reservoirs	1,489,806	3.2	1,755,053	3.9	3,244,859	3.4
Urban and industrial	1,415,691	2.9	1,810,858	4.0	3,226,549	3.4
Wetlands	624,088	1.3	1,681,479	3.7	2,305,567	2.4
Evergreen forest and woodland	1,301,366	2.6	64,911	0.1	1,366,277	1.4
Deciduous forest and woodland	332,295	0.7	715,022	1.6	1,047,317	1.1
Semi-arid shrubland	739,652	1.5	470,422	1.0	1,210,073	1.3
Riparian woodland	272,131	0.6	325,159	0.7	597,290	0.6
Badlands	413,481	0.8	106,650	0.2	520,130	0.5
Total acres	49,351,325		45,251,093		94,602,418	

Source: Landsat satellite data for 2016 used to produce figure 1.4.

Fig. 1.5. Flat land is common on the "bottoms" of former glacial lakes, such as near the Red River of the North, west of Grand Forks. Glacial Lake Agassiz covered this area from about 13,800 to 10,900 years ago and again from about 9,900 to 9,300 years ago. Similar flatlands, known as *glacial lake plains*, are found in the lowlands of the Souris River and James River where Glacial Lakes Souris and Dakota existed at about the same time (see fig. 1.2). There are numerous smaller lake plains. Such places are highly valued for croplands because there are essentially no rocks and the soils are fertile. Shelterbelts are planted for various benefits (see chapter 6).

Fig. 1.6. Flat land also is common on glaciated land away from the location of ancient glacial lakes, such as west of the Red River lake plain and east of the Missouri Coteau. The soils in such places have a coarser texture than those that developed on former lake beds, but they still are highly valued for crop production, in this case wheat. In contrast to the hilly moraines of Turtle Mountain and the Missouri and Prairie Coteau, the gravel, sand, silt, and clay in the ice sheets were deposited more or less evenly.

Fig. 1.7. Much of the land north and east of the Missouri River is hilly or undulating, with many depressions that enabled the formation of lakes and marshes in the Prairie Pothole Region (see chapter 11). Glaciated land, which comprises about two-thirds of North Dakota and half of South Dakota, is highly valued for waterfowl habitat as well as agriculture (wheat and sunflower in this photo). Shoreline trees are plains cottonwood and peachleaf willow.

Fig. 1.8. The Missouri Coteau and Prairie Coteau often are too hilly and the soils too rocky for cultivation. Mixed-grass prairie predominates on the hilltops, tallgrass prairie on wetter slopes and in swales. Granitic boulders are common. Photo by Justin Meissen.

Fig. 1.9. Buttes and escarpments are remnants of sedimentary strata deposited millions of years ago and are characteristic of the unglaciated western Dakotas. Ponderosa pine is common, such as in northwestern South Dakota and southwestern North Dakota. The mixed-grass prairie and occasional shrublands with silver sagebrush or big sagebrush are used primarily for livestock grazing.

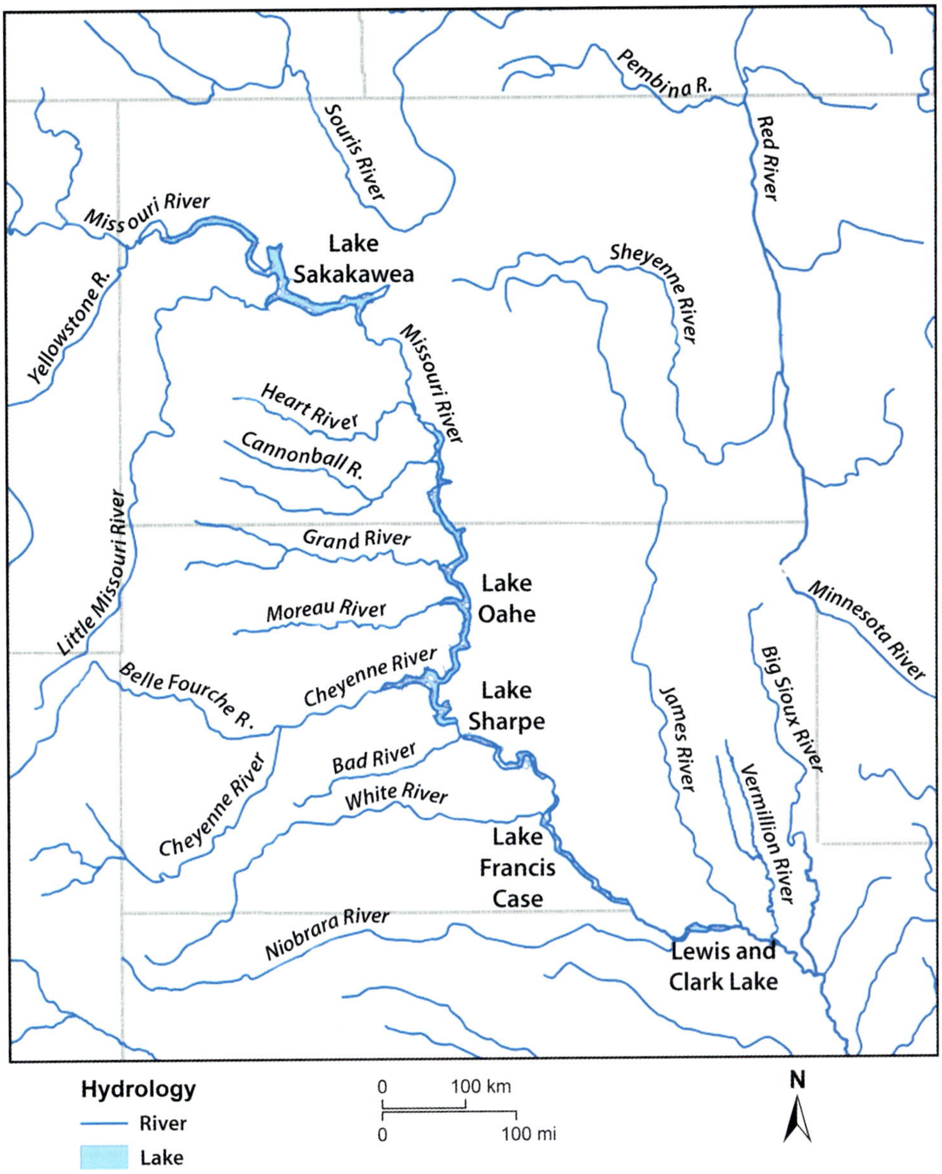

Fig. 1.10. The rivers in the eastern Dakotas flow mostly north or south, whereas rivers in the west generally flow eastward to the Missouri. A continental divide (977 feet above sea level) occurs near where the east corners of the two states come together, with the Red River of the North flowing north into Hudson Bay via the Nelson River (as do the Sheyenne, Pembina, and Souris Rivers) and the Minnesota River flowing south to the Gulf of Mexico via the Missouri and Mississippi Rivers (as do the Big Sioux, James, and Vermillion). Cartography by Malia Volke.

recreational opportunities they provide, but they destroyed much of the riparian ecosystem (see chapter 10).

For various reasons the Dakotas now support more people than ever before, with wood for construction, wells and reservoirs for water, and fossil fuel that powers machinery and provides light and heat for homes. Free-roaming bison, grizzly bears, black bears, and wolves are gone, and elk, prairie chickens, and sharp-tailed grouse are less common, but white-tailed deer have expanded their range. Fishing is excellent for native species such as walleye, northern pike, catfish, perch, and crappies. Waterfowl hunting is as popular today as it was in the 1800s; and the ring-necked pheasant, introduced to the Dakotas from China in the early 1900s, provides recreation for hunters and an important source of revenue for many others. This colorful, delectable bird has thrived in farmed landscapes.

Land management objectives in the Dakotas vary with ownership. Most land is privately owned farms and rangeland (fig. 1.11), where staying in business calls for practices that sustain economic returns without degrading the soil and other resources. There is no single recipe for success because all parcels of land have different soils, climate, and water availability. Moreover, markets for grain and livestock are notoriously variable and unpredictable. Tribal councils face similar challenges on the land they oversee. Other lands are managed by state or federal agencies and are part of the public domain, such as the Missouri River reservoirs, the Black Hills National Forest, several national grasslands and national parks, and many state parks, forests, historical sites, and wildlife management units. Much of the financial support for these lands is from taxation and fees paid for hunting and fishing licenses. Another category of land is managed and sometimes owned by nongovernmental organizations, for example, The Nature Conservancy, Ducks Unlimited, Pheasants Forever, and other organizations that work collaboratively to conserve wildlife and biological diversity in general. All management strategies benefit from the traditional knowledge of people from all walks of life as well as from research by scientists associated with colleges and universities, agricultural experiment stations, and other government agencies. Numerous fields of study now inform decision making, including the science of ecology.

Ecology and Ecosystem Services

Ecology is a diverse science that can be viewed as the systematic study of ecosystems—areas in which plants, animals, microbial organisms, and humans interact with one another and their environment (fig. 1.12). Waterfowl scientists focus on ecosystems with wetlands, such as prairie marshes; fish biologists work on lake and stream ecosystems; and agricultural scientists study cropland and rangeland ecosystems, commonly referred to as agroecosystems. The boundaries of an ecosystem depend on the kinds of plants or animals being considered and the objectives of a study. To illustrate, the ecosystem of spiders is much smaller than the ecosystems of wide-ranging animals such as deer or bison. Some ecologists work to understand why different plants and animals are restricted to certain habitats; others study how crop production can be improved, how ecosystems might be restored following disturbances, how they are likely to change as the climate changes, and how various factors affect plant growth rates or animal population sizes. In addition to studying specific kinds of plants or animals, some ecologists concentrate on understanding food webs, nutrient cycling, and the movement of water or pollutants—both above the ground and in the soil. This information enables more informed management decisions.

The terms *ecosystem* and *landscape* are used frequently in this book, sometimes interchangeably. Ecosystems may be large or small, from Earth as a whole to the microbial interactions associated with a decomposing root. A landscape is commonly viewed as a miles-wide mosaic of different ecosystems. For example, a watershed, farm, ranch, national park, or topographic feature such as Turtle Mountain is a landscape with patches of woodland, prairie, and aquatic ecosystems. One purpose of this book is to describe transformations to those landscapes brought about by immigrants of European descent.

The ecosystems of the Dakotas have been developing and changing for millions of years. During this time, several thousand species of plants and

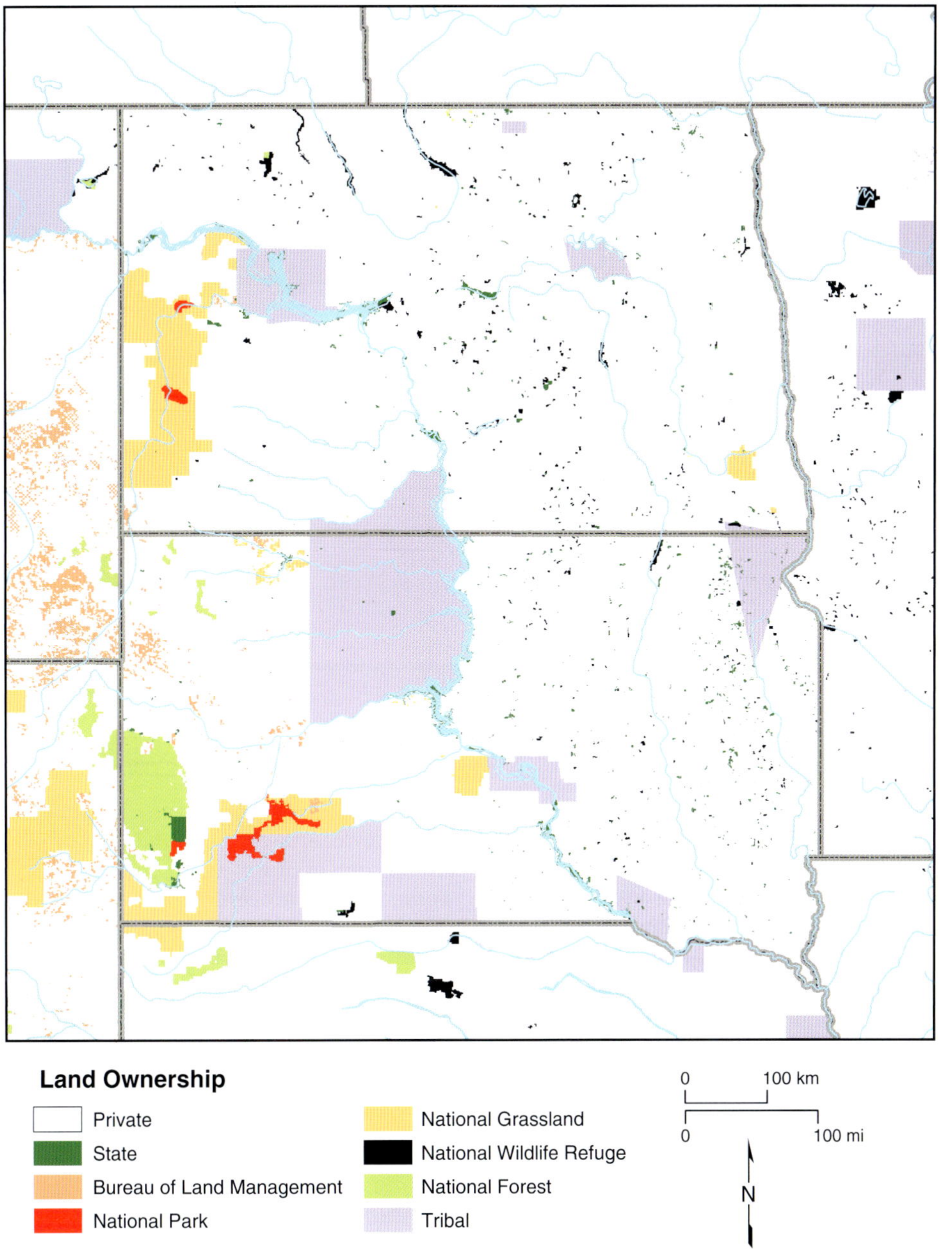

Fig. 1.11. About 10 percent of the Dakotas are public lands. They include the Little Missouri National Grassland, Buffalo Gap National Grassland, Fort Pierre National Grassland, Sheyenne National Grassland, and Black Hills National Forest (all managed by the U.S. Forest Service); Theodore Roosevelt National Park, Wind Cave National Park, and Badlands National Park (managed by the National Park Service); 42 National Wildlife Refuges and Districts (managed by the U.S. Fish and Wildlife Service, 9 in South Dakota and 33 in North Dakota); and lands managed by the Bureau of Land Management (all rangelands in the far west). Tribal lands include the Cheyenne River, Crow Creek, Flandreau Santee, Fort Berthold, Lake Traverse, Lower Brulé, Pine Ridge, Rosebud, Spirit Lake, Standing Rock, Turtle Mountain, and Yankton Reservations. Some public and tribal lands are small and not shown on this map. State lands are in numerous small parcels, mostly along the Missouri River, in the Black Hills, and in the Prairie Pothole Region north and east of the Missouri River. The largest tract of state land is Custer State Park, in the Black Hills adjacent to Wind Cave National Park. Cartography by Malia Volke.

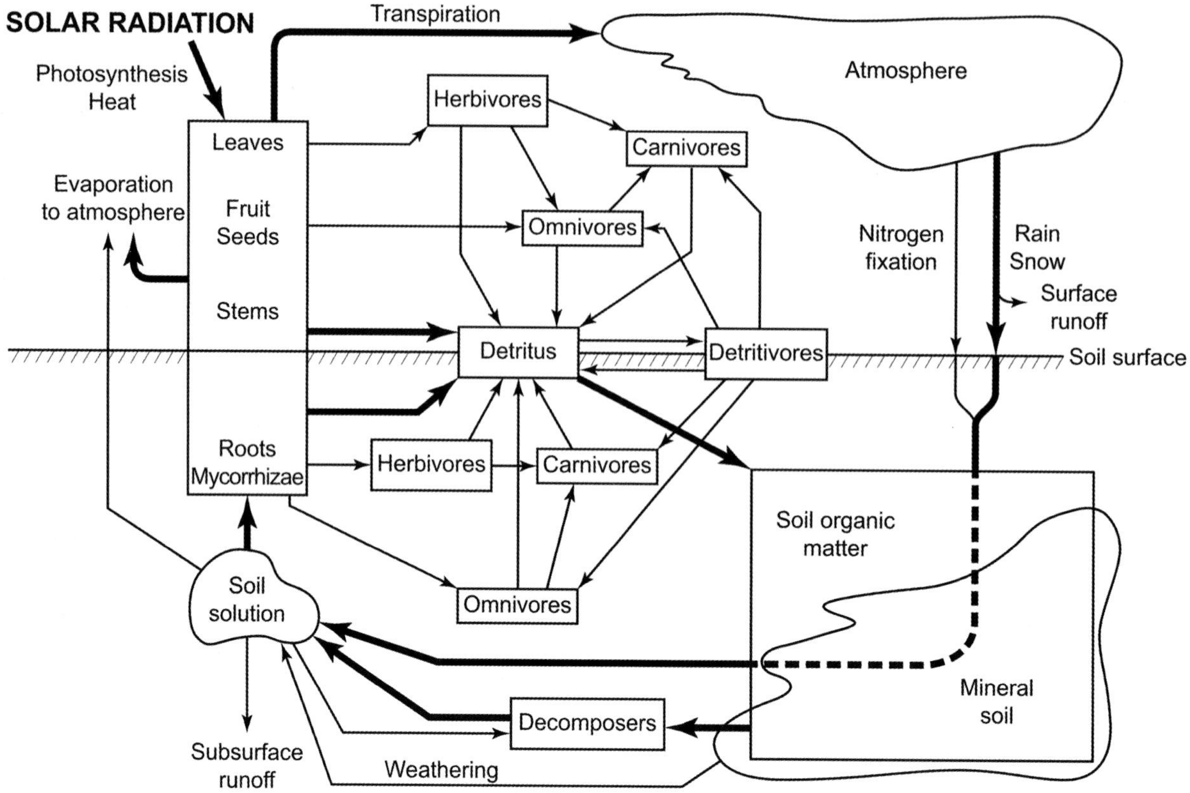

Fig. 1.12. Ecosystems are composed of components (boxes) that interact through various ecological processes (arrows), as illustrated here for a terrestrial ecosystem. Each box represents numerous species; arrow width is a measure of the relative amount of energy or water moving along a pathway. Food webs exist above- and belowground. The irregular shapes indicate sources of water and nutrients. A change in one component or process, for whatever reason, leads to changes in the others. From Knight et al. (2014), with permission from Yale University Press.

animals, and unknown numbers of microbial organisms, evolved special adaptations for surviving in the various environments found in the region. As the climate changed, due to continental drift and other factors, plants and animals better adapted to the new environments became more common. Those that could not adapt emigrated to more favorable environments—or they became extinct, such as the camels, mammoths, giant ground sloths, and horses that thrived during the Pleistocene, and the dinosaurs that prevailed long before them. These ancient ecosystems produced fertile soils, filtered water, and habitat for many highly valued native species that facilitate ecosystem recovery after disturbances—benefits that are still provided in places where native plants grow in abundance. Powered by solar energy, these benefits have come to be known as *ecosystem services* (table 1.2).[3] In modern croplands, large quantities of fossil fuel are required to make and apply inorganic fertilizers that previously were provided by the grassland ecosystem.[4] Knowledge about nutrient cycling informs decisions about how to minimize the costs of fertilization, as discussed in subsequent chapters.

Notably, the economic benefits of grain, meat, and other commodities are more easily monetized than those of native ecosystems. Considerable research is now under way on this topic, but today, when land is under consideration for conversion

Table 1.2. Ecosystem services provided by assemblages of native organisms

Maintenance of soil fertility and reduced soil erosion
Provision of edible protein, primarily from grasslands
Water filtration and the provision of clean water
Mitigation of the adverse effects of droughts and floods
Habitat for insects that pollinate crops and native plants
Detoxification and decomposition of waste materials
Sequestration of carbon dioxide
Inhibiting the spread of invasive plants
Biological control of insect pests
Habitat for fish, wildlife, and the conservation of biological diversity
Genetic resources for restoration and the production of food and other commodities

to some other use, it is reasonable to ask about the costs of replacing ecosystem services that would be lost or diminished by the conversion. Having made these calculations, New York City restored watersheds in the Catskill Mountains, a major source of drinking water, after concluding that doing so would be much cheaper for improving water quality than constructing and maintaining a filtration plant.[5] In another example, the Des Moines Water Works in Iowa sued three upstream drainage districts in 2015 to recover the costs of removing nitrates from the Raccoon River, a primary source of water for the city.[6] The source of the polluting nitrates is drainage from heavily fertilized cropland. The utility lost the suit in 2017 after judges ruled that the drainage districts could not be held responsible for the runoff, but these examples illustrate pragmatic reasons for maintaining and restoring native ecosystems in some places. Moreover, remnants of ancient grasslands and other kinds of ecosystems are appreciated for aesthetic, recreational, and educational values.

Traveling across the Dakotas, it is clear that humankind is now the dominant species. Native ecosystems have become rare, and because of that they are highly valued. Conserving them, where that is still possible, seems prudent considering the high rate of habitat fragmentation and the prospect of relatively rapid climate change. In some areas the restoration of native ecosystems is proving to be an economically sound decision. Likewise, concern over the long-term consequences of the historical deterioration of soil quality on Dakota farmland has stimulated some managers to promote the practices of *regenerative agriculture* (see chapter 5). In general, the results of ecological research, such as those reviewed in the chapters that follow, provide insights on how and where innovative practices can be implemented or expanded in the ongoing quest to achieve the sustainable production of the commodities on which so many people now depend.

Chapter 2 Landscape History

Although difficult to comprehend, terrestrial ecosystems are now found where marine life once flourished in a vast inland sea. The climate was much warmer also. Tropical swamps provided an environment conducive to the formation of coal, oil, and natural gas, and bottom sediments were compressed into rocks that eventually would contribute to the development of fertile soils. After the sea retreated, long periods of erosion, deposition, and glaciation created hills, valleys, flatlands, and wetlands. Understanding this history helps explain why some areas are excellent for crop production and others less so, why wetlands are not evenly distributed in the region, and why the badlands of North Dakota are different from those in South Dakota.

Other significant changes occurred when humans arrived about 13,500 years ago.[1] Though few in number, the populations of these itinerant people eventually became large enough to contribute to the extinction of mammoths and some other large mammals—animals that may have become vulnerable because of their confinement to marginal habitats by glaciers. Thousands of years passed before some of these people established villages and planted gardens of squash, beans, and corn on the rich soils of floodplains. They also hunted bison.[2] Eventually, in the mid-1800s, farmers of European descent cultivated prairie soils with flax, wheat, barley, and rye. They also raised cattle and sheep. The bison were nearly eliminated from the Great Plains, wetlands were drained, rivers were impounded, and groves of trees were planted for windbreaks and fuel. Some of the introduced plants and animals became invasive species, but agriculture prospered. Much of this book focuses on the ways that generations of Dakotans have modified native ecosystems, but let us first consider what happened during the several hundred million years before people arrived, starting with the Paleozoic era (fig. 2.1).

Millions of Years Ago

Sedimentary rocks with fossils of marine organisms are found throughout the Dakotas, evidence that seawater once covered this part of North America.[3] Fish were common, as were insects and amphibians. Tidal flats had swamps of tree-sized liverworts, ferns, and horsetails. The climate during part of the Paleozoic was tropical, much different from today. Tall conifers and now-extinct seed ferns grew on the uplands as the water level subsided, but flowering plants and dinosaurs would not appear in the fossil record for millions of years.

As the seas expanded and receded, successive

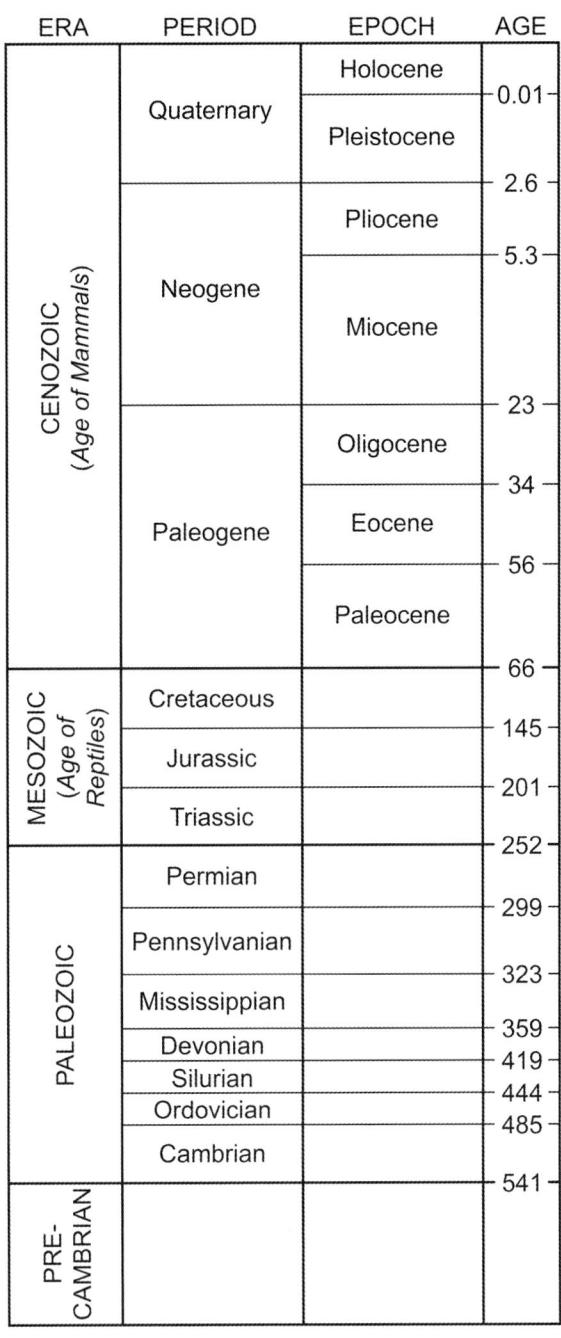

ERA	PERIOD	EPOCH	AGE
CENOZOIC (*Age of Mammals*)	Quaternary	Holocene	0.01
		Pleistocene	2.6
	Neogene	Pliocene	5.3
		Miocene	23
	Paleogene	Oligocene	34
		Eocene	56
		Paleocene	66
MESOZOIC (*Age of Reptiles*)	Cretaceous		145
	Jurassic		201
	Triassic		252
PALEOZOIC	Permian		299
	Pennsylvanian		323
	Mississippian		359
	Devonian		419
	Silurian		444
	Ordovician		485
	Cambrian		541
PRE-CAMBRIAN			

Fig. 2.1. Geologic timetable with ages in millions of years. The Paleogene and Neogene periods of the Cenozoic era were previously known as the Tertiary period. Paleontologists emphasize that some reptiles were found late in the Age of Fishes and some mammals were found late in the Age of Reptiles. Adapted from the International Commission on Stratigraphy (http://www.stratigraphy.org). Bluemle (2000) provides a more detailed timetable (pp. 138–144) that is specific for the Northern Great Plains.

layers of sand accumulated in shallow water near shorelines and were eventually cemented into sandstones. At greater depths, fine-textured silt, clay, and volcanic ash were deposited, where they would eventually become shales, siltstones, and claystones. These rocks were exposed much later—all sedimentary, formed during the Cretaceous period 145 million to about 66 million years ago. Plate-tectonic movement caused a regional uplift about 65 million years ago, which accelerated erosion and exposed much-older rocks, especially in the Black Hills and the Rocky Mountains to the west. A conspicuous example is the colorful Spearfish Formation that had formed during the Triassic Period, 200 million to 250 million years previously. Rich in iron oxides and commonly known as redbeds, this formation was once deeply buried under younger sedimentary strata. Even older granites and gneisses, 2.5 billion years old, were exposed, mainly in the central part of the Black Hills (see chapter 8). To the east these ancient Precambrian rocks were buried a mile deep or more in the central Dakotas, sloping upward from there to the Great Lakes region. Small outcrops of this granite can be found on the eastern border of South Dakota near Milbank, where it is quarried for monuments and building construction. Somewhat younger but still ancient, the 1.6 billion-year-old Sioux quartzite forms palisades and rapids on Split Rock Creek and the Big Sioux River in southeastern South Dakota, near Sioux Falls, Garretson, and Dell Rapids (fig. 2.2). Though exposed only on a small portion of the surface, these igneous and metamorphic outcroppings add diversity to the landscape. In North Dakota the oldest surface rocks are 90 million-year-old Greenhorn Formation shales found in the bottom of the Pembina River.[4]

Of the younger rock formations, the most widespread is the Cretaceous Pierre Shale, which began to form about 80 million years ago. More than 1,000 feet thick in some places, it covered most of central North America, including land that would become the Dakotas. Light to dark gray in color, Pierre Shale

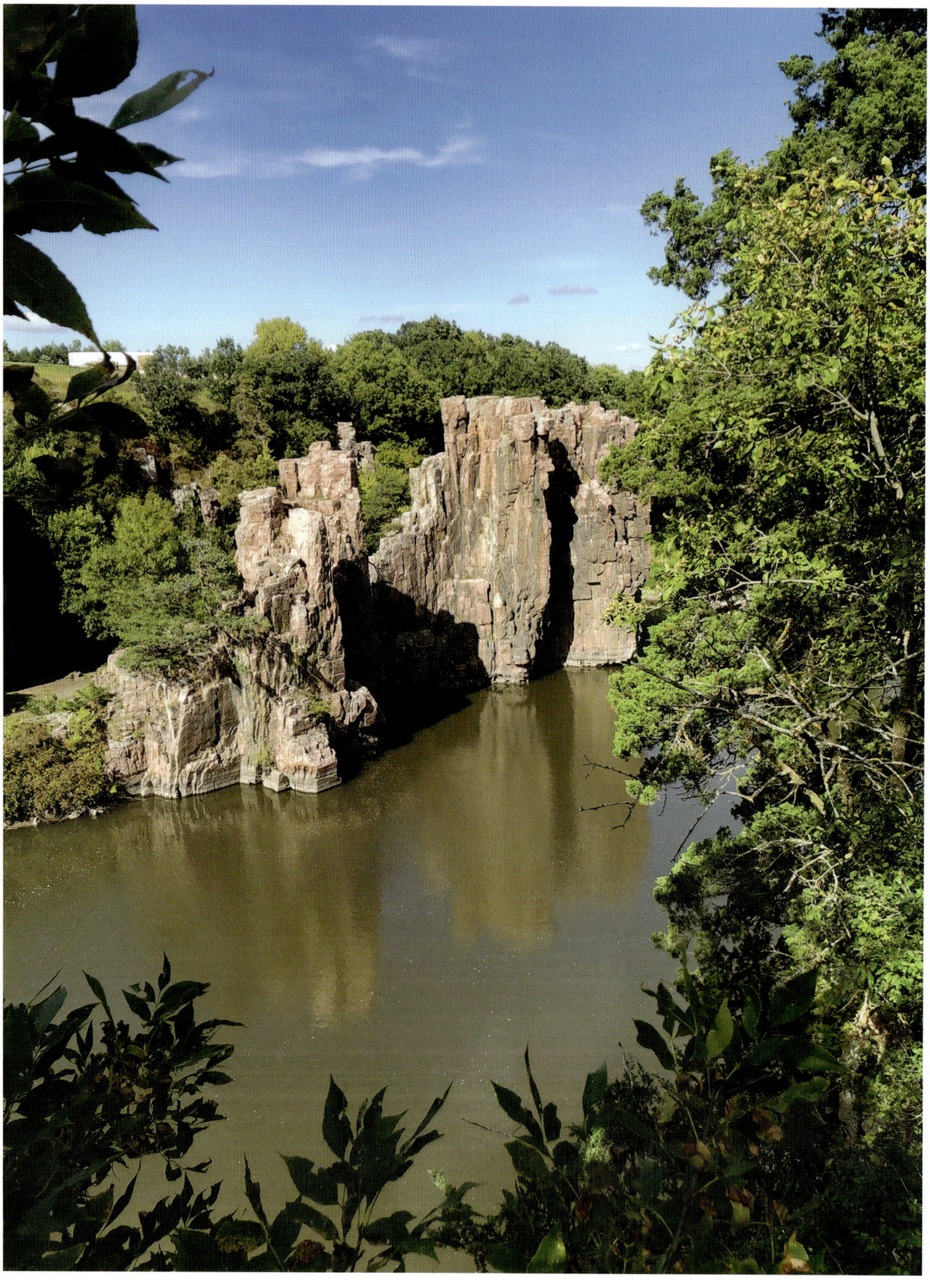

Fig. 2.2. Palisades along Split Rock Creek and the Big Sioux River in southeastern South Dakota were carved by water flowing across Precambrian Sioux quartzite after the relatively thin mantle of glacial deposits in this area was eroded away. Common trees and shrubs in the valley are green ash, bur oak, hackberry, basswood, American elm, boxelder, plains cottonwood, eastern redcedar, chokecherry, Virginia creeper, and poison ivy.

developed from mud with varying amounts of silt, clay, volcanic ash, and organic material. Much of the Pierre Shale has now eroded away, but where it remains, the soils are best known for the infamous gumbo they form when wet—a sticky, slippery mud that can make travel impossible.

Fossils of ammonites, oysters, clams, crabs, snails, fish, sharks, and turtles are common in the Pierre Shale. Also found in the shale and in the younger, overlying Niobrara Chalk are the fossils of 40- to 50-foot-long swimming reptiles—the mosasaurs and plesiosaurs (fig. 2.3). Much of the seawater they depended on drained eastward during the late Cretaceous and early Cenozoic.[5] Rivers originating in mountains meandered widely across the gently sloping plains of the Dakotas, depositing overlapping alluvial fans as the channels shifted. With alluvial sediments came pebbles and cobbles that included chert, quartzite, and agate. Periods of deposition over the Pierre Shale and other strata created the Fox Hills and Hell Creek Formations, which are famous for fossils of dinosaurs, turtles, crocodiles, and large trees of sequoia (a conifer) and katsura (an angiosperm). Birds had evolved much earlier, in the Paleozoic era, but are not well represented in the fossil record because of their hollow, lightweight bones and typical terrestrial habitat.

Fig. 2.3. Plesiosaurs were common in the late Cretaceous sea that covered the Dakotas. Like Mosasaurs, they were 40- to 50-foot-long carnivorous reptiles that are known from Pierre Shale fossils. They became extinct at the end of the Cretaceous, the same time as dinosaurs. Painting by Roger Harris/Science Source.

Fig. 2.4. Fossil tree stump in the Little Missouri Badlands of North Dakota. This tree, probably dawn redwood (*Metasequoia*), lived during the Paleocene, about 60 million years ago. Other trees in the area at that time could have included ancestors of modern palms, pines, ginkgo, bald cypress, magnolia, oak, maple, birch, walnut, hickory, and sycamore. Photo by John Bluemle.

The end of the Cretaceous period is marked in the fossil record by the extinction of dinosaurs and many other forms of life—plants and animals. The most probable cause was an abrupt episode of intense heating and subsequent cooling attributed to the colossal impact of a small asteroid on the north shore of the present-day Yucatan Peninsula. Fires would have burned wherever there was sufficient fuel. Astonishingly, evidence from all around Earth suggests that this global extinction occurred within a few hours or days. Working near Bowman in southwestern North Dakota, the paleontologist Robert DePalma and his associates concluded that most organisms were extirpated there in about an hour. Small, warm-blooded mammals in burrows were largely spared from the heating and cooling and, without dinosaurs as competitors, their abundance and diversity increased rapidly. The Age of Mammals, known as the Cenozoic era, began in this way.[6]

In the Cenozoic, long periods of alluvial deposition were followed by episodes of further uplifting and subsequent erosion, which created ravines and valleys. Some low-lying flatlands included swamps and marshes that provided an environment favorable for the growth and fossilization of tree-sized clubmosses, ferns, and horsetails—all of which are well represented in the fossil record of the Paleocene (66 million to 56 million years ago). On the uplands were forests of cycads, ginkgo, and conifers (including pine, spruce, fir, and sequoia). Petrified wood that formed during this time can still be found in the badlands (fig. 2.4). Subsequent uplifting led to the drying of carbon-rich sediments that became lignite, a soft coal that was ignited from time to time by lightning or spontaneous combustion. In some places the lignite burned for many years, and still does today, sometimes underground and hot enough to bake the overlying shale and other sediments into a hard, erosion-resistant rock known as *scoria* or *clinker*. Some hills and buttes persist because of beds of clinker on the top (see chapter 12).

Lakes and seas were less widespread during the more recent Cenozoic era, but they still covered some of the land that would become the Dakotas. The Fort Union group of sedimentary rocks formed on top of Cretaceous strata and included additional coal beds. Some exposed shales weathered to colors of red, yellow, and purple, such as in the South Dakota badlands (see chapter 12). Plant fossils suggest that the vegetation was subtropical woodland. Trees at that time included ferns, palms, pines, metasequoia, redwood, ginkgo, magnolia, oak, elm, boxelder, maple, birch, alder, walnut, horse chestnut, hickory, hazelnut, and sycamore.

About 23 million to 5 million years ago, during the Miocene, another episode of regional uplifting raised the Rocky Mountains and western Great Plains to their approximate present-day elevations above sea level. Erosion was again accelerated, and large amounts of ash blew in from newly erupting volcanoes in the Rocky Mountains and Great Basin, much of it settling to the bottom of large lakes and contributing to the hardness of sandstones and limestones. Such rocks, along with clinker, would eventually form the erosion-resistant tops (caps) of buttes, including the Killdeer Mountains. The rising Rocky Mountains to the west reduced the amount of precipitation over the plains—the rain-shadow effect. Forests became patchy and grasslands more widespread as the climate became drier, and herbivores adapted to eating difficult-to-digest grasses and forbs became more abundant.[7] By this time the North American continent had drifted northward to near its present location and the flora and fauna had evolved to tolerate cold winters. Fossil evidence indicates that the plant life was similar to that seen in the Dakotas today, although the mammals were still very different, with giant ground sloths, saber-toothed cats, a piglike animal known only as *Archaeotherium*, and now-extinct ancestors of horses, camels, hyenas, and rhinoceroses (fig. 2.5).

Toward the end of the Pliocene epoch, about 3 million years ago, most of the woodlands probably

Fig. 2.5. An artistic rendition based on fossil evidence of an Oligocene landscape, about 25 million years ago, on what is now the Northern Great Plains of North America. Among the mammals that existed at this time were hoofed pantodonts that included the ancestors of odd-toed perissodactyls, such as rhinos and horses, and even-toed artiodactyls, which included the ancestors of cattle, bison, deer, camels, and pigs. These ancestors have unfamiliar names, such as the piglike *Archaeotherium* featured in this photographic mural at the North Dakota Heritage Center and State Museum in Bismarck. Courtesy of the State Historical Society of North Dakota.

were restricted to ravines and along rivers, with grasslands more common on the uplands. Erosion continued, wearing away most of the younger rocks as well as much of the underlying Cretaceous formations, including the Pierre Shale (especially at higher elevations in the west). The terrain became similar to what is seen today in western South Dakota and southwestern North Dakota, with scattered buttes (see fig. 1.9), broad river valleys, and relatively flat alluvial fans flanked by low escarpments. The rivers flowed mainly from west to east or northeast, but unlike today, all of them merged with other rivers flowing into the Arctic Ocean. By this time the Red River Valley existed and the Paleozoic and Precambrian rocks of the Black Hills were exposed.

Most of the landforms at the end of the Pliocene were the result of upward and downward warping of Earth's crust and the exposure of different kinds of bedrock with varying resistances to wind and water erosion. The magnitude and time frame of these geological processes are difficult to comprehend, but perspective is gained by considering, for example, that the tops of isolated buttes seen today often have rounded, polished stones produced by thousands of years of tumbling in rivers.

The Ice Age

About 2.6 million years ago, at the beginning of the Pleistocene epoch, most of the land had prairies and woodlands with soils derived from Paleogene and Cretaceous sedimentary rock. Continued cooling led to the formation of ice sheets at northerly latitudes, which advanced and retreated at least six or seven times. In some places these glaciers were more than a half-mile thick—but they disappeared during warm interglacial periods that would last for thousands of years, time enough for new grasslands and woodlands to develop on the glacial material left behind. The last 11,700 calendar years (approximately 10,000 radiocarbon years), since the last glaciers retreated, are the Holocene epoch.[8]

The most recent glacial advance reached North Dakota about 26,000 years ago. Known generally as the Late Wisconsin, this advance persisted for about 14,000 years and eventually covered the eastern half of South Dakota and all but the southwestern quarter of North Dakota (see fig. 1.2). As the glaciers expanded southward, the climate just beyond the ice became colder and the vegetation changed, first to woodlands dominated by spruce and poplar and later to tundra. The patterned ground typical of Arctic tundra is still apparent on the tops of several buttes in the western Dakotas today; permafrost would have been common. Only cold-hardy plants and animals persisted in the vicinity of the glacial lobes, some on pockets of soil on the ice itself. There would have been seasonal temperature changes, with short periods of melting during the summer, but the mean annual temperature was below freezing for thousands of years at a time.[9]

The ice sheets transported large amounts of soil, sand, gravel, crushed rocks, and boulders that would be deposited in various landforms when the ice melted. Some of this material formed *end moraines*, that is, elongated hills still visible today that indicate the farthest advances or pauses of a glacier. End moraines might be a few tens of feet thick, several miles wide, and tens of miles long. The maximum advance of these ice sheets also can be approximated from isolated boulders, known as *erratics*, that could have been transported only by glaciers (fig. 2.6). Nearly flat *ground moraines* formed directly under melting ice sheets and, with multiple glaciations, reached thicknesses of up to 150 feet or more. Rivers draining from the ice carried gravel, sand, silt, and clay away from the glacier, which, when deposited downstream, created alluvial fans commonly referred to as *outwash* or *outwash plains*. The headwaters of some glacial rivers formed tunnels or valleys through the ice, which led to the deposition of sinuous landforms known as *eskers* (fig. 2.7).[10] Windblown silt (loess) covered much of the land and contributed significantly

Fig. 2.6. This granitic boulder was transported 400 miles or more by glaciers, from the Canadian Shield to what is now the northern unit of Theodore Roosevelt National Park, south of the Missouri River. Known as a glacial erratic, its presence indicates that glaciers extended this far south at one time, perhaps 80,000 years ago (early Wisconsin). Except for similar boulders here and there, all other evidence of the ice sheets in this area has been eroded away or become hard to find. The plants of the mixed-grass prairie seen in this photo include little bluestem (straw colored), patches of western snowberry (buckbrush) just beyond the boulder, and chokecherry in the distant ravines.

Fig. 2.7. The Dahlen Esker, one of the most spectacular eskers in North America, is a ridge formed by a meltwater river that flowed through an ice-walled valley or tunnel in a glacier that once covered this part of North Dakota. Cone-shaped hills, known as *kames*, are formed in much the same way as eskers (Bluemle 2016). Photo by Tom Bean.

to soil development on the Great Plains. In some places silt dunes formed that were 200 feet high, such as in southeastern South Dakota and northwestern Iowa. Once stabilized with grasslands and woodlands, they became known as loess hills.[11]

Present-day landscapes were influenced by several periods of ice advance and retreat, each lasting thousands of years and separated by equally long interglacial periods. More recent glaciers buried the vegetation and soils that had developed since the last retreat, and they sometimes overtopped portions of older moraines. Former stream channels were filled and new deposits were mixed with the old. If huge blocks of ice were buried, lakes or ponds formed hundreds of years later when the ice melted. The resulting land, known as *hummocky collapsed glacial topography*, is characteristic of the Missouri Coteau, Prairie Coteau, and Turtle Mountain, which comprise a large portion of the Prairie Pothole Region. When glaciers advanced over some aquifers, the pressurized water helped force the overlying material upward into the path of the advancing ice. This process, labeled *ice thrusting* by John Bluemle, resulted in characteristic landforms, sometimes consisting of stacked layers of sediment, like fallen dominoes.[12]

Large lakes were formed where rivers were dammed by the ice sheets. Icebergs probably were common. Glacial Lake Agassiz was by far the largest, much larger than Lake Superior. Other glacial lakes (Regina, Souris, Mose, and Dakota) were comparable in size to some of the smaller Great Lakes

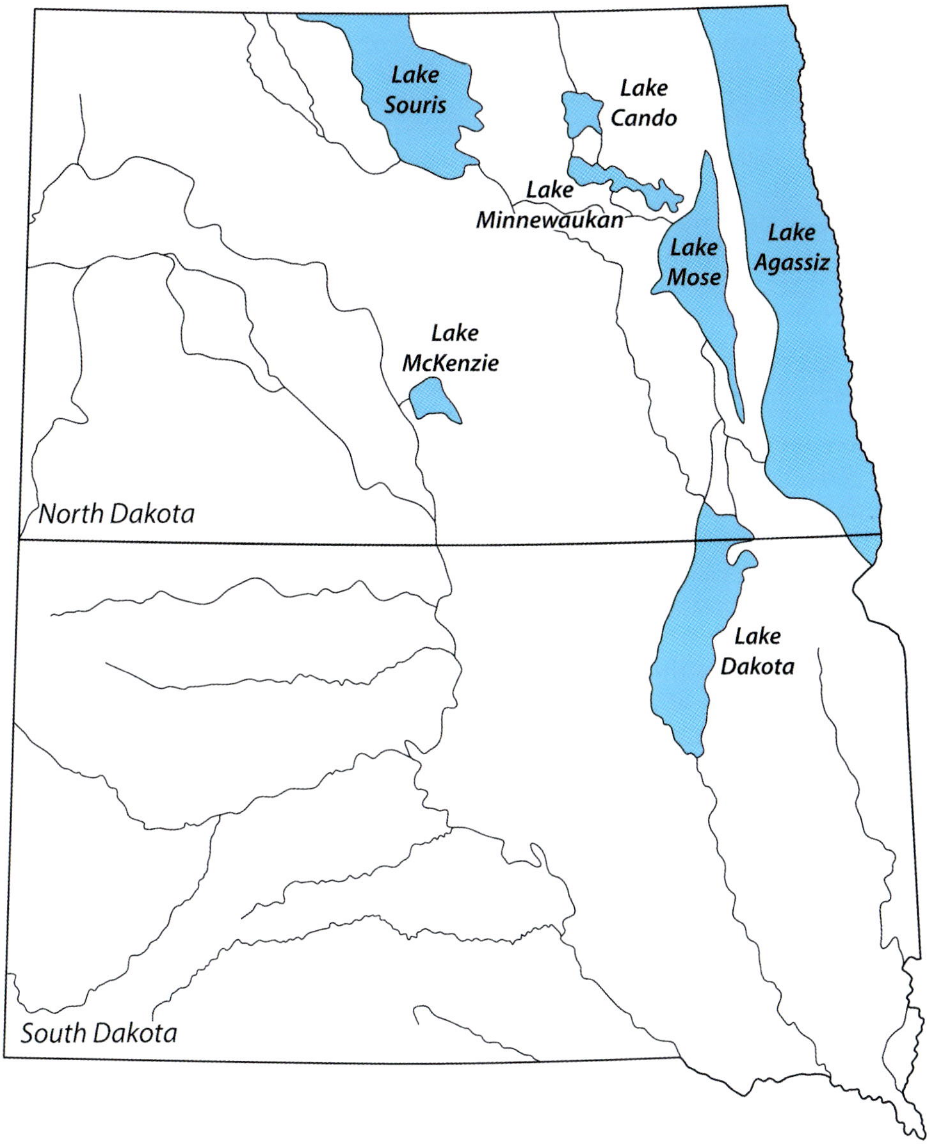

Fig. 2.8. Some of the glacial lakes that existed at the end of the Pleistocene, including the four largest: Agassiz, Dakota, Mose, and Souris. Lake Agassiz was mostly in Canada, with less than 10 percent of the lake in North Dakota and northern Minnesota. Lake Mose predates Lake Agassiz, having formed on higher ground just to the west when the present-day Red River Valley was still essentially filled with glacial ice. As described in chapter 1, flat land over large areas is an indication of former postglacial lakebeds or, in some places, glacial outwash plains (see figs. 1.5 and 1.6). Adapted from Bluemle (2016) and Gries (1996); Bluemle provides a detailed map of glacial lakes for North Dakota.

as they exist today (fig. 2.8, box 2.1). Many smaller lakes were created as well. As lake levels fluctuated, beaches of sand were formed, some that are now stranded on dry land but still visible.[13] The volume of water in the lakes increased greatly at the beginning of warm interglacial periods, overflowing spill points and causing huge volumes of cascading water that formed much broader valleys than could have been created by the small rivers that remain today (fig. 2.9). Good examples are the

Box 2.1. Icebergs in the Dakotas

Long after most of the Red River Valley's fertile prairie soils had been plowed and cultivated, University of North Dakota geologist Lee Clayton observed that shelterbelts planted west of Grand Forks sometimes have a "beaded" appearance, that is, segments of tall trees alternating with segments of shorter trees. This was unusual because all the trees in a single row were the same species, same age, and planted at the same time. After flying over the area and making on-the-ground measurements, he and his colleague John Bluemle found that the tall trees occur in shallow "grooves" in the landscape that are up to six miles long, 75 to 100 feet wide, and three to seven feet deep. Except from the air, the linear depressions are nearly imperceptible. The geologists concluded that the grooves were formed when large pieces of ice, possibly icebergs from retreating glaciers, gouged shallow bottom sediments as they were blown across Glacial Lake Agassiz during a specific spring thaw. Many of the grooves on the land are parallel and trend from north-northwest to south-southeast, probably the direction of prevailing winds at the time. Similar grooves are formed today near the north shore of Canada and Alaska, carved by drifting ice.

The cause of shorter trees on the upland is uncertain, but several factors appear to be involved. Clayton and Bluemle found beaded shelterbelts only where the groves coincide with saline artesian water seeping to the surface. Thus, the higher ground could have been drier, which might have slowed tree growth, and perhaps tree growth also was influenced by differences in salinity. Another factor could be soil erosion from fields on the higher ground prior to tree planting. Whatever the cause, this landscape pattern was first revealed to modern-day scientists by trees planted for shelter. Previously, Native Americans probably knew that prairie grasses and forbs in the grooves were different from those on the adjacent upland—another example of how geologic history affects modern plant distribution patterns.

valleys of the Souris, Pembina, Sheyenne, James, and Minnesota Rivers. The terraces that can be seen in these valleys are remnants of floodplains created when the glacial rivers were much wider and deeper (fig. 2.10).

The flat Red River Valley existed before glaciers arrived, but Glacial Lake Agassiz flooded and deposited sediments over a much larger area to the north. Over a period of about 4,400 years (12,900 to 8,500 years before present), the lake drained in various directions, first to the southeast, creating Glacial River Warren that shaped the broad valley through which the rather narrow Minnesota River now flows. Some of Lake Agassiz's overflow may have drained southward into Glacial Lake Dakota. Drainage to the east and north began about 9,000 years ago, most likely first to the Great Lakes and then to Hudson Bay.[14]

One of the most dramatic effects of glaciation was the creation of the Missouri River as it exists today. Previously, a few rivers flowed northward to Hudson Bay, such as the Little Missouri, while others traversed most of the Dakotas before bending northward, such as the Cannonball and the Cheyenne. However, the ice sheets and end moraines of the Missouri Coteau diverted flows to the east for the Little Missouri and to the south for others (see figs. 1.2 and 1.10). The resulting confluence of numerous rivers formed the Missouri.[15] Erosion along the Little Missouri River was accelerated because its new route was steeper than before, which exposed the colorful Paleocene sedimentary rocks of the Little Missouri Badlands in North Dakota. Badlands in South Dakota began to develop at about the same time but cannot be attributed to glaciation.

A second dramatic effect of the glaciers was the deposition of finely ground material rich with plant nutrients. Combined with windblown loess, these minerals contributed to the eventual formation of more fertile soils, compared to those derived from sedimentary bedrock alone. Agriculture would not be as prosperous in the Dakotas had glaciers not covered much of both

Fig. 2.9. This broad valley was created when the Sheyenne River was much larger than it is today, at a time when the river was more than a mile wide and over 200 feet deep and had torrents of water draining Glacial Lake Souris. Floodplain terraces can be found on the sides of the river valley, far above the elevation of present-day floodwaters (fig. 2.10). This photo was taken south of Esmond, North Dakota.

Fig. 2.10. This road reveals terraces created when the Sheyenne River was much larger as a result of drainage from Glacial Lakes Souris and Regina. In some areas the terraces are no longer visible (see fig. 2.9). Photo by John Bluemle south of Esmond.

states. Erosion by water and by wind continues to shape fields and pastures, as it has for millions of years—but now it occurs more rapidly as a result of human activities.

Contributing to the hilly topography of some glaciated land is the unevenness of the underlying Cretaceous and Paleogene sedimentary rocks. Where glacial deposits were only a few feet thick, the hills and ravines seen today existed long before the ice arrived, but where the deposits are hundreds of feet thick, the older topography is obscured. The highlands of Turtle Mountain, the Missouri Coteau, and the Prairie Coteau are attributable in part to underlying preglacial escarpments that forced some of the advancing ice sheets upward (box 2.2).

Box 2.2. Les Coteaux des Prairies

Named Les Coteaux des Prairies by French explorers, the Prairie Coteau is the most conspicuous topographic feature in eastern South Dakota. This massive highland is a flatiron-shaped plateau that rises to a height of 800 feet above the surrounding plains and extends 200 miles from the North Dakota–South Dakota border to northwestern Iowa (see figs. 1.1 and 1.2). Before glaciation, hills underlain by Cretaceous shales existed where the Coteau is located now. The first glacial advance deposited sediments that buried the shale and began to raise the elevation of the Coteau. Later advances built the Coteau higher, eventually forcing, in the most recent advance, a split of the ice sheet into the western James lobe and eastern Des Moines lobe.[a] Because the advancing lobes pushed ice blocks and debris aside onto the Coteau surface, rather than passing over it, the surface geology of the Coteau is often called a *dead ice* or a *stagnant ice* glacial feature. The topography and soils created by the advance and retreat of glaciers now support a wide range of ecosystems, including clear-water lakes, prairie wetlands, a mosaic of mixed and tallgrass prairie, and woodlands with green ash, bur oak, and, on the north end, aspen and sugar maple.

[a]Flint 1955.

After the Glaciers

The glaciers had nearly disappeared when the first people arrived in the land that would become the Dakotas, about 13,500 years ago.[16] Little is known about those first Americans, as most traces of their activities have disappeared, but archaeologists report that they arrived from the west and might have encountered large blocks of ice on cooler north slopes. Smoke might have been seen billowing upward on the horizon, leading them to wonder whether the fires had been started by lightning or by people who had arrived before them. Occasionally archaeologists find a dart or spear point in or near the skeleton of a mammoth, indicating their coexistence, but only one mammoth kill site has been found in the Dakotas.

Over thousands of years, numerous cultures developed on the Great Plains. Others emerged more recently, such as the Mandan, Hidatsa, and Arikara, who arrived about a thousand years ago. They lived in semipermanent villages of earthen lodges along rivers (see chapter 10). Nomadic tribes like the Dakota, Lakota, Assiniboine, Cheyenne, Plains Cree, Blackfeet, and Crow began to spend more time in the area about 300 to 400 years ago.[17] The first immigrants of European descent did not arrive until the early 1700s, first on foot but soon on horseback. Some were intent on visiting tribal communities, seeking guidance about travel, and finding sources of furs.

The wildlife encountered by the first Americans (the Clovis culture) included many species that soon would become extinct, including large herbivores such as mammoths, mastodons, camels, horses, and giant ground sloths (figs. 2.11 and 2.12). Mammalian carnivores that became extinct included saber-toothed cats, dire wolves, and short-faced bears. Most scientists think that humans and large carnivores were the primary cause of

Fig. 2.11. Mammoths and the long-horned bison coexisted on the Northern Great Plains during the Pleistocene, along with numerous other now-extinct mammals. Painting by Karen Carr; printed with permission of the artist and the Sam Noble Museum at the University of Oklahoma, Norman. A photographic mural with this image can be seen at the North Dakota Heritage Center and State Museum in Bismarck.

extinction. They surmise that the herbivores were pushed south by the expanding ice sheets and cold climate, where many of them became vulnerable when concentrated by topographic bottlenecks and cul-de-sacs. Additional factors may have been extreme weather and the warm, dry climate following deglaciation.[18] Some meat eaters became extinct when their prey became scarce.

Dramatic changes in plant life occurred during the past 12,000 years, as is well documented by the kinds of pollen, seeds, and cones preserved in sediments at the bottoms of the most permanent lakes. Radiocarbon dating can be used to identify the time period when the sediments were deposited. Deeper sediments are older. Trained observers identify the species from which the plant material originated in each layer, providing a basis for determining how vegetation in the vicinity has changed. Based on this kind of evidence, it is clear that forests and woodlands dominated by spruce were widespread from 12,000 to 10,000 years ago while much of the land was still covered with ice (fig. 2.13). The nearest spruce trees now, except where they were planted, are where the climate

Fig. 2.12. The skulls of the modern bison (*Bison bison*) on the right and the now-extinct *Bison latifrons* (center) and *Bison antiquus* (left). The relatively small modern bison is still the largest native land mammal in North America today. Exhibit in the North Dakota Heritage Center and State Museum, Bismarck. Courtesy of the State Historical Society of North Dakota.

is still relatively cool, such as in the Black Hills, southern Manitoba, and northern Minnesota. Notably, several layers of buried spruce forest, with some tamarack, have been identified in the Red River Valley—the result of fluctuating water levels in Lake Agassiz caused by climate change.[19]

Research on lake-bottom sediments indicates that, as the climate warmed, conifers were replaced by hardwood trees such as oak, elm, ironwood, hazel, black ash, birch, and aspen. Meadows, shrublands, and grasslands were interspersed among the woodlands. Pine trees—probably jack pine—persisted in the Sheyenne River delta until about 9,000 years ago (see fig. 1.2) but were more common in Canada and northern Minnesota. Grasslands would have been more widespread in the western Dakotas, especially in South Dakota, where the rain-shadow effect of the Rocky Mountains was stronger. Further warming caused aridity and more frequent fires that killed many trees. Prairie plants became more common, with woodlands persisting where grassland fires were less likely to spread, such as on the leeward sides of lakes or steep ridges. Trees also survived in areas with more precipitation and consequently fewer fires, such as the Black Hills and Turtle Mountain. Aspen parklands were common in the cooler environment of northeastern North Dakota, northwestern Minnesota, and Manitoba (see chapter 7).[20]

In general, the Dakotas were drier during the early parts of the Holocene, but evidence from pollen and other sources suggests that severe, multidecadal droughts alternated with wet periods during the mid-Holocene. There were times when Glacial Lake Agassiz, the Devils Lake basin, and much of the Waubay Lake basin were dry.

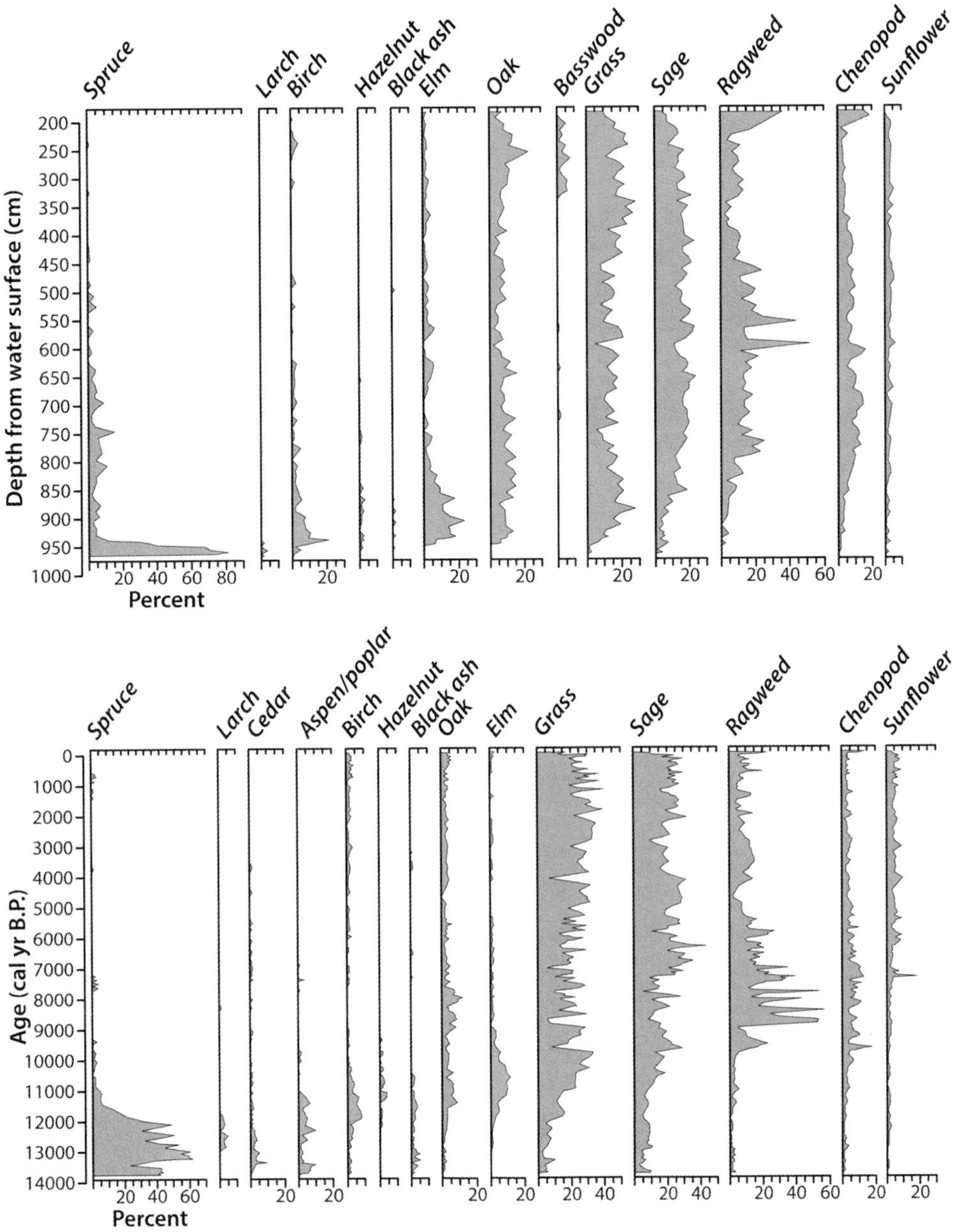

Fig. 2.13. Pollen profiles derived from lake sediments reveal how common plants on the surrounding land have changed during the past 10,000 years. The horizontal axis is the percentage of pollen in a sample belonging to each species or species group. Some pollen grains can be identified to genus, others cannot. In both diagrams, note how spruce was very common 14,000 to 10,000 years ago, and to a lesser extent larch (tamarack), birch, and elm. Spruce and larch soon disappeared and oak, grasses, and some other prairie plants became more common. The abundance of weedy species such as ragweed and chenopods probably was the result of disturbances, drought, the creation of mudflats, and other factors that reduced the competitive ability of prairie and woodland species. Top profile: Pickerel Lake in the Prairie Coteau of northeastern South Dakota; data for basswood are exaggerated five times to illustrate more clearly that this tree is a recent arrival. Depth in the sediment core is used as a surrogate for age because only one radiocarbon date was obtained for this site, from the bottom (10,670 years ±140 years). Sediments closer to the surface are assumed to be younger. Adapted from Grimm (2001) and Watts and Bright (1968). Bottom profile: Pollen from Moon Lake near Bismarck, in the glacial drift prairie of North Dakota, illustrate vegetation changes similar to those at Pickerel Lake. Basswood pollen was not found in Moon Lake sediments. The vertical axis is sediment age, made possible because radiocarbon dates were obtained at several depths. Adapted from Grimm (2001) and Clark et al. (2001).

Box 2.3. Prehistoric Droughts Had Dramatic Effects

Learning about the past has become highly relevant to discussions of climate change in the twenty-first century, but how is it possible to learn about the climate of ancient ecosystems? Fossilized bones, shells, wood, and leaves provide evidence for long-term changes, as discussed in this chapter, but now scientists can detect fluctuations in mean annual temperature and precipitation during recent centuries using tree rings. Also, pollen grains and other materials preserved in sediments at the bottoms of some deep lakes are useful because their age increases with depth, and often their source is identifiable for 15,000 years or more (see fig. 2.12). Shifts in species represented in the layers from bottom to top are a reliable reflection of changes in climate. The approximate ages of organic materials in each layer can be determined with radiocarbon dating methods.

Duke University professor James Clark and his associates studied bottom sediments in Kettle Lake in northwestern North Dakota.[a] Their research revealed fluctuations during the past 10,000 years (the Holocene) in the abundance of grass and weed pollen, charcoal fragments, dust, and other climate indicators. Low amounts of grass pollen coincided with low amounts of charcoal and high amounts of dust and weed pollen. Less charcoal was formed, they reasoned, because the climate was too dry to produce enough grass to fuel the prairie fires required to form charcoal. At the same time, more dust was likely the result of increased wind erosion from the drought-stricken prairie and the creation of erodible mudflats as water levels dropped. Weed pollen would have increased because mudflats created more space for weeds. The scientists documented multidecadal droughts that are much longer than the drought of the 1930s and that, they concluded, would have been too long to support the kind of agriculture practiced in the region today.

[a]Clark et al. 2002.

Also, stabilized sand dunes that formed during wet periods, such as downwind from the Sheyenne River delta, became active again as dune plants died for lack of water. Woodlands and prairies changed dramatically during droughts that were much more severe than during the Dust Bowl years of the 1930s (box 2.3).[21]

Centuries Ago

> *"Ancient sea floors, coral reefs, shorelines, coastal swamps, tropical river systems, melting ice sheets, and a variety of other environments have combined to produce the framework that forms our land. . . . We tend to think of the landscape as permanent. But our hillsides, floodplains, buttes, badlands, broad rolling prairies, lake plains and floodplains are dynamic, always changing."*[22]

This observation penned by former North Dakota state geologist John Bluemle is a reminder of the astonishing changes that led to the development of modern landscapes—and how that change continues. Throughout geologic time, to the late Holocene, long-term change occurred so slowly that it would have been imperceptible to members of any single generation. But new machinery fabricated during the Industrial Revolution changed the land rapidly. Most important were combustion engines that could pull steel plows and facilitate the transportation of crops to distant markets.

What was the Dakota landscape like just prior to that time, two or three centuries ago?

- There were various kinds of grassland ecosystems almost everywhere (see fig. 1.3), but with deciduous woodlands of aspen, bur oak, and green ash on Turtle Mountain and forests of ponderosa pine, white spruce, and aspen in the Black Hills. Patchy savannas or parklands with bur oak and green ash were found in the transition between forests and grasslands. Ponderosa pine and Rocky Mountain juniper grew on the rocky, coarse-textured soils of some buttes and escarpments in the western Dakotas.

- Patches of hardwood trees and juniper would have been found in river breaks (ravines, draws, coulees), on the leeward side of some lakes, and on a few leeward hill slopes—wherever rainfall was sufficient and prairie fires burned less frequently, such as on the Missouri Coteau and Prairie Coteau.
- Cottonwood and other riparian trees would have grown along some creeks and rivers, especially larger rivers.
- There were lakes, potholes, marshes, and fens on glaciated terrain, surrounded by prairie, many without trees.
- Aspen parklands were common near Turtle Mountain and Pembina.

The first explorer of European descent known to walk on land that would become the Dakotas was a Frenchman, Philip Renault.[23] In 1719—more than 300 years ago—he and his party traveled northwest to the headwaters of the Minnesota River in present-day northeastern South Dakota. From there they crossed a continental divide and continued northward down the Red River to present-day Canada. Regrettably, no reports from that expedition have been found, but historians assume Renault was exploring the French-owned Louisiana Territory. About 19 years later, a French-Canadian explorer, Pierre Gaultier de la Vérendrye, traveled through the area with his sons and wrote the first known journal. On their first trip, in 1738, they walked from Fort La Reine west of Winnipeg to a Mandan village on the Missouri River near present-day New Town, crossing the Souris (Mouse) River twice along the way—a distance of about 275 miles. Horses apparently had not yet arrived in Manitoba, although they were in the southern part of present-day South Dakota by 1725.[24] Five years later, in 1743, two of the Vérendrye brothers left a lead plate inscribed with their names along the Missouri River near present-day Fort Pierre. The plate was found by school children 179 years later, in 1913. Numerous other explorers, trappers, and fur traders worked in the Dakotas during the 1700s and early 1800s.[25] Some of their journals express appreciation for assistance from tribal members.

The best-recorded expedition through land that would become Dakota Territory was led by Captain Meriwether Lewis and Second Lieutenant William Clark. Commissioned by President Thomas Jefferson, they traveled up the Missouri River in search of a route to the Pacific Ocean and to make observations about natural resources along the way. Reaching southeastern South Dakota in August 1804, one of their camps was near the mouth of the Vermillion River. From there Lewis and Clark hiked north about six miles and climbed to the top of Spirit Mound, once covered by glaciers (figs. 2.14 and 2.15). Clark wrote, "We beheld a most butiful [*sic*] landscape; numerous herds of buffalo were seen feeding in various directions." He added that the plains to the northwest and northeast extend "without interruption as far as can be seen." Trees were infrequent, except along rivers and in some ravines, where they would have seen plains cottonwood, green ash, American elm, boxelder, hackberry, basswood, willow, and numerous shrubs, such as wild rose, snowberry, currant, chokecherry, wild plum, serviceberry, and hawthorn.[26]

Farther upriver Lewis and Clark reported gray wolves and grizzly bears, and they visited an Arikara village with gardens of corn, tobacco, and beans. Large flocks of geese, ducks, and swans were observed migrating southward as the days shortened and temperatures cooled. Their trip through the Dakotas was the first to document the existence of numerous animals and plants already familiar to indigenous people but new to scientists of European descent. Examples include pronghorn, gray wolf, grizzly bear, black-tailed prairie dog, prairie chicken, black-billed magpie, channel catfish, prairie rattlesnake, plains cottonwood,

Fig. 2.14. An artistic rendition of Meriwether Lewis, William Clark, and some of their crew on August 25, 1804, after they climbed Spirit Mound, about six miles north of Vermillion on the Missouri River in southeastern South Dakota. They wrote in their journals about seeing herds of bison and elk. Indigenous tribes were helpful to the expedition, but Lewis and Clark reported that they avoided this area. The vast prairie without trees was most likely a mosaic of mixed-grass prairie on hilltops and upper slopes and tallgrass prairie below. With permission from the artist, Ron Backer (photo by Mark Wetmore).

Fig. 2.15. A view to the south from the top of Spirit Mound in 2018 (see fig. 2.14). Croplands and trees planted in shelterbelts are now common in all directions. With support from the Spirit Mound Trust, the Spirit Mound Historic Prairie was established as a state park in 2001 and the Spirit Mound Summit Trail was designated a national recreation trail in 2004. Work is under way to restore the native grasslands.

silver buffaloberry, silver sagebrush, and purple coneflower—among many others.

While Lewis and Clark were on their expedition, a new culture was developing in the northern Red River Valley, where French and Scottish fur traders intermarried with Native Americans. Their descendants became known as the Métis, and their southernmost colony would become Pembina, established as a trading post in 1819 and the oldest settlement in the Dakotas.[27] Many Métis were fur traders, but others became buffalo hunters and farmers. Buffalo hides had become an important export item, and their acquisition was greatly facilitated by acquisition of horses and rapid-fire rifles. The hides were loaded onto two-wheeled "Red River carts" made entirely of wood and taken to ports on the Mississippi River. For various reasons, bison numbers declined precipitously and were extirpated by 1888 throughout the land that would become the Dakotas.[28] Beavers were trapped nearly to extinction as well (see chapter 9). With horses, firearms, and traps, combined with markets for buffalo hides and beaver pelts, Native Americans and Euro-Americans alike brought about socioeconomic and ecological changes more quickly and over larger areas than ever before.

A tragic development was caused by the introduction of smallpox, which killed about two-thirds of the Mandan, Hidatsa, and Arikara people 20 years before Lewis and Clark arrived. Another wave of smallpox mortality occurred in the early 1830s when steamboats traveled up the Missouri to western North Dakota. An estimated 17,200 Plains Indians died during this epidemic. For this and other reasons, some of the tribes became hostile to European explorers, which led to the establishment of military forts to protect newcomers. Perhaps the best known is Fort Abraham Lincoln, south of Bismarck-Mandan, from which Lieutenant Colonel George Armstrong Custer began an expedition in 1874 to verify the existence of gold in the Black Hills—which he did. Two years later he departed from Fort Lincoln again, this time on his ill-fated expedition to the Little Big Horn River in Montana.

The motivations for this early exploration by Europeans varied but included finding furs, hides, other resources, and a place for a better life. The vast prairie appeared inhospitable to many of the first immigrants, but essentially free land provided by the Homestead Act of 1872 was a strong incentive. During a period of about ten years (1878–1887), several hundred thousand farmers and others arrived in the area from Northern Europe, a period that became known as the Great Dakota Boom (fig. 2.16).[29] They were encouraged by the knowledge that travel to the Dakotas by

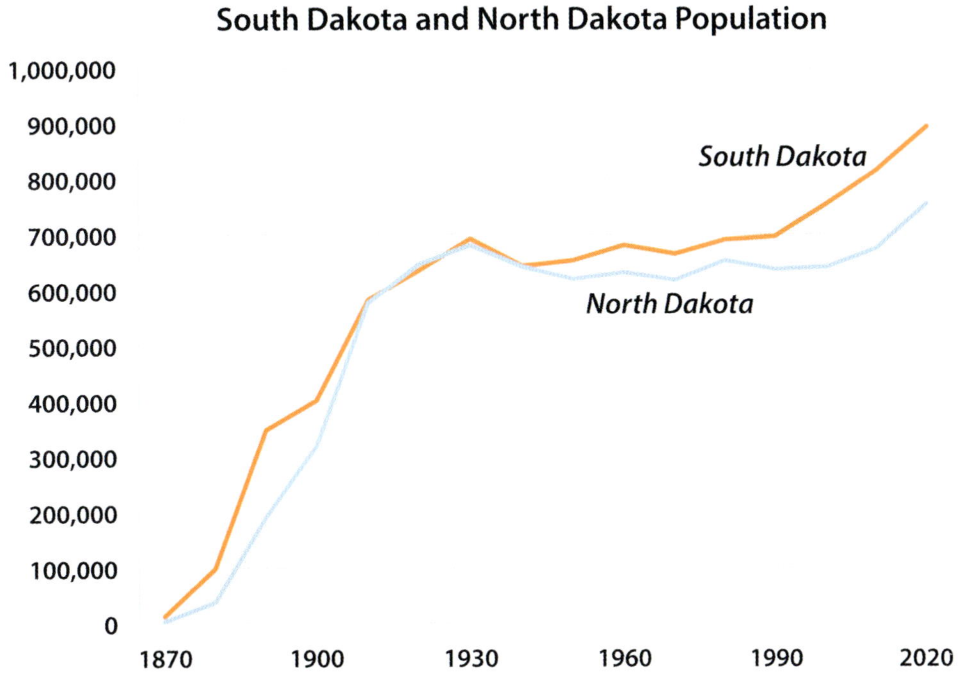

Fig. 2.16. Human population growth in North and South Dakota from 1870 to the present. The rapid rise in the late 1800s was due largely to immigration from Northern Europe.

railroad was relatively easy, that military forts would protect them, and that the oxen and draft horses required to pull sod-breaking plows were available. Over much of the Dakotas, water could be obtained from rivers, wetlands, springs, and artesian wells; and wood could be harvested along river bottoms and the leeward side of lakes. For those in the western Dakotas, where annual precipitation was too low for grain crops, cattle were driven north from Texas. A timber industry developed on Turtle Mountain and in the Black Hills. As the land filled with immigrants, engineers began to think about how Missouri River floods could be contained and how the river could provide hydroelectric power and water for municipalities and industry as well as agriculture.

Now, early in the twenty-first century, most native ecosystems have been greatly altered or lost. Introduced plants are more abundant than native plants over much of the region; and the remaining native grasslands, wetlands, woodlands, and forests have become more highly valued. Subsequent chapters focus on the nature of these ecosystems, and how ecological information can lead to improved land management practices. First, however, consider how temperature and precipitation are expected to change during the next 80 years—the subject of the next chapter.

Chapter 3 Climate

Climate is the daily weather averaged over a long period, commonly 30 years, and is described in terms of temperature, precipitation, solar energy, wind, and length of growing season. The effects of these environmental variables are well known. Soil characteristics also are influential, as are periodic disturbances caused by floods, fires, windstorms, and insect or disease epidemics. But if the climate changes, everything else changes as well. The extinction of innumerable species was triggered by climate change in the past; the very nature of our homes, farms, and cities is now governed by climate—and will be in the future.

Because of long distances to oceans and the Rocky Mountain rain-shadow effect, the mean annual precipitation in the Dakotas is relatively low—from 14 to 18 inches in the west and 24 to 28 inches in the east (fig. 3.1). Average annual precipitation is highest in the Black Hills and southeastern South Dakota. Usually, June is the wettest month and January the driest (fig. 3.2), with 70 percent or more of the annual precipitation falling during the spring, summer, and fall (April–September). Extended droughts occur at irregular intervals and have lasted up to six years during recorded history. Before that, information gleaned from tree rings and lake sediments indicates that severe droughts lasted for 20 years or more.[1] The warmest temperatures occur in southern South Dakota, but summers are hot in both states during occasional heat waves (fig. 3.3). The average maximum temperature from 1980 to 2015 at six weather stations was 105.6° F (Pierre), 100.4° (Dickinson), 99.2° (Mandan), 95.6° (Grand Forks), 95.1° (Brookings), and 93.5° (Lead).

High variability in weather is typical of the Northern Great Plains. From one year to the next, summers can range from drought to deluge and winters from very cold with abundant snow to relatively warm with little snow.[2] Unusually wet and cool years followed by warm drought years explain the episodic, often dramatic changes in crop production, wildlife population sizes, and water levels in rivers and lakes. The most severe drought in recorded history occurred from 1933 to 1939. Known as the "dirty thirties," the soil laid bare by extensive plowing caused devastating dust storms throughout the Great Plains. Many farmers abandoned their land; the importance of climate became eminently clear.

Aside from droughts, early explorers wrote about great variation in daily temperature. On a summer day near Grand Forks—August 2, 1823—William Keating wrote: "[We] suffered much from cold. The thermometer, which had stood at 83°

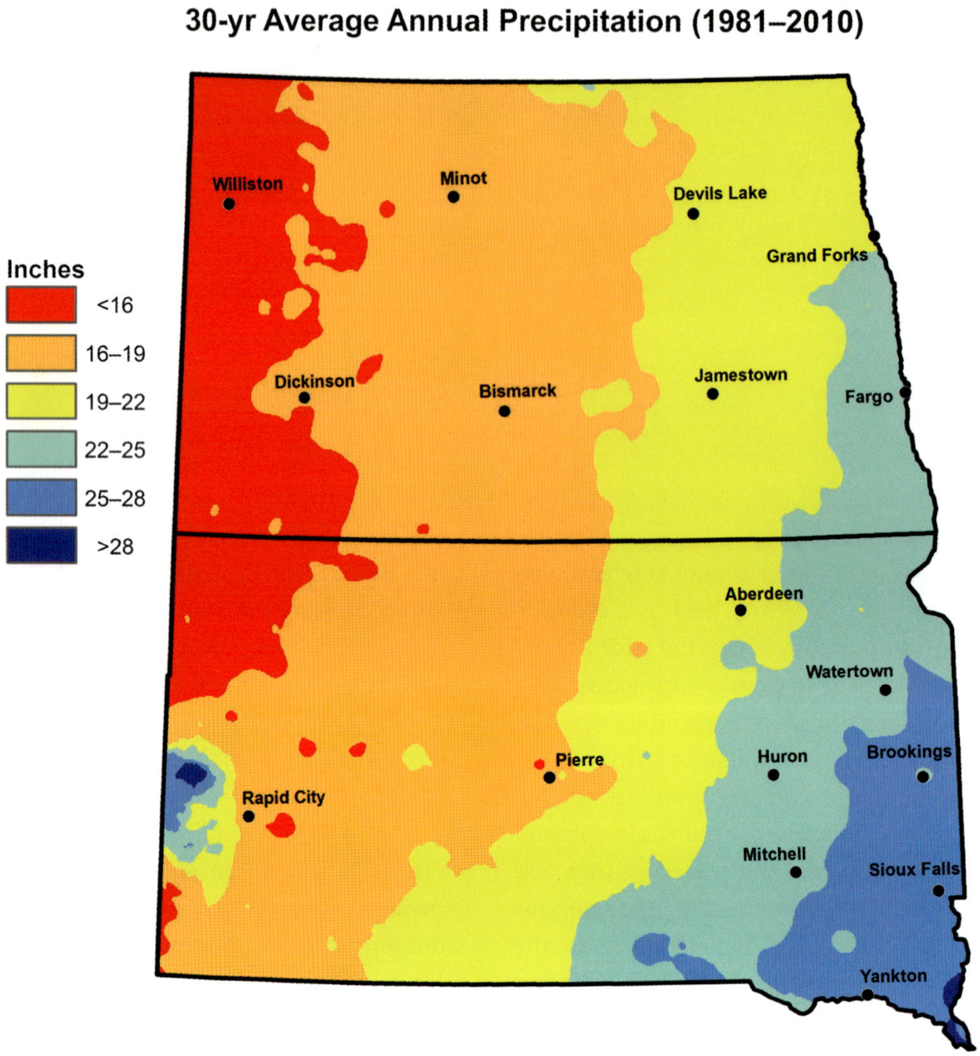

Fig. 3.1. Average annual precipitation in North and South Dakota for the period 1981–2010. The Black Hills in western South Dakota receives more precipitation than the surrounding prairie. Source: PRISM Climate Group data, Oregon State University (http://prism.oregonstate.edu). Cartography by Christopher Nicholson.

the preceding day at noon in the shade, had sunk to 43° at sunrise." In the winter, blizzards are common in the Dakotas. On December 20, 1826, Alexander Ross was near Pembina and described a storm that "lasted for several days, drove the buffalo beyond the hunters reach, and killed most of their horses." Hunters were so scattered that "they could render each other no assistance . . . thirty-three lives were lost."[3] Storms would kill large numbers of cattle, such as during the winter of 1886–1887. People still die during blizzards, usually when becoming lost after abandoning a vehicle stuck in the snow.

Until about 30 years ago, most observers thought that climate-human interactions were essentially one-way, that is, climate affected the lives of people but people did not affect the climate. However, abundant research indicates that industrialization has altered the climate significantly because of large increases in greenhouse

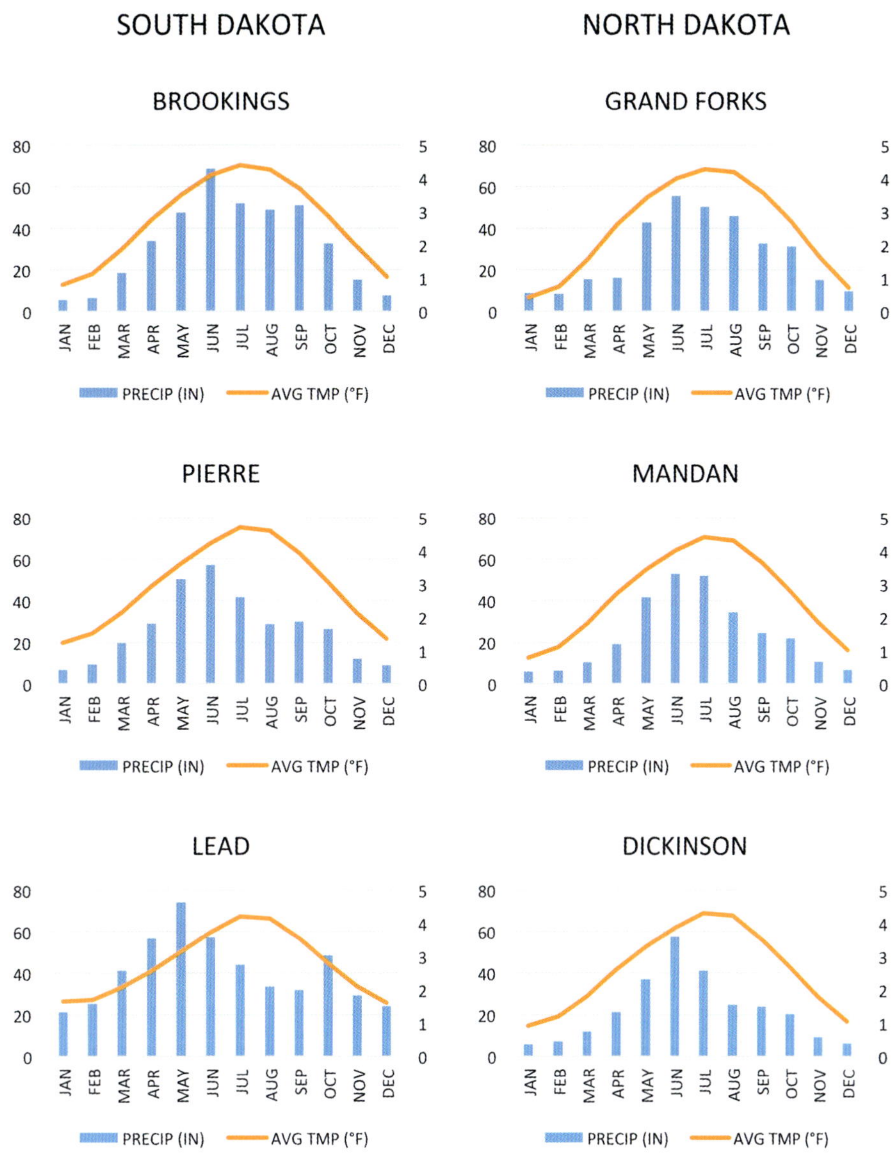

Fig. 3.2. Climographs that illustrate seasonal temperature and precipitation changes during the year at six weather stations, arranged from east to west across North and South Dakota. The lines are for mean monthly temperature; the bars are mean monthly precipitation from 1981 to 2010. Source: PRISM Climate Group data, Oregon State University (http://prism.oregonstate.edu).

gases—mostly carbon dioxide but also methane, nitrous oxides, and fluorinate gases—through the burning of fossil fuels, and because of a reduction in the capacity of Earth's ecosystems to withdraw carbon dioxide from the atmosphere and store it in plant biomass and soil organic matter.[4] After extensive peer-reviewed analysis, the U.S. Global Change Research Program reached the following conclusions in 2017:

- The carbon dioxide concentration of the atmosphere has increased by 43 percent during the past 200 years, since the start of the Industrial Revolution, from 280 to over 400 ppm.[5]

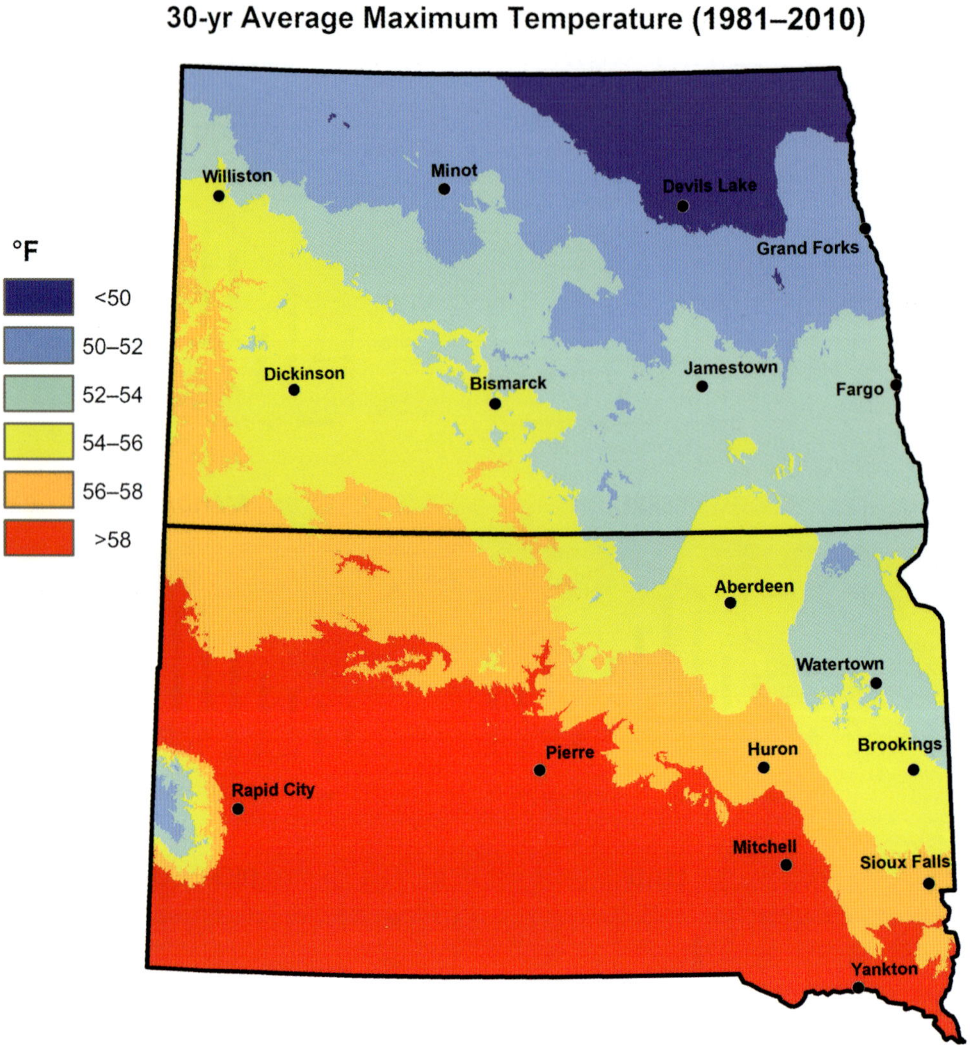

Fig. 3.3. Average maximum daily temperature during a year in North Dakota and South Dakota for the 30-year period 1981–2010. Mean annual temperature has essentially the same geographic pattern. Source: PRISM Climate Group data, Oregon State University (http://prism.oregonstate.edu). Cartography by Christopher Nicholson.

- Ocean acidification is occurring.
- Global average annual surface air temperature has increased by about 1.8° F over the past 115 years (1901–2016). Sixteen of the warmest years on record for the globe occurred in the last 17 years (1998 was the exception).
- Human activities, especially emissions of greenhouse gases, are the dominant cause of the observed warming since the mid-twentieth century. There is no convincing alternative explanation supported by scientific observations.
- Global average sea level has risen by about seven to eight inches since 1900, with almost half of that rise occurring since 1993, caused by thermal expansion and meltwater from land-based glaciers. As a result, coastal flooding and saltwater intrusion into freshwater supplies occur more often.

- Atmospheric water vapor is higher and heavy rainfall is increasing in intensity and frequency.
- Mountain glaciers have been retreating everywhere, sea ice at the poles is at record low levels, and mountain snow often disappears earlier in the spring.
- Heat waves and forest fires are becoming more frequent.

Detecting regional trends is difficult but mean annual temperature has increased slightly throughout the Dakotas since 1895—perhaps more in the west than the east (fig. 3.4). In contrast, long-term precipitation trends have been variable, declining, or remaining the same in the west while the eastern Dakotas have become slightly wetter (fig. 3.5). The Palmer Drought Severity Index (PDSI), calculated by combining data on precipitation and temperature, indicates that droughts have been increasing in frequency and severity in the western Dakotas, including the Black Hills, but not in the central and eastern Dakotas (fig. 3.6).

Notably, summer rain in a single day is intensifying in the Dakotas (fig. 3.7). To illustrate, only seven rains of four inches or more were recorded during the first half of the period of record (1905–1960), but 16 events of this magnitude have occurred since that time, including the official record for both states—8.74 inches in north-central South Dakota. Flooding and erosion are more likely and planting and harvesting more difficult with more intense rainfall.[6] In general, climate change leads to changes in the seasonal timing of events such as flowering and wildlife migration. Conservation biologists are concerned that, unlike in the past, climate change is occurring more rapidly than many native species can evolve adaptations for survival or disperse to more favorable environments.

Anticipating the Future

The current rate of climate warming worldwide seems slow, but small changes can lead to significant, measurable effects. Satellite images have documented a steady decline in sea ice in Antarctica and the Arctic, and photos have shown how mountain glaciers are retreating everywhere on Earth. With even slightly warmer temperatures, the atmosphere holds more moisture, thereby providing water and energy for the more frequent extreme weather events illustrated in figure 3.7. Warming is not uniform everywhere on Earth, hence climatologists often write about climate change rather than climate or global warming. They also emphasize that short-term changes in the weather should not be confused with climate change. Extreme cold events or high rainfall can occur during periods of warming or drought.

Complex mathematical models and advanced computers have been used to project how climate is likely to change, an approach that has been used with confidence for space exploration and national defense.[7] Only in this way can pertinent geophysical, atmospheric, and ecological data be combined quantitatively. The most comprehensive climate modeling was started in 1988 for the United Nations Intergovernmental Panel on Climate Change (IPCC). Led by the world's best climate scientists, the first report was published in 1990, and subsequent updates have appeared as new information became available, the most recent in 2019. The models are routinely tested, peer-reviewed, and improved at every opportunity. Numerous scientific organizations have endorsed the IPCC reports, including the National Academy of Sciences, the American Association for the Advancement of Sciences, the American Meteorological Society, and the Geological Society of America.[8]

Forecasting the future is challenging and requires making assumptions. Therefore, several plausible scenarios are produced, enabling users of the models to have a basis for describing a range of future climates. Reviewing all IPCC scenarios is beyond the scope of this chapter but, assuming no significant change in the rate of fossil fuel

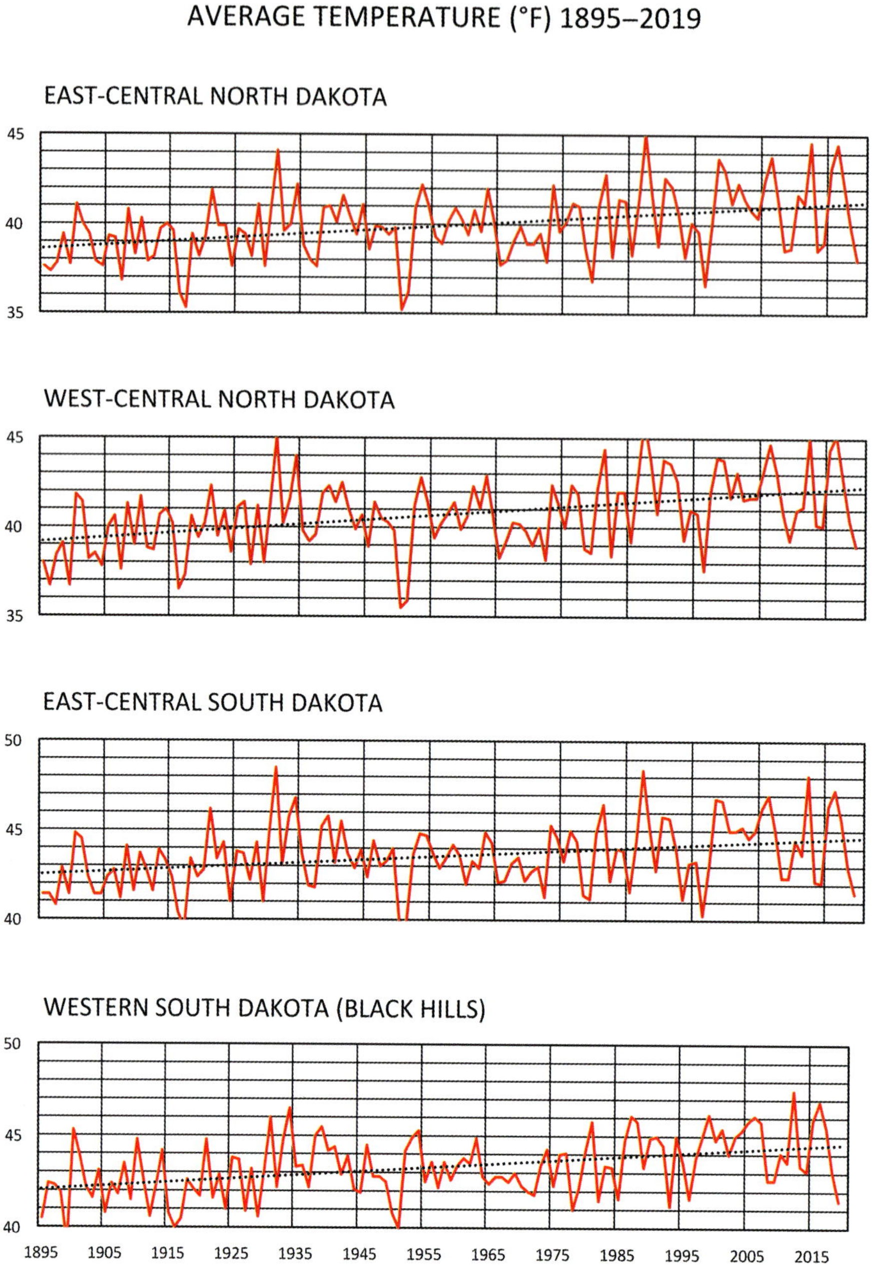

Fig. 3.4. Average annual temperature for each year from 1895 to 2017 for each of four climate regions in North and South Dakota. Each point on the graph is an average of several weather stations. Note the year-to-year fluctuations and gradual increase for each region. The temperature increases about 0.3° and 0.2° F per decade in the west and the east, respectively. The trend line is based on regression analysis. The data are from the U.S. National Oceanic and Atmospheric Administration (NOAA). West-central and east-central North Dakota are NOAA's North Dakota climate divisions 4 and 6, respectively, and western South Dakota (Black Hills) and east-central South Dakota are NOAA's climate divisions 4 and 7, respectively. For more detail, see https://www.ncdc.noaa.gov/cag/.

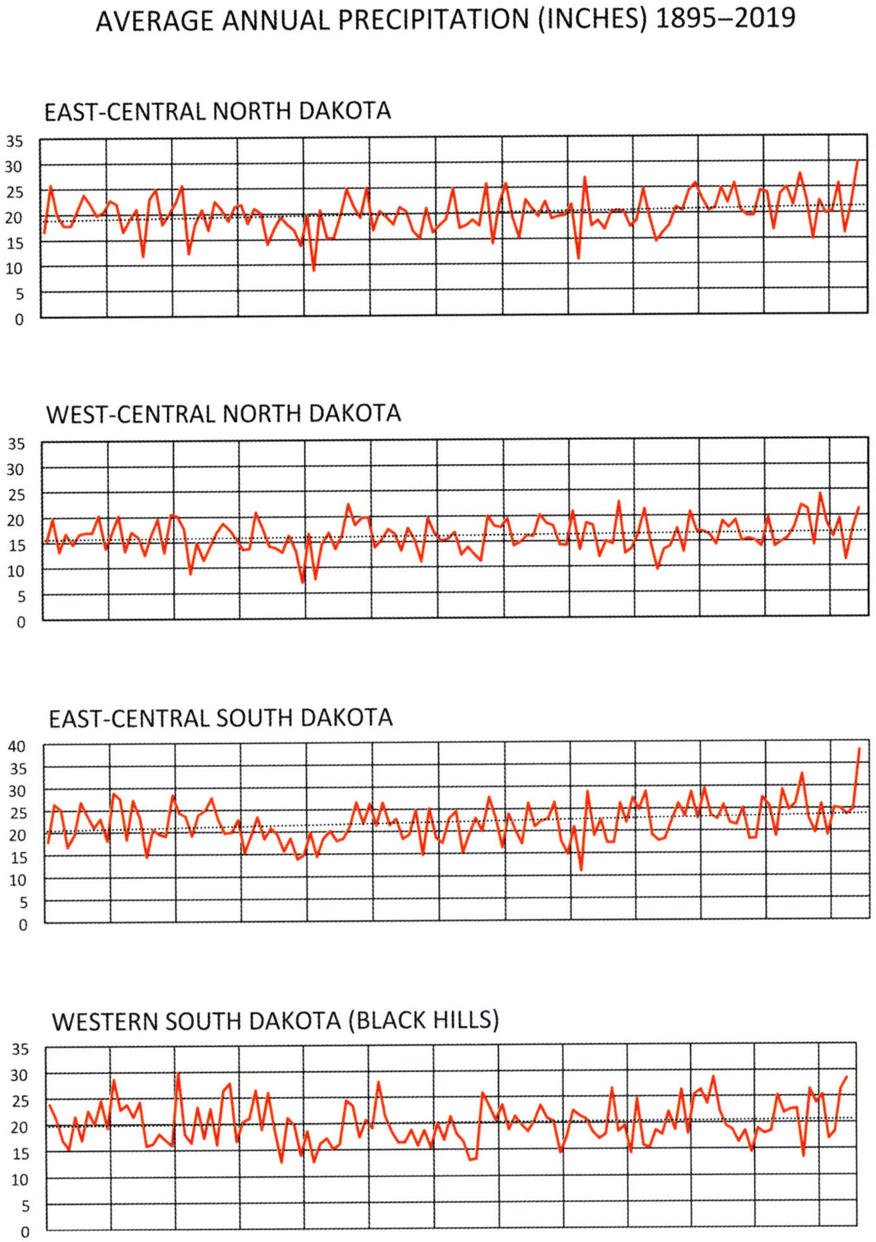

Fig. 3.5. Average annual precipitation for each year from 1895 to 2017 for each of four climate regions in North and South Dakota. The data are averages of several weather stations located in each region. The trend line is based on regression analysis. Note the year-to-year fluctuations and the subtle trend upward in the east. Not shown is 2019, which was the wettest year on record in both states. Data are from NOAA, and climate divisions are the same as for fig. 3.4. For more detail, see https://www.ncdc.noaa.gov/cag/.

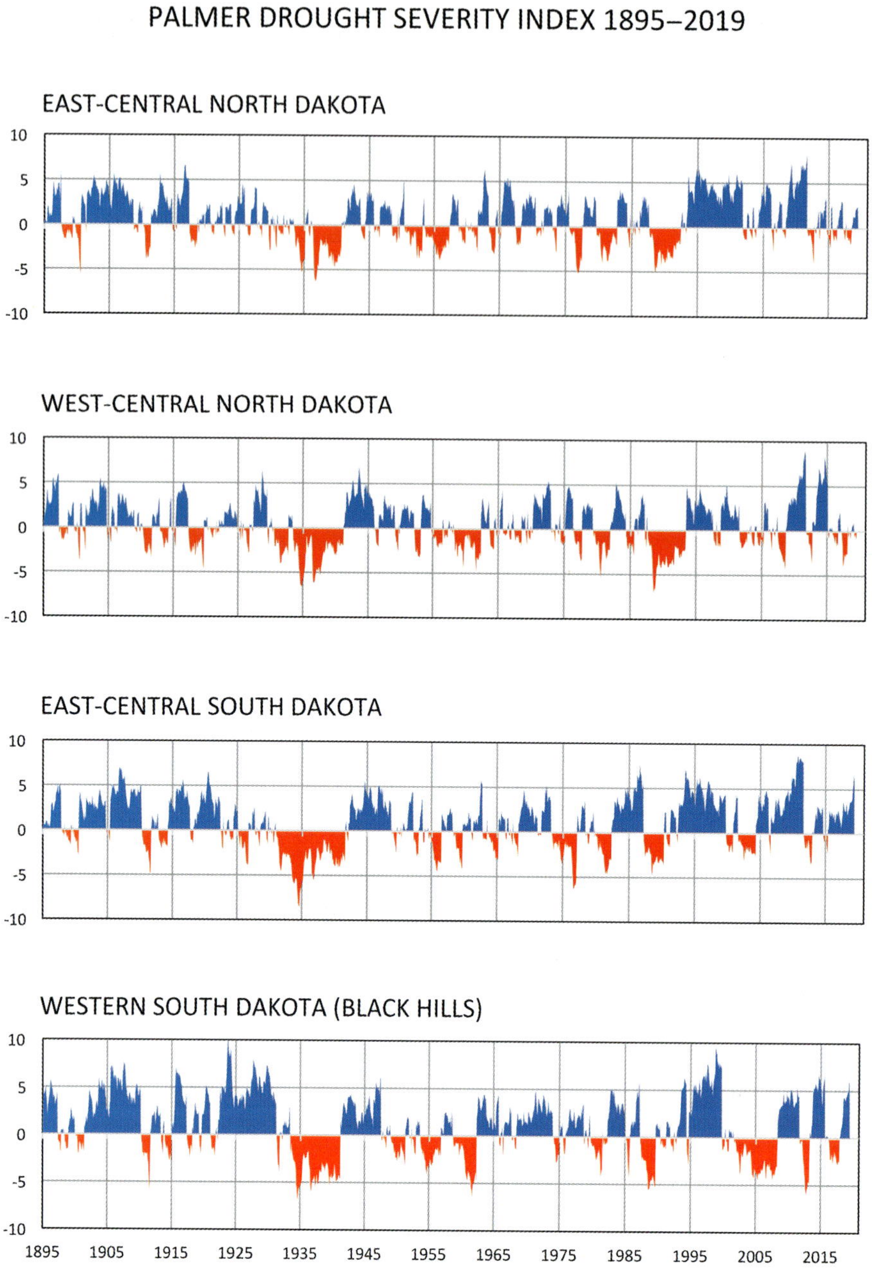

Fig. 3.6. Relatively wet (blue) and dry (red) periods from 1895 to 2017 for each of four climate regions in North and South Dakota, as estimated using the Palmer Drought Severity Index (PDSI). Calculated using mean annual temperature and precipitation data, the PDSI values are based on averages from several weather stations located in each region. The temperature and precipitation data used for calculating the PDSI are from NOAA. Climate divisions are the same as for fig. 3.4. For more detail, see https://www.ncdc.noaa.gov/cag/.

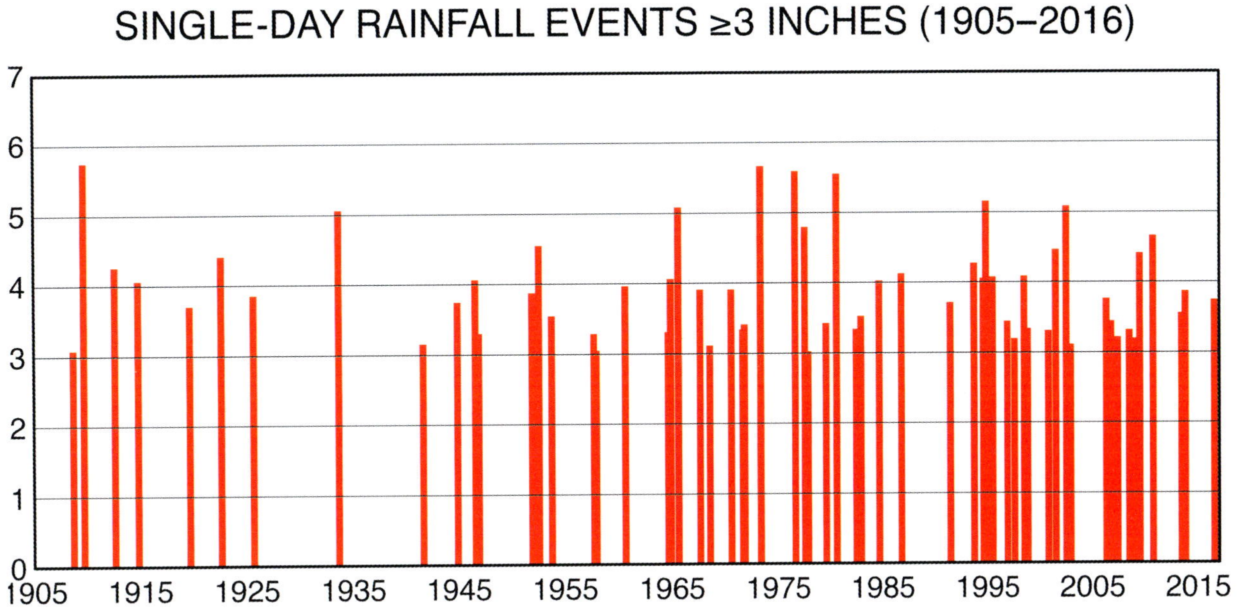

Fig. 3.7. Note the gradual increase in the number of high-intensity rainfall events from 1905 to 2016 at six weather stations: Grand Forks, Mandan, and Dickinson, North Dakota; and Brookings, Pierre, and Lead, South Dakota. Each line represents one event. The record rainfall during a 24-hour period in South Dakota is 8.74 inches (Groton, May 6, 2007); in North Dakota, 8.1 inches (Litchfield, June 29, 1975). An unofficial 14-inch rain was reported near the Black Hills. The data are from NOAA. For more detail, see https://www.ncdc.noaa.gov/cag/.

consumption (the IPCC A2 scenario), the following climate changes for a region encompassing North and South Dakota have been projected:[9]

- Warming of the mean annual temperature by as much as 9° F by 2085. Seasonal temperature increases will be greatest in the summer, possibly leading to more heat waves and water scarcity. Evapotranspiration will become higher as the climate warms, drying the soil more quickly, slowing groundwater recharge, and hastening the decline in groundwater availability. The frost-free season across the Great Plains will become 20 to 30 days longer by the middle of the century.
- Declines in precipitation will occur in the summer, a time of highest water demand, though an increase in annual precipitation is projected. Heavy precipitation events could increase in South Dakota by 30 percent, a state where heavy rains and costly floods already have increased in the past two decades. Extreme rainfall events will delay spring planting, increase soil erosion, and reduce crop production in some areas, though they may possibly increase production in traditionally dry areas.
- Warmer winters are likely to reduce the snowfall and spring runoff that recharges rivers, lakes, and wetlands. Alpine ice and snow in the Rocky Mountains will decline, an important source of water for rivers on the Great Plains. Earlier spring runoff from the mountains is occurring now, causing greater dependence on ground water and spring and summer rainfall.
- More extreme weather in the form of droughts and deluges is expected, often in back-to-back years. Extreme weather events such as floods, tornadoes, and hailstorms will become more frequent and more intense.

Climatologists have been warning that climate change is accelerating. A 2007 analysis of climate trends by paleoecologists John W. Williams and Stephen T. Jackson concluded that the characteristics of the rapidly emerging climates will be novel. How will Dakota ecosystems change? How can municipalities, farmers, ranchers, wildlife biologists, and others adapt? Jackson discusses some of the options.[10] Preparing for the effects of climate change seems urgent—a recurring topic in the chapters that follow.

Chapter 4 Grasslands

Grasslands, often referred to as prairies, began to develop in central North America as the climate became drier about 5 million to 10 million years ago. Trees slowly succumbed over large areas to increasing aridity and more frequent fires, which enabled the spread of herbaceous plants. The number of large herbivores adapted for eating grasses increased and included camels, horses, mammoths, and giant ground sloths, in addition to the ancestors of bison and pronghorn. Prairie birds became more common, including the ancestors of meadowlarks, bobolinks, lark buntings, prairie chickens, upland sandpipers, grasshopper sparrows, predatory raptors, scavenging vultures, and others. And there were prairie dogs, ground squirrels, various kinds of small reptiles, and dozens of species of ants, grasshoppers, beetles, and other insects—all dependent on the astounding abundance of microbial life in the soil.

Prairie plants are as diverse and well adapted to their environment as the animals are. Numerous species of grasses and sedges produce new leaves within a few days after being grazed. Some grow most rapidly in the spring, others in the summer. Wilting during dry periods is minimal because of rigid cell walls. Wind-pollinated grasses account for most of the grassland biomass, but over half of the species are forbs, which are plants with wider leaves than grasses and that have conspicuous, insect-pollinated flowers—blazing star, purple coneflower, black-eyed Susan, sunflower, prairie turnip, western prairie fringed orchid, and many others (fig. 4.1). Where these plants grow, their nectar sustains hundreds of different kinds of pollinators, and they create a scene comparable to colorful meadows. Indeed, French explorers in the 1700s first used the word *prairie*, which means "meadow," to describe this landscape.

Grasslands and other kinds of vegetation were obliterated as the glaciers advanced. However, many of the species survived just beyond the ice sheets and new grasslands developed during the next interglacial period. Since the last retreat, about 10,000 years ago, the prairie ecosystem has formed fertile soils that are extremely important for the region's economy today. Most of the eastern prairies have been plowed for agriculture, except on hills or where stones and boulders are abundant. Cultivation expanded into some western grasslands that were drier. Native prairies, where they are still found, are valued for livestock grazing, wildlife habitat, and watershed protection.

The vast prairies would have been both fascinating and perplexing to the first Europeans, who were more accustomed to forests. The author and naturalist John Madson wrote that "the newcomers

Prairie sunflower

Gayfeather

Purple coneflower

Fig. 4.1. Numerous forbs with colorful, insect-pollinated flowers are common in prairies, including prairie sunflower (photo by Justin Meissen), purple coneflower (photo by Neil Shook), and gayfeather (photo by Justin Meissen). Common grasses in the photos are big bluestem, little bluestem, and sideoats grama. See table 4.1 for the names of other prairie plants.

brought no real knowledge of grasslands. . . . They had no basis for even imagining wild fields through which a horseman might ride westward for a month or more, sometimes traveling for days without sight of trees. . . . Most arrivals ventured timidly into the edges of the grass and clung to the outriders of forest like mice hugging a wall."[1]

While he was in southeastern South Dakota, Meriwether Lewis commented in 1804 that "the country in every direction around us was one vast plain in which innumerable herds of Buffalo were seen attended by their shepherds the wolves."[2] In 1839, Judge James Hall wrote that "the scenery of the prairie is striking and never fails to cause an exclamation of surprise. . . . The verdure and the flowers are beautiful . . . a gaiety which animates the beholder."[3]

For John C. Van Tramp in 1860, the unbroken prairies inspired "feelings so unique, so distinct from anything else, so powerful, yet vague and indefinite, as to defy description."[4] Theodore Roosevelt commented in 1884, "Nowhere, not even at sea, does a man feel more lonely than when riding over the far-reaching, seemingly never-ending plains."[5] In 1821 Major Stephen Long labeled the region west of Bismarck and Pierre as the Great American Desert, but Lt. Col. George A. Custer wrote in 1874 that the western Dakotas presented "an almost unbroken sea of green, luxuriant, wavering grass, from six inches to one foot in height."[6]

The Effects of Climate, Topography, and Soil

There are strong climatic gradients across the Dakotas, with tallgrass prairie in the eastern quarter of the two states where the annual precipitation is relatively high, 20 to 28 inches per year (figs. 4.2, 4.3, and 4.4). More rainfall and more fertile soils enable more plant growth. Mixed-grass prairies to the west have a drier climate, receiving about 14 to 20 inches per year, and consequently the plants are shorter. Shortgrass prairies, found only in a small area in southwestern South Dakota, are even drier.[7] The term *mixed* implies that short, medium-height, and tall species often grow together.

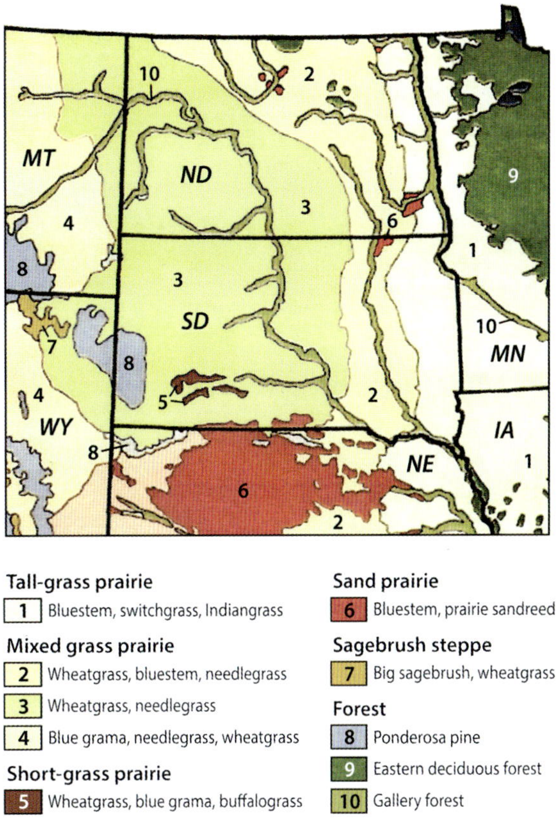

Fig. 4.2. Map of major kinds of grassland in the Dakotas with the names of grasses common in each one. Gallery forests and woodlands occur along creeks and rivers (see fig. 1.3). Adapted from Küchler (1964) and Grimm (2001).

Fig. 4.3. Oakville Prairie, west of Grand Forks, is a rare remnant of once-widespread tallgrass prairie on the Red River glacial lake plain. It is maintained by the University of North Dakota. There is considerable variation in the kinds of native and introduced plants found in the prairies of this region (e.g., Hadley and Buccos 1967; Hadley 1970; Redman 1972; Facey et al. 1986). The invading tree is Russian olive.

Fig. 4.4. Mixed-grass prairie with bison on the slopes of the Cheyenne River valley in western South Dakota. Cattle have replaced bison as the largest herbivore in most grasslands, although they differ somewhat in their grazing behavior. Even with large numbers of bison or introduced livestock, most of the herbivory is belowground, where most of the plant biomass is located (fig. 4.6). Photo by Jill O'Brien.

Within each kind of grassland there is considerable variation in terms of height, the amount of biomass produced, and the abundance of different plant species (table 4.1). Characteristic tallgrass species are big bluestem, little bluestem, Indiangrass, switchgrass, and porcupine grass. Mixed-grass prairie usually has western wheatgrass, green needlegrass, blue grama, sideoats grama, needle-and-thread grass, junegrass, little bluestem, and threadleaf sedge. Shortgrass prairie, where it occurs, is dominated by blue grama, buffalograss, and western wheatgrass. Each kind of grassland can be subdivided into numerous categories. For example, a shrub known as big sagebrush grows in the far-western Dakotas at higher elevations, forming a distinctive variety of mixed-grass prairie that verges on being a shrubland sometmes referred to as sagebrush steppe. Also, the presence of plains rough fescue in northern North Dakota and southern Canada identifies *fescue mixed-grass prairie*.[8]

Topographic and soil characteristics vary greatly across the prairie landscape and influence the kinds of plants that are most common. Hilltops and south-facing slopes in the tallgrass prairie region are relatively dry because they receive more direct sunlight than north slopes do, and therefore mixed-grass prairie species are common there. Plants near the bottom of hillslopes benefit from deeper, often wetter soils and are taller (fig. 4.5). These *low prairies* include wet-meadow species such as prairie cordgrass (see chapter 5). Other species are common where much of the soil is composed of sand, such as prairie sandreed,

Table 4.1. Some characteristic plants of three kinds of grassland common in North Dakota and South Dakota[a]

Common name[b]	Latin name	Tallgrass prairie	Mixed-grass prairie	Sand prairie[c]
GRASSES & SEDGES				
Barley, foxtail[H]	*Hordeum jubatum*	X	X	–
Bluegrass, Kentucky[I]	*Poa pratensis*	X	X	X
Bluegrass, Sandberg	*Poa secunda*	–	X	–
Bluestem, big	*Andropogon gerardi*	X	X	–
Bluestem, sand	*Andropogon hallii*	–	–	X
Bluestem, little	*Schizachyrium scoparium*	X	X	X
Brome, smooth[I]	*Bromus inermis*	X	X	–
Buffalograss	*Bouteloua (Buchlöe) dactyloides*	–	X	–
Cordgrass, alkali[H]	*Spartina gracilis*	X	X	–
Cordgrass, prairie[W]	*Spartina pectinata*	X	X	–
Dropseed, prairie	*Sporobolus heterolepis*	X	–	–
Dropseed, sand	*Sporobolus cryptandrus*	–	X	X
Dropseed, tall	*Sporobolus compositus*	X	X	–
Fescue, plains rough[ND]	*Festuca hallii*	–	R	–
Grama, sideoats	*Bouteloua curtipendula*	X	X	–
Grama, blue	*Bouteloua gracilis*	X	X	–
Indiangrass	*Sorghastrum nutans*	X	X	X
Junegrass	*Koeleria macrantha*	X	X	X
Muhly, plains	*Muhlenbergia cuspidata*	X	X	–
Needle-and-thread grass	*Hesperostipa comata*	X	X	X
Needlegrass, green	*Nassella viridula*	X	X	–
Needlegrass, porcupine	*Hesperostipa spartea*	X	X	–
Ricegrass, Indian	*Achnatherum hymenoides*	–	X	X
Sacaton, alkali[H]	*Sporobolus airoides*	–	X	–
Saltgrass[H]	*Distichlis spicata*	X	X	–
Sandreed, prairie	*Calamovilfa longifolia*	–	X	X
Sedge, threadleaf	*Carex filifolia*	–	X	–
Sedge, needleleaf	*Carex duriuscula*	–	X	–
Switchgrass	*Panicum virgatum*	X	X	–
Three awn, purple	*Aristida purpurea*	–	X	–
Wheatgrass, slender	*Elymus trachycaulis*	X	–	–
Wheatgrass, thickspike	*Elymus lanceolatus*	–	X	–
Wheatgrass, western	*Pascopyrum smithii*	X	X	–

(continued)

Table 4.1. (*continued*)

Common name[b]	Latin name	Tallgrass prairie	Mixed-grass prairie	Sand prairie[c]
FORBS				
Anemone, meadow	*Anemone canadensis*	x	–	–
Aster, Geyer's	*Symphyotrichum leave*	x	X	X
Aster, white heath	*Symphyotrichum ericoides*	x	X	–
Beardtongue, slender	*Penstemon gracilis*	–	X	–
Black-eyed Susan	*Rudbeckia hirta*	x	X	X
Blanketflower	*Gaillardia aristata*	–	X	–
Blazingstar	*Liatris aspera*	x	–	–
Cactus, pricklypear	*Opuntia polycantha*	–	X	X
Cinquefoil, prairie	*Potentilla pennsylvanica*	x	–	–
Clover, purple prairie	*Dalea purpurea*	x	X	–
Compass plant[SD]	*Silphium laciniatum*	x	–	–
Coneflower, prairie	*Ratibida columnifera*	x	X	–
Death camas	*Zigadenus elegans*	–	X	–
Echinacea (purple coneflower)	*Echinacea angustifolia*	x	X	–
Gaura, scarlet	*Gaura coccinea*	–	X	–
Gayfeather, dotted	*Liatris punctata*	–	X	–
Globemallow, scarlet	*Sphaeralcea coccinea*	–	X	–
Goldenrod	*Solidago missouriensis*	x	X	X
Groundplum	*Astragalus crassicarpus*	–	X	–
Larkspur, prairie	*Delphinium carolinianum*	x	–	–
Lily, prairie	*Lilium philadelphicum*	x	–	–
Licorice, American	*Glycyrrhiza lepidota*	x	X	–
Locoweed, purple	*Oxytropus lambertii*	–	X	–
Milkvetch, plains	*Astragalus gilviflorus*	–	X	–
Milkweed	*Asclepias syriaca*	X	X	–
Onion, wild	*Allium textile*	x	X	–
Orchid, prairie fringed	*Platanthera praeclara*	x	–	–
Paintbrush, downy	*Castilleja sessiliflora*	–	X	–
Phlox, downy	*Phlox pilosa*	x	–	–
Phlox, Hood's	*Phlox hoodii*	–	X	–
Prairie smoke	*Geum triflorum*	x	–	–

Table 4.1. (*continued*)

Common name[b]	Latin name	Tallgrass prairie	Mixed-grass prairie	Sand prairie[c]
Pasque flower	*Pulsatilla patens*	x	X	–
Sagewort, cudweed	*Artemisia ludoviciana*	x	X	X
Scurfpea	*Pediumelum argpphyllum*	x	X	–
Spurge, leafy[I]	*Euphorbia esula*	x	X	X
Sunflower, Nuttall's	*Helianthus nuttallii*	x	X	X
Sweetclover, yellow[I]	*Melilotus officinalis*	x	X	–
Tansyaster, hoary	*Machaeranthera canescens*	–	X	–
Thistle, Canada[I]	*Cirsium arvense*	x	X	X
Turnip, prairie (Indian breadroot)	*Pediomelum esculentum*	–	X	–
Vetch, American	*Vicia americana*	x	X	–
Wormwood	*Artemisia absinthium*	X	X	–
Yarrow	*Achillea millifolium*	–	X	–
Yucca	*Yucca glauca*	–	X	–
SHRUBS				
Buckthorn[I]	*Rhamnus cathartica*	x	–	X
Buffaloberry, silver	*Shepherdia argentea*	–	X	–
Ivy, poison	*Toxicodendron rydbergii*	x	X	X
Leadplant	*Amorpha canescens*	x	X	X
Rabbitbrush, rubber	*Ericameria nauseosa*	–	X	–
Rose, prairie	*Rosa arkansana*	x	X	X
Sagebrush, silver[W]	*Artemisia cana*	–	X	–
Sagewort, fringed	*Artemisia frigida*	X	X	–
Snakeweed, broom	*Gutierrezia sarothrae*	–	X	–
Snowberry, western	*Symphoricarpos occidentalis*	x	X	–
Sumac, skunkbush	*Rhus trilobata*	–	X	–

[a]Some species occur in more than one kind of grassland. Abundance is indicated as follows: X = common, x = occasional, and R = rare; a dash indicates that the plant is absent or not common.

[b]Superscripts provide additional information for some species: H = halophytes, I = introduced, W = wet meadow, ND = North Dakota only, SD = South Dakota only. See Johnson and Larson (2016). Marriott and Faber-Langendoen (2000b) describe grasslands in the vicinity of the Black Hills.

[c]See also table 12.2. Sand prairies are found primarily on sandhills and sandplains in eastern North Dakota and south-central South Dakota (figs. 1.2 and 1.3).

Fig. 4.5. A mosaic of mixed-grass and tallgrass prairie exists on hilly terrain of the Missouri Coteau, such as here north of Bismarck. Hilltops have mixed-grass prairie plants (high prairie) and the lower slopes have tallgrass species (low prairie). Wet meadows occur adjacent to wetlands. Parts of the Prairie Coteau are similar. Photo by Justin Meissen.

sand bluestem, sand dropseed, and Indian ricegrass. Saline soils, which develop in depressions where salts are deposited as surface water evaporates, are prime habitat for salt-tolerant plants known as halophytes, such as saltgrass, alkali grass, foxtail barley, seablite, and salicornia.[9] Additional water in depressions normally enhances plant growth, but high soil salinity has the opposite effect. Because of this variation from place to place, the vast prairies first seen by early explorers are best thought of as a mosaic of different plant communities.

Compared to tallgrass prairie, a smaller portion of mixed-grass prairie has been cultivated. Still, farms of wheat, corn, soybeans, sunflowers, and planted hayfields are planted wherever the soil is favorable and precipitation or irrigation water is adequate, especially on the relatively flat land on the glaciated north and east sides of the Missouri River (see fig. 1.4). Mixed-grass prairies are used for livestock production where water or soil characteristics limit the potential for row-crop production, such as on portions of the Prairie Coteau, Missouri Coteau, and most of the Missouri Plateau (see fig. 1.2).

Most Grassland Herbivory Occurs in the Soil

Extensive roots are required to obtain sufficient water to maintain prairie plants during the entire year. Consequently, only about 20 percent of prairie plant biomass is visible aboveground (fig. 4.6). Most carbohydrates produced by photosynthesis are transported belowground through the stems, where they provide the energy needed to produce roots or are stored until needed for the growth of new leaves and stems in the spring. The energy of stored carbohydrates is also used for replacing plant parts that are consumed by herbivores or burned by fires. With most plant biomass in the soil, it is not surprising that most herbivory occurs there also. Ants, mites, nematodes, insect larvae, and other invertebrates consume more than aboveground mammals and other herbivores do—even where aboveground grazing is heavy. This is possible because every cubic yard of soil has miles of plant roots and fungal filaments (hyphae) that sustain millions of bacteria and thousands of invertebrates, each group represented by numerous species (table 4.2). Black, fertile soils known as *mollisols* are maintained as aging roots, rhizomes, fungi, mycorrhizae, bacteria, and belowground animals die and become part of the soil's organic matter, which is eventually mineralized into the inorganic nutrients that plants need for more growth and photosynthesis.

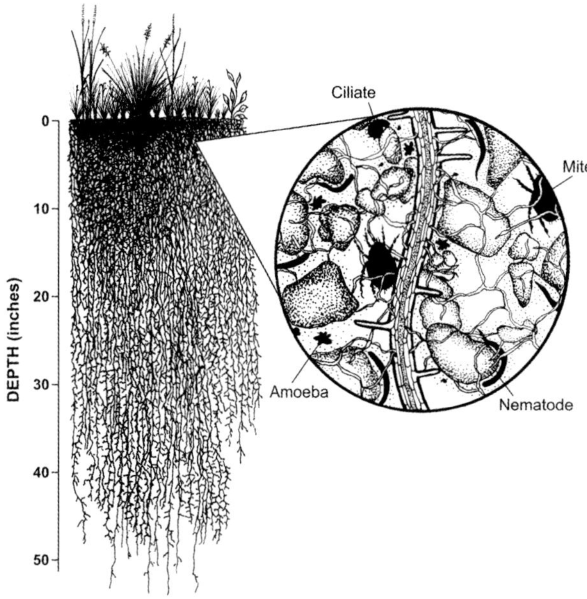

Fig. 4.6. As this diagram illustrates, most of the grassland biomass is in plant roots that extend downward 50 inches or more. The enlargement shows a single root with root hairs and the filaments (hyphae) of mycorrhizal fungi. A film of water (stippled area) on each soil particle provides habitat for nematodes and numerous protozoans, including amoebae and ciliates. Mites, insect larvae, and other invertebrates are found in air spaces (table 4.2). Bacteria are abundant but too small to include at this scale. Magnification about 15 times. Source: Knight et al. 2014, with permission from Yale University Press.

Table 4.2. Examples of organisms of different sizes found in the soils of grasslands and other ecosystems

Microflora (microscopic, <100 micrometers): Bacteria, fungi
Microfauna (microscopic, <100 micrometers): Nematodes, protozoa, rotifers
Mesofauna (100 micrometers–2 mm): Mites (Acari), springtails (Collembola), two-pronged bristletails (Diplura), garden centipedes (Symphyla), pot worms (Enchytraeae), termites and ants (Isoptera/Formicidae), fly larvae (Diptera)
Macrofauna (>2 mm): Isopods (Isopoda), millipeds and centipedes (Myriapoda), spiders (Arachnida), beetle larvae (Coleoptera), earthworms (Oligocaeta)
Megafauna (soil vertebrates): prairie dogs, pocket gophers, ground squirrels, badgers, snakes, toads, burrowing owls

Source: Wall 2012 (chapters by Lavelle and Wurst et al.)

Box 4.1. Earthworms

Long before the concept of soil health was introduced, children and their parents knew that earthworms were good for a garden. The worms created friable, aerated soil that absorbed water like a sponge. Also, numbering hundreds of thousands per acre, they mix a large portion of the soil with beneficial bacteria, fungi, and nutrients—which improves fertility. Like prairie dogs, the worms have been labeled *ecosystem engineers* because they improve the environment for numerous other organisms—protozoans, nematodes, arthropods such as ants and beetles, and other invertebrates. To the surprise of many, however, the earthworms that most people see were introduced to North America from Europe, most likely in potted plants. About 45 exotic species are now found here, including the common red worms and nightcrawlers used by anglers and biology teachers. Native earthworms are less conspicuous.[a]

In general, all native and introduced species of earthworms have been classified into three categories according to where they obtain most of their food. Surface feeders eat decomposing litter or mulch. Consequently, they must be adapted to survive high temperatures and periods of soil desiccation. Well below the surface are species that consume organic matter already incorporated into the soil. They burrow through the soil by eating it, depositing further decomposed organic matter (castings) behind them. Perhaps best known is a third group that feeds on the surface but burrows down into the soil three feet or more. This group includes the introduced nightcrawler (*Lumbricus terrestris*) and red worm (*L. rubellus*), both of which are very abundant and widespread in cultivated soil. Larger than the native species, they pull fragments of dead leaves into the soil, creating a food cache that soon becomes part of the soil organic matter. They are commonly seen on the soil surface after rainstorms, especially at night. Conventional tillage of cropland, the application of pesticides, and frequent mowing or burning can cause declines in most soil invertebrates.

[a]Henshue et al. 2018; Hendrix 2006.

Belowground plant parts also are consumed by burrowing mammals. The best-known examples are prairie dogs, pocket gophers, ground squirrels, and voles. Their burrows provide them access to food, but they also are shelter from summer heat, winter cold, and aboveground predators. Of course, the soil is no barrier for some predators, such as badgers and black-footed ferrets, but the burrowing lifestyle has proved relatively safe for many animals. Their waste products fertilize the soil. Also, their digging mixes and aerates the soil, thereby contributing to its fertility and water-holding capacity. Soils made friable in this way absorb more water, which enables more plant growth (box 4.1). Such biological activity has had a strong influence on soil development and how soils are classified, but just as important are the effects of climate and whether the underlying substrate was laid down by glaciers or originated as sediments in ancient seas.

Plant Adaptations for Drought, Fire, and Grazing

Many grassland plants have rigid stems and leaves that do not readily wilt during inevitable dry periods. By not wilting, the leaves remain upright, poised to resume photosynthesis when rain replenishes soil moisture. Such plants also use water efficiently; that is, they produce a relatively large amount of plant biomass for every ounce of water absorbed by their roots. Some, known as cool-season plants, begin their growth early in the spring when water is less likely to be limiting, thereby minimizing water stress. Others, the warm-season species, begin growing later, are more drought tolerant, and use their water more efficiently.[10] The growing seasons for both groups of plants begin at different times, but growth stops when water becomes limiting—sometimes before mid-August in the western Dakotas, when the stems and leaves lose their chlorophyll and turn tan or light brown. Some of the dead shoots remain vertical until pressed to the soil surface by wind, heavy rain, and snow, which forms a natural mulch that is typical of grasslands after a period without grazing or fire.

Standing dead biomass and mulch are easily ignited flash fuels. Prairie fires were frightening events for early settlers, especially in tallgrass prairie, where the loud, crackling flames moved rapidly across the landscape and were difficult to put out (fig. 4.7). One teenager wrote in 1886, "There is a prairie fire in sight every night almost and sometimes five or six."[11] Fires might burn for weeks until reaching a lake, river, or heavily grazed area, or until a substantial thunderstorm occurred. Nearly every acre of tallgrass prairie burned during a three- to five-year period, sometimes annually. In contrast, mixed-grass prairie had less aboveground biomass and burned less frequently, every 5 to 25 years.[12]

With rare exceptions, prairie plants are not killed by fire. Nearly all are perennials with root crowns and rhizomes that are not burned and that have buds capable of producing new stems and leaves within weeks. Burning may cause changes in the relative abundance of different species, depending on the season in which the fire occurs and how many fires occur in a period of time (fire frequency), but essentially all the prairie species survive. Some may even require fire for their persistence, as fire suppression often leads to the invasion of shrubs, trees, or introduced weeds, especially in the tallgrass region. Arguably, preventing fires is the disturbance, not the fire itself. Successful prairie restoration requires prescribed burning.

Grassland plants survive grazing in about the same way as they do fire: buds and roots in the soil persist even if the leaves and stems are consumed. Also, the leaves of grasses have tissues (intercalary meristems) capable of replacing the leaf material that has been eaten, in the same manner that lawn grass grows after being mowed. Thus, the grassland ecosystem is maintained if individual plants are not defoliated too often during a growing season. In general, grazing itself, whether by bison, elk, pronghorn, livestock, or grasshoppers, does not cause the demise of grassland plants unless there is insufficient time for the plant to produce the carbohydrates needed to replace the leaves and stems that were lost. Common indicators of ecosystem disturbance, such as soil erosion accompanied by the invasion of weeds and declines in native plant biomass and species diversity, are observed only when too many grazing animals are confined by fences or other barriers for too long a time or at the wrong time of year. Notably, bison have been observed to graze heavily in some areas, as observed in 1800 by Alexander Henry. He wrote that willows along the Red River were "entirely trampled and torn to atoms, even the bark of the smaller trees are in many places totally rub'd off by the Buffalo. . . . The numerous paths . . . and the vast quantity of dung . . . gives this place the appearance of . . . where Cattle have been kept for many years."[13] Such reports about unfenced bison are rare and, as will be discussed, prairie ecosystems commonly are thought to benefit from strategic grazing.

North American prairie plants are so well adapted for drought, fires, and grazing that ecologists often do not think of them as disturbances. All the plant biomass may be burned, leaving a blackened soil surface, or most of the plants may have been heavily grazed—some might say overgrazed—but within a year or two the prairie commonly regains or exceeds its prefire biomass of native species. Similarly, after the Dust Bowl years of the 1930s, some prairie species recovered quickly after the drought ended.[14]

What, then, does it take to disrupt a grassland ecosystem? First, recall that most grassland biomass is in the soil, in the form of roots, rhizomes, corms, and bulbs. Therefore, the most significant disturbances require disturbing the soil down to a depth of a foot or more. Fires and grazing, even by large herds of bison, do not do that. For thousands of years the primary cause of disturbances

Fig. 4.7. A well-controlled prescribed fire in a tallgrass prairie. Note the effective firebreak in the foreground, created by a backfire. Lightning ignites fires more often than people do in some places. Prairie fires spread rapidly on dry, windy days and often would burn for weeks, if not months. Photo by Pete Bauman.

in grasslands was burrowing animals—all small, such as prairie dogs, pocket gophers, ground squirrels, and badgers, and all disrupting the soil to some degree. Colonies of prairie dogs and pocket gophers were common, and sometimes large, but they are not found everywhere. Prairie plants have evolved with burrowing mammals.

Buffalo wallows also are sometimes considered a natural disturbance. Early explorers frequently observed bison rolling in these small, dusty depressions, presumably for the purpose of gaining some relief from flies and skin parasites—as they still do today. Created by many different animals pawing the soil in the same place, the wallows became vegetated depressions after the bison disappeared. Early journalists suggest that there were numerous wallows per square mile, each one about three to six yards in diameter.[15] Considering that bison have existed in the Northern Great Plains for thousands of years, some wallows could have been used for at least several hundred years. And some became ephemeral wetlands after heavy rains, favoring the local abundance of other kinds of plants and animals. Like burrowing animals, bison wallows were a part of the landscape, a natural part of the ecosystem.

Immigrants, however, plowed the prairie sod over large areas to eliminate competition for their crops. About 98 percent of the tallgrass prairie and about half of the drier mixed-grass prairie were destroyed in a surprisingly short time. Significantly, the cultivation of crops required disrupting the ecosystem that produced the soil on which crop production depends.

The nature of prairie ecosystems is further illustrated by considering the effects of removing carcasses of large mammals, primarily bison and cattle, and eliminating wolves and grizzly bears. With regard to carcasses, early explorers saw numerous dead animals and often they traveled through "bone fields" where large numbers of bison and other large mammals had died—whether by winter storms, hunting, or some other cause.[16] For thousands of years no one removed the carcasses. A host of scavengers came to depend on them, including the grizzly bear, gray wolf, coyote, turkey vulture, raven, eagle, magpie, and numerous insects and other decomposers. Today bison have been eliminated in most places, and state laws mandate that dead cattle, horses, and other livestock be removed to avoid attracting scavengers, spreading disease, and producing the odor of decay. In fact, livestock normally are sold before they are likely to die from natural causes; a dead animal on a pasture is a sign of bad luck, if not poor husbandry. Thus, carcass removal has become a disturbance to the community of scavengers. As with all systems, if one feature of an ecosystem is changed, other changes occur as well. It can be argued that grasslands are healthy only if they have sufficient carcasses to sustain viable populations of scavengers, just as healthy forests have enough dead trees to sustain cavity nesters.

In addition to the essential absence of large mammalian carcasses, grasslands now have lost the gray wolf and grizzly bear. Native Americans had coexisted with these predators for thousands of years, but European immigrants feared close encounters with such animals. For many years they shot them on sight, and the animals soon disappeared.[17] Removing top predators from any ecosystem has come to be viewed as one of the most significant disturbances of our time. Although they are few in number compared to their prey, carnivorous predators now are recognized for their beneficial effects on grasslands:

- Hazing herds of bison, elk, and pronghorn so grazing is spread more evenly across the landscape, thereby maintaining a higher level of plant and animal diversity, above- and belowground.
- Consuming weaker individuals of all ages, thereby maintaining wild genomes that enable large herbivore survival for millions of years and fostering the evolution of new genomes as environmental conditions change.
- Providing food for scavengers, which consume those portions of the carcasses not eaten by predators.
- Keeping the populations of some herbivores in check, at least locally, so that a greater diversity of herbivorous species can coexist, whether the predators be large wolves and bears or smaller coyotes, swift fox, black-footed ferrets, or various raptors (kestrels, falcons, hawks, or eagles).

Understanding the role predators played in the development of grasslands can inform decisions about how to achieve sustainable land uses, as discussed in chapter 13.

Benefiting from the Interaction of Fire and Grazing

Before fences, bison moved freely across the landscape, grazing some areas more intensively than others and thereby creating patches of relatively short plants that persisted until the bison moved on and regrowth could occur. Bison were not abundant everywhere; for example, none were seen by the 1874 Custer expedition from Fort Abraham Lincoln on the Missouri River to the Black Hills, a distance of about 200 miles. Pronghorn were common. In contrast, Charles Frémont wrote in 1839 about bison in the Red River Valley:

"For three days we were in their midst, traveling through them by day and surrounded by them at night. We could not avoid them."[18] On the same expedition, south of Devils Lake, Joseph Nicollet wrote, "We see here bison grazing so tranquilly that it is clear no hunting party has disturbed them for some time."[19] Bison consumed mostly grasses, which facilitated the growth of the forbs preferred by pronghorn. Without grazing by bison or other large mammals, grasses increase, forbs decrease, plant species diversity declines, and the ecosystem becomes more homogeneous.[20]

Prairie fires created patches also. Once ignited, whether by lightning or humans, the fires often burned for weeks through the dead biomass that had accumulated since the last fire or last episode of heavy grazing. But the flames would move in different directions as the winds shifted, and consequently not all the grassland would burn. Also, the flames sometimes would not burn the leeward side of steep ridges or in ravines, which accounts for occasional groves of trees that remain in such places (see chapter 7).

Regrowth after fire is often rapid because the blackened soil surface, devoid of mulch, warms up earlier in the spring than unburned areas—at a time when water usually is available. The heat of combustion volatilizes some of the nitrogen, but this is a minor loss considering that most nitrogen is stored in unburned soil organic matter and continues to be released slowly during decomposition. Research has shown that the foliage of young shoots growing back after a fire is more nutritious than older plants, a fact learned by large herbivores as well.

Recognizing this interaction between fire and grazing, range managers sometimes recommend a *patch-burn grazing system*, by which a portion of the grassland is burned.[21] The livestock benefit from the more palatable and nutritious forage that grows following the fire, and some birds and small mammals thrive in the unburned grassland that remains. Prescribed fires of this nature are easily controlled if the amount of fuel accumulation is moderate. Over a period of five to ten years, all parts of the grassland are burned—but not all at once. The appropriateness of adopting the patch-burn grazing system depends on various factors, including the management objectives of the landowner, the size of the management unit, the nature of adjacent land, and the feasibility of controlling the prescribed fire. With so many roads and much more cultivated land, prairie fires are more manageable today than they were before.

Curiously, studies of western Dakota butte tops with no large-mammal grazing and no fires during the past few decades have led to the following observations: mulch on the soil surface is much greater than on the plains below and can weigh six or seven times more than the green forage, a highly flammable situation; and western wheatgrass and dryland sedges have a higher, more uniform cover on the buttes, with less needle-and-thread and blue grama (see chapter 12). Such observations illustrate how large mammals and fire can have an important influence on the grassland ecosystem.

Prairie Dogs, Other Burrowing Animals, and Black-Footed Ferrets

Prairie dogs are sometimes referred to as ecosystem engineers and keystone species.[22] Their communal behavior creates large patches in the prairie landscape, and their intricate burrows, extending downward three to nine feet, provide habitat for other animals—burrowing owls, rattlesnakes, black-footed ferrets, cottontail rabbits, mice, crickets, and beetles. Preying on these animals are badgers, coyotes, swift fox, and various raptors. Thus, observing prairie dog colonies, known as *towns*, can be an exciting experience. The rodents instinctively clip plants around their burrows, nearly down to the soil surface, not for food but presumably to improve visibility and reduce the success of their predators (figs. 4.8 and 4.9). Most

Fig. 4.8. Mixed-grass prairie in Wind Cave National Park with bison grazing on a prairie dog town where the rodents have already clipped the plants short. New sprouts from burned or clipped plants are more nutritious than standing dead biomass.

of their food consists of insect larvae, roots, and rhizomes.[23] Grazing by other herbivores, whether bison, cattle, or sheep, also can improve visibility, thereby making the establishment of new towns more likely.

In addition to providing wildlife habitat, prairie dog digging also creates patches of disturbed soil where less competitive, disturbance-dependent plants can grow.[24] Examples of these native species are scarlet globemallow, fetid marigold, fringed sagewort, and dwarf horseweed. Introduced plants may benefit as well, such as white horehound and yellow sweetclover. Prairie dogs do not clip some of these robust forbs, which conceal predators to the detriment of the prairie dogs and other animals in the colony. Wildlife biologists have used herbicides to control such plants, to maintain prairie dog populations for the benefit of the endangered black-footed ferret.[25]

As noted previously, all burrowing animals contribute to the mixing and development of grassland soils, thereby improving the infiltration of water into the soil and enabling plant growth later in the summer. Also, their waste products and carcasses provide readily available nutrients for plant growth. The new sprouts from

Fig. 4.9. A prairie dog town in the Conata Basin of Buffalo Gap National Grassland, adjacent to Badlands National Park, with a black-tailed prairie dog in the background and a black-footed ferret in the foreground. Photo by Michael Forsberg.

clipped plants are more palatable and nutritious than the standing dead biomass adjacent to the town, which explains why bison and pronghorn spend more time grazing on prairie dog towns than would be expected by chance, considering the land area that the towns occupy. Research has shown that weight gain by domestic livestock can benefit from burrowing animals.[26]

Thus, while there may be a reason to control prairie dog populations in some rangelands, they provide significant benefits, and their populations overall are already low in many areas, mostly the result of poisoning, shooting, and habitat loss. Once widespread across the Great Plains, the black-tailed prairie dog has been proposed for protection afforded by the Endangered Species Act. That protection has already been given to the black-footed ferret (fig. 4.9), primarily because it became rare as prairie dog populations declined.

Sustaining ferrets has become a formidable challenge.[27] The ferret was declared extinct in 1979, but two years later a few dozen were found on a ranch in western Wyoming. Unfortunately, epidemics of sylvatic plague and canine distemper killed all but 18 ferrets, and plague killed most nearby prairie dogs. In desperation, the surviving ferrets were captured by conservation biologists, quarantined, and successfully bred in captivity. Now reintroduced to the wild, there are currently about 300 ferrets living in 30 large prairie dog towns in 16 western states. One town is in Wind Cave National Park on the south end of the Black Hills, and there is another in the Conata Basin, a part of Badlands National Park and Buffalo Gap National Grassland. The constant threat of disease has led to vaccination programs for the ferret. Also, sylvatic plague, spread from one ferret or prairie dog to another by fleas, is suppressed by dusting colonies with an insecticide. A long-term conservation strategy depends on several colonies in an area so that the prairie dogs and ferrets can survive even if one colony is eliminated by an epidemic.

Pollinators and Grasshoppers

Watching bees, butterflies, and moths flit from one prairie flower to another is a fascinating experience. The flowers of each species have evolved shapes, colors, and fragrances that attract specific insects, and often only a few kinds of insects have a body size and mouthparts specific for each kind of plant—an example of coevolution (figs. 4.10 and 4.11, box 4.2). Many species of plants have persisted for millions of years because of these pollinators, as they are essential for the sexual reproduction required to produce an abundance of genetically diverse seed. Genetic diversity has been critical for plant survival during periods of climate change in the past, as it will continue to be in the future. This is true because, curiously, most prairie plant seedlings die because they cannot compete with dense, well-established plants. However, from time to time seeds land in

Fig. 4.10. The monarch is one of hundreds of insects that pollinate prairie plants. Though dependent on various kinds of milkweed for egg laying, this butterfly pollinates numerous other prairie plants, such as the meadow blazingstar shown here. Photo by Justin Meissen.

Fig. 4.11. The western prairie fringed orchid, with spectacular flowers that are about two inches across, is a rare tallgrass prairie plant that is pollinated by several species of hawkmoth (also known as the sphinx moth). See box 4.2. Photo by Jim Fowler.

Box 4.2. A Prairie Orchid, Hawkmoths, and Plant Diversity

One of the most spectacular native orchids in North America grows in the Red River Valley. Known as the western prairie fringed orchid and found in tallgrass prairie, it is sometimes three feet tall and produces prominent sprays of up to 20 white flowers, each two inches across (fig. 4.11). The flowers are especially fragrant in the evening and at night, to attract night-flying pollinators like moths. Also, each flower has a two-inch-long spur with nectar in the bottom, indicating that the pollinator most likely has a tongue that is about that long—as hawkmoths do. Those that have been captured often have small packets of orchid pollen, known as *pollinia*, attached to their head. Hawkmoths are fairly common, but the orchid has become so rare that it is protected by the Endangered Species Act.

The most fundamental cause of rarity is the conversion of over 95 percent of tallgrass prairie to cropland and pasture, but the explanation is more complicated.[a] Five kinds of hawkmoths are known pollinators in the Red River Valley, four natives and one introduced from Europe as a biocontrol agent for leafy spurge. Curiously, the spurge moth is the one that is most frequently trapped with attached pollinia. This species of hawkmoth has a tongue slightly shorter than the length of the flower spur. Thus, to access as much of the nectar as possible, the moth must touch the flower, thereby collecting pollen, delivering pollen, or both. Two of the native species have tongues longer than the spur and reach the nectar while hovering near the flower—not good for pollination. Dubbed *nectar thieves*, they get what they want with no benefit for the plant.

There seem to be enough hawkmoths to effectively pollinate the fringed orchid. Biologists emphasize, however, that hawkmoths need energy from other kinds of flowering plants when the orchid is not in bloom. Moreover, the hawkmoths lay their eggs on other plants, not the orchid, and their host plants depend on other kinds of pollinators. Grasslands with very few forbs, whether due to herbicides, invasive grasses, or fire suppression, do not provide good habitat for pollinating insects of any kind, including those needed by farmers and ranchers. North Dakota scientists Steven Travers, Marion Harris, and their associates wrote in 2011 about the "hidden benefits" of pollinator diversity. Such benefits depend on a diversity of native plants and other organisms—an old principle that is strengthened by new studies pertinent to removing the western prairie fringed orchid from the endangered species list.

[a]Travers et al. 2011.

a suitable seedbed and a new plant survives, perhaps on soil cleared of plants by burrowing animals. Although nearly all of new spring growth is vegetative, that is, from sprouting buds in the soil, sexual reproduction and subsequent seedling establishment—though infrequent—is the key to maintaining the genetic diversity required for long-term survival.

Early farmers benefited greatly from pollinators, as they still do now. Sunflowers, squash, beans, flax, alfalfa, and canola soon became important—all insect pollinated. Also, farmers set out handmade beehives to capture honey, a practice that was common in Europe. Honey was a sweet treat that could be sold. Today there is great concern about declining populations of pollinating insects, most likely caused by the loss of habitat for the prairie forbs on which the pollinators depend and the widespread use of insecticides. Pollination is an ecosystem service powered by solar energy that would be very difficult and expensive to replace. State departments of agriculture now promote the maintenance of pollinator populations. Notably, the pollinator now most important for farmers is an introduced species from Europe—the honey bee (fig. 4.12). Many are concerned by how rare it has become in some places.[28]

Grasshoppers also are of great importance (fig. 4.13). Over a hundred native species have been identified in the Northern Great Plains, all

Fig. 4.12. The western honey bee, introduced from Europe, is the best known of the various kinds of honey bees in North America. Commonly thought of as a domestic insect and raised on farms, these bees are extremely important economically for the production of honey and the pollination of crops. However, they also can displace native pollinators. Photo by Maja Dumat, Flickr, CC BY 2.0.

Fig. 4.13. The red-legged grasshopper (top, photo by Tom Murray) and the plains lubber grasshopper (bottom, photo by Robert Pfadt) are two of more than 100 species of Orthoptera found in the Northern Great Plains. The lubber grasshopper image is from *Grasshoppers of the Western U.S., Edition 4*, USDA APHIS Identification Technology Program.

herbivores ranging in size from one to four inches at maturity. Most species were always present in low numbers and were dispersed among the different kinds of grassland according to their food preferences. However, the population eruptions of a few species can become dark clouds of noisy insects—often soon after a period of drought. Dry conditions apparently reduce the likelihood of population-limiting grasshopper diseases, but there is much to be learned about other factors that keep populations at low levels, such as predation and parasites.

Grasshoppers often consume as much or more of the plant biomass aboveground as any other herbivore. And when their populations are large, they consume nearly all of it. With 50 or more on every square foot in mixed-grass prairie, a rangeland scientist observed in 1941 that they "destroyed all edible vegetation." Much earlier, in 1868, another traveler through Dakota Territory wrote, "These cursed insects hit us in the face, [got] caught in our eyes or in our beards, got into our clothes, and jumped from the ground in swarms at our every step." Some years had more outbreaks than others and some tracts of grassland were not affected when the outbreaks did occur, adding further patchiness to a landscape already patchy because of topographic variation, changes in soil salinity, prairie dog colonies, fires, and grazing by large mammals. One of the most notorious grasshoppers was the Rocky Mountain locust, which from time to time swarmed over hundreds of square miles on the Great Plains in the eighteenth and nineteenth centuries, darkening the sky and consuming crops and native plants alike. Surprisingly, this locust became extinct in the early 1900s, apparently because of plowing and irrigation—a happy outcome for some but also alarming because of how rapidly it occurred.[29]

The high cost of controlling grasshopper outbreaks can be considerable because insecticides have to be purchased and applied. Such expenditures are rationalized by expectations that the populations can be reduced for several years. Research suggests, however, that there is little basis for such multiyear benefits and that working to conserve natural predators could be more economical.[30] Notably, grasshoppers play the same role as other herbivores do in grassland ecosystems: facilitating the cycling of nutrients by consuming plant tissue and excreting more easily decomposed, nutrient-rich waste products. They also are an important food source for birds, small mammals, and even fish.

Bobolink

Bird Diversity

As with all other forms of life in grassland ecosystems, birds have evolved to occupy a diversity of ecological niches, that is, different portions of the habitat where specific species find the shelter and food they need in different ways (fig. 4.14). Each species has adaptations that minimize competition

Burrowing owl

Dickcissel

Greater prairie chicken

Upland sandpiper

Fig. 4.14. Birds are common in Dakota prairies. To minimize competition, the various species have different preferences for food and nesting habitat. Representatives shown here are the dickcissel, bobolink, burrowing owl, greater prairie chicken, and upland sandpiper. The prairie chicken is a year-round resident, but most birds migrate hundreds of miles southward to a warmer climate for the winter. The upland sandpiper and bobolink often migrate to South America. Photos by Doug Backlund.

from other species. Some are scavengers, like turkey vultures, crows, and ravens, which feed on the carcasses of other animals; others are predators, like hawks, falcons, and owls, which kill and consume live birds, mammals, and reptiles. These carnivores often nest on nearby high points, such as trees or escarpments, although the ferruginous hawk, burrowing owl, and short-eared owl nest on the ground. Many prairie birds are small and consume seeds, insects, and other invertebrates. Examples are the eastern meadowlark, lark bunting, dickcissel, bobolink, and various kinds of sparrows in the tallgrass prairie—colorful enough or sufficiently vociferous to attract mates but with the instincts to build hidden, well-camouflaged nests that minimize egg and fledgling predation by other birds, snakes, and small mammals. Mixed-grass prairies with shorter plants have some of the same birds, but others become common as the climate changes, for example, the western meadowlark, McCown's longspur, chestnut-collared longspur, and prairie horned lark. The much-larger greater prairie chicken and sharp-tailed grouse also may be seen. Some grassland birds appear out of place, like the upland sandpiper (other species of sandpipers are seen more often along the shores of wetlands and lakes). Ducks such as the mallard, pintail, gadwall, and teal can be considered prairie birds because they build their nests on the upland, sometimes a mile or more away from prairie wetlands (see chapter 11).

It is common for upland prairie birds to frequent nearby swales, where water and food are sometimes more readily available, or they are found in nearby shrubs or trees in the ravines of creeks and rivers. Similarly, some species gravitate to burned or heavily grazed prairies. A striking example occurred when, during a prescribed prairie fire, several dozen migrating Swainson's hawks were observed feeding on the insects, rodents, snakes, and frogs that had just lost their cover.[31]

In general, prairie bird species diversity in an area is dependent on the patchiness of the landscape mosaic, whether caused by topography, fire, grazing, wetlands, burrowing animals, or the presence of trees and shrubs. During the winter, birds migrate to better habitat, a few only a short distance away, possibly to a nearby frozen cattail marsh with sufficient thermal cover to keep warmer when food is scarce, thereby minimizing the expenditure of body fat. Most birds migrate southward to warmer climates, even to other continents. For example, upland sandpipers, bobolinks, and Swainson's hawks commonly migrate to South America.

Unfortunately, much of the habitat for prairie birds has been degraded at both ends of their migration routes, where so much of the prairie has been converted to croplands and many wetlands have been drained to provide still-larger fields. Fencing can lead to more uniform grazing and less patchiness in pastures, and fires usually are suppressed, burning only small areas. Prairie dog colonies provide habitat for various kinds of birds but are now much less common than they were previously. Trees have been planted for shelterbelts, which may add a new kind of patch to the landscape that has some benefits, but the trees also provide perches for raptors that prey on rare prairie birds below. All these developments have led to a widely recognized conservation crisis—the rapid, widespread decline in the abundance of grassland birds.[32] Various solutions to this problem have been proposed, as discussed in chapters 5, 6, and 13. Management strategies that benefit birds are likely to increase the diversity of other organisms as well.

Invasive Plants

A fundamental principle of ecology is that individual species, whether plant or animal, tend to expand their distribution ranges whenever possible. Often such migrations are not successful, perhaps because a better-adapted competitor already occupies the same ecological niche, or because plant seeds are dispersed to a site that is not favorable for seedling establishment. From time to time, however, individuals of new species become established and grow to maturity, thriving if the newcomer escapes excess herbivory and historic predators and diseases. Such species can become *invasive*, that is, capable of expanding rapidly throughout an ecosystem dominated by native species, changing it dramatically in the process. Native species become less common during such invasions, biological diversity usually declines, and expensive control measures are often advised (box 4.3).

Box 4.3. North America's Lawn Grass Gone Wild

There are numerous native species of bluegrass in the Northern Great Plains, but the most loved and problematic is an exotic. Despite its common name, Kentucky bluegrass was introduced to North America in the mid- to late 1600s. A rhizomatous, sod-forming, cool-season perennial, it was cultivated in pastures, lawns, and golf courses across the continent and was often the first to bring a much-appreciated touch of green after a long winter. But this grass became a difficult-to-control invasive plant in prairies. After so many years, some consider it to be naturalized, a designation that may be a copout for failed attempts to minimize the problems it causes.

Arriving in the Dakotas by the mid- to late 1800s, Kentucky bluegrass most likely gained a foothold in disturbed mixed and tallgrass prairies. Excess grazing over long periods would have weakened the native plants, enabling the establishment of this newcomer. Mature plants thrived, benefiting from early spring growth when many native plants are still dormant. Smooth brome was introduced to the Great Plains about 1900 and became invasive after the Dust Bowl years. Fire suppression favored both invaders.

How serious is the problem? North Dakota ecologists Shawn DeKeyser, Lauren Dennhardt, and John Hendrickson, in a study published in 2015, found that Kentucky bluegrass was often the most abundant species in remnants of native prairie. Smooth brome was close behind. Native plant growth and diversity declined significantly as these invasive plants increased. Some native grasses common in 1984, such as blue grama and junegrass, had disappeared 23 years later from half of their 25 study areas. And the problem is worsening, in part because of longer growing seasons and higher atmospheric carbon dioxide concentrations, both of which are known to benefit the bluegrass. Restoration ecologists are developing strategies to resolve this predicament.

Invasive plants have become one of the most important problems associated with global change, primarily because intercontinental commerce has skyrocketed.[33] Some plants were brought to North America because of their known value as crops and ornamentals, and most of them grew only where they were planted. But others, now known as weeds, escaped and became abundant where they were not wanted (table 4.3). Invasive animals also have become problems, such as the starling, English sparrow, and Eurasian collared dove.

Notably, native species also can be invasive. This happens when management practices create novel environments that are more favorable for one or a few species. The most obvious example is fire suppression, which enables the invasion of ponderosa pine in the foothills of the Black Hills, aspen in the foothills of Turtle Mountain, eastern redcedar along the breaks of some large rivers or where planted for wildlife habitat, and green ash or bur oak in prairies near a woodland seed source (see chapter 6).

The control of invasive species—introduced or native—is difficult. One promising method for plants is burning the prairie intentionally at a time when weedy species are most vulnerable, often in the spring. Another approach is to introduce grazers that prefer eating target plants, a form of biological control. For example, leafy spurge has been contained in some places using introduced goats, flea beetles, stemborer beetles, and hawkmoths (fig. 4.15). Herbicides can be used, but they often kill many of the desirable forbs that sustain pollinators. Unfortunately, weed and pest control districts sometimes feel obliged to spray prairie preserves to control species that could invade neighboring pastures and cropland. Invasive plants can contribute to some ecosystem services, such as erosion control and wildlife habitat, but the long-term value of the grassland usually declines.

Active management is required to minimize the success of invasive plants in grasslands. This

Table 4.3. Some introduced plants that are considered problems in the grasslands of North Dakota and South Dakota[a]

GRASSES	
Bluegrass, Kentucky	*Poa pratensis*
Brome, Japanese	*Bromus japonicus*
Brome, smooth	*Bromus inermis*
Cheatgrass/downy brome	*Bromus tectorum*
Foxtail, creeping meadow	*Alopecurus arundinaceus*
Quackgrass	*Elymus repens*
Timothy grass	*Phleum pratense*
Wheatgrass, crested	*Agropyron cristatum*
TREES AND SHRUBS	
Buckthorn[b]	*Rhamnus cathartica*
Russian olive	*Eleagnus angustifolia*
Siberian elm	*Ulmus pumila*
FORBS	
Bindweed, field	*Convolvulus arvensis*
Henbane, black	*Hyoscamus niger*
Knapweed, diffuse	*Centaurea diffusa*
Knapweed, Russian	*Centaurea repens*
Knapweed, spotted	*Centaurea maculosa*
Mullein, common	*Verbascum thapsus*
Sickleweed	*Falcaria vulgaris*
Sowthistle, perennial	*Sonchus arvensis*
Spurge, leafy	*Euphorbia esula*
Sweetclover, yellow	*Melilotus officinalis*
Thistle, Canada	*Cirsium arvense*
Wormwood, absinthe	*Artemisia absinthium*

[a]Additional species invade croplands, gardens, and wetlands.
[b]Found in sandhill prairie swales but more common in woodlands.

Fig. 4.15. Leafy spurge is a noxious invasive plant in some tallgrass prairies and relatively moist mixed-grass prairies. Control measures include grazing by goats and the introduction of beetles and hawkmoths that consume this plant. The deep root system is difficult to kill with herbicides or fire. Cattle do not normally eat the plant. Source: Montana Statewide Noxious Weed Awareness and Education Program, Montana State University, www.bugwood.org.

can be accomplished by the following: allowing periods of rest from grazing so that the energy reserves in native plant roots are restored, controlling weeds while they are restricted to small areas, and taking steps to prevent the inadvertent introduction of other potentially invasive species. The collaboration of neighboring landowners increases the effectiveness of this approach. In some cases, land managers have decided that containment is a more reasonable goal than total elimination, which is expensive and could require steps that lead to further ecosystem degradation. Invasive species tend to be a bigger problem in relatively moist tallgrass prairie and meadows, where such grasslands still exist, and are less of a problem on well-managed, drier prairies (fig. 4.16).

As the climate and other environmental variables change, maintaining ecosystem benefits may lead to accepting some introduced plants. Still, an advantage of restoring native species is that they persist year after year, are more likely to survive extended droughts, and they greatly reduce the amount of herbicide needed. Native plant growth may at times be less than the introduced species, but the expense of weed control is much less as well.

The Probable Effects of Climate Change

Grassland plants are capable of tolerating periods of high temperature and drought, but will they survive the anticipated changes described in chapter 3? Will survivors be the species that are most appreciated today, whether for their forage value or for the role they play in sustaining certain kinds of wildlife, or will they be undesired invasive species or pathogens? In general, successful plants and animals will either evolve adaptations to their new environment or disperse to favorable environments. Regrettably, for native tallgrass prairie species, the option of dispersal to more favorable environments has been greatly reduced. Only small remnants remain, usually many miles apart. Tallgrass prairie reconstruction has become a priority for many managers (see chapters 5 and 13).

In contrast to tallgrass prairie, there are still extensive mixed-grass prairies on the Prairie Coteau in the east and some parts of the Missouri Coteau and Missouri Plateau to the west (see fig. 1.4). How will they respond as the Dakotas become warmer and generally more arid in the twenty-first century?[34] Some research indicates that total average plant growth could remain the same or increase

Fig. 4.16. Smooth brome, an introduced invasive grass, is the dominant plant in the roadside ditch to the right of the fence and extends a short distance into the adjacent mixed-grass prairie. The plant is more abundant where it was planted, where runoff is greater (e.g., from road pavement), and where there is little or no grazing.

because of earlier spring emergence, when water is more likely to be available. Also, at such times, when water and nutrients are not limiting, higher levels of atmospheric carbon dioxide are likely to stimulate the growth of some species.[35]

The interaction of such factors is complicated. Average annual precipitation is not expected to decline in the Dakotas during this century, but there is now a consensus that climate change will lead to the amplification of precipitation events; that is, rainfall and snowfall will produce more water on average when they occur. Such downpours could wet the soil to greater depths on relatively flat land, which could be beneficial to deeply rooted prairie species.[36] Thus, even with longer summer droughts, more water could be available for plant growth if more of the annual precipitation infiltrates to depths where it is less likely to be evaporated before plants can use it. On steeper slopes, a larger proportion of the precipitation would flow over the soil surface, especially if the soil is frozen or the amount of plant cover is low. Droughts of only one or two years can greatly reduce the amount of plant growth.[37]

The most noticeable long-term effect of climate change on grasslands in the Northern Great Plains is likely to be shifts in the relative abundance of different species, the result of changes in the proportion of rainfall received during the spring, summer, and fall. Cool-season species could become more common if a multiyear sequence of summer droughts is too severe for warm-season species to survive. In general, climate change will facilitate the expansion or immigration of species, native or introduced, that have adaptations for tolerating the new climatic conditions.[38] For example, species now common in Nebraska may be preadapted to warming climates in the Dakotas. Comparable shifts in the prevalence of animals, including soil organisms, could occur as well.[39] Wet prairies will respond differently than drier prairies. Another variable affecting ecosystem response to climate change is biological diversity. There is more genetic diversity where species diversity is higher, which increases the likelihood that some native species will survive. Prairies with low diversity are likely to be affected more dramatically.[40]

Overall, the combined effects of fire suppression, invasive species, habitat fragmentation, and climate change have had, and will continue to have, significant effects on the grasslands that remain.[41] Different kinds of grasslands will be affected differently. To adapt, ranchers have been advised to adjust traditional management practices.[42] As prairies have become less common, their biological diversity, wildlife habitat, and ecosystem services have become highly valued by society as a whole. Consequently, grasslands are being restored, such as those enrolled in the Conservation Reserve Program (CRP) of the U.S. Department of Agriculture. CRP lands and pastures are much different than the native grasslands that once existed on the Great Plains, but year-round plant cover has significant benefits, as described in the next chapter.

Chapter 5 Agriculture and Agroecology

The First Farmers

About 13,500 years ago, as the massive continental glaciers were receding, the first people set foot on land that would become the Dakotas.[1] Members of the itinerant Clovis culture hunted large mammals and gathered wild plants for thousands of years before the first farmers arrived—sedentary, woodland tribes from the south and east. These farmers needed specific natural resources to maintain their relatively permanent villages, including perennial rivers with wooded valleys, wild game, and fertile floodplains on which to grow crops. In the Dakotas they included members of the Oneota tribe along the Big Sioux River and near Devils Lake, the Cheyenne tribe along the Sheyenne and James Rivers, and the most numerous and best-known farmers, the Mandan, Hidatsa, and Arikara tribes along the Missouri River.[2]

For nearly a thousand years the three Missouri River tribes cultivated gardens to supplement their diet of meat and wild plants. Riparian forests provided lodge posts, willow mats, grain-drying racks, and fuel for cooking and heating, as well as space for crops once the trees were cleared (fig. 5.1). Garden plots were established on fertile, lower floodplain sites in spring. Their crops were of Mexican origin—corn or maize, squashes, pumpkins, gourds, melons, and beans.[3] The tribes started their gardens in the spring by using hoes and digging sticks to plant six to nine corn seeds in small mounds. Nearby mounds were planted with beans and squash. Together, these three crops formed what is now referred to as a *symbiotic polyculture*: nodules on the bean roots fixed nitrogen in the soil that improved production of companion crops; the twining beans supported corn stalks in strong winds and intercepted more sunlight by growing vertically; and the squash vines with large leaves covered much of the soil, reducing weed growth and evaporation. Fields were weeded during the growing season. Many of them were bordered by a row of sunflowers likely derived from wild species. One or two large trees were left in garden plots, perhaps to shade workers. Indian farmers gradually produced varieties of corn better adapted to the cooler climate and shorter growing season of the Northern Great Plains. At least 15 different types of corn and beans were grown near the Mandan villages.[4]

Most families farmed from three to five acres of floodplain and river islands.[5] Villages as a whole cultivated 300–500 acres. The combined production of corn from the Mandan and Hidatsa villages was considerable. Numerous pits capable of storing a total of approximately 70,000 bushels of corn were dug by the Mandans. These caches

Fig. 5.1. Depiction of Mandan farm fields on the Missouri River floodplain. Women are carrying garden produce on their backs in burden baskets. Painting by Robert Evans. Courtesy of the State Historical Society of North Dakota.

were sealed with skins, soil, and grass; some were six or more feet below ground, requiring ladders to access them. Stacked ears of corn lined the pit; loose kernels of corn and dried squash were placed in the middle (fig. 5.2). After overwintering with the Mandans in 1804–1805, Lewis and Clark's 33-man party left with about 70–90 bushels (about 2 U.S. tons) of food for their journey upriver.[6]

The Mandans were the master marketers of corn in a trading region that extended from the Great Lakes to the northern Rockies. Beginning in about AD 1400 and lasting for nearly five centuries, central North Dakota was the "granary" for the nomadic Crow, Shoshone, Kiowa, Assiniboine, Lakota (Dakota), and Blackfeet—and later for Euro-American traders. Both sedentary gardeners and nomadic hunters benefited from trading. While many sedentary tribes had gardens, the Mandans made farming into a regional business enterprise by producing large surpluses.[7]

Euro-American Agriculture

Agriculture has been the major industry in the Dakotas since large numbers of Euro-Americans arrived late in the nineteenth century. Gross income from crops and livestock has greatly exceeded other revenue in both states, except in North Dakota during 2012, when income from agriculture was essentially tied with oil and gas, and in 2014, when oil and gas income was considerably higher.[8] The first form of agriculture was "open range" livestock grazing on large unfenced tracts of grassland with cattle and sheep brought in by large companies, some funded by foreign investors. This romanticized period in American frontier history was short lived because of overgrazing, the fencing of open rangeland by farmers, and better profits from row crops where rainfall was higher. One of the last open-range cowmen in the Dakotas was Bert "Cap" Mossman,

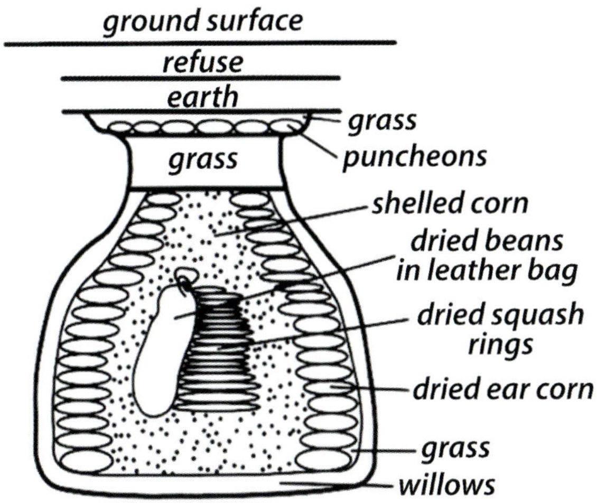

Fig. 5.2. Diagram on left of a cache pit dug into the ground and used to store durable food such as dried corn, beans, and squash in Mandan villages (adapted from Handy-Marchello and Swenson 2018, State Historical Society of North Dakota). The photograph on the right shows a reconstructed cache pit on display at the Knife River Indian Villages National Historic Site, Stanton, North Dakota

a former Arizona Ranger and general manager of the Colorado-based Diamond A Cattle Company. He managed up to 50,000 cattle and horses on about 1 million acres in north-central South Dakota from about 1902 to 1914.[9]

Grazing was an obvious use of the expansive grassland once populated by bison, elk, and pronghorn, but plowing and cultivating grassland was new. Immigrants from the eastern United States or Europe had carved out farmland by clearing forests adapted to a wetter climate, but they were unsure how to "break" the seemingly impenetrable prairie sod and then fence their land, where wood for posts and rails was scarce. And if they could, would the plowed grassland produce good crops? Fortunately for them, the timely invention of the steel, moldboard plow solved the first problem, at least for those who owned oxen or could pay for their services (fig. 5.3). With only less-effective horses available, it took Oscar Micheaux all summer to break his farm of 120 acres, the most among his neighbors. He estimated that, to do this, he had walked 1,400 miles. Even with the plow, breaking the tough prairie sod required skill, strength, and endurance. The second problem was eventually solved by the invention of barbed wire, although trees for the fence posts often were scarce. Over the years, the more successful farms were enlarged by buying adjacent land when it became available.[10]

The tallgrass prairie in the eastern Dakotas was homesteaded first during a period that became known as the Great Dakota Boom (1870–1890).[11] This area had a favorable climate for agriculture, and wood and water were sufficient. With approximately 40 million acres of grassland, farmers had plenty of land to cultivate. Wheat farms that were thousands of acres in size and owned by corporations, known as *bonanza farms*, were soon established in the Red River Valley (fig. 5.4). Immigration from the eastern United States and Europe increased rapidly, spurred on by the Homestead Act

Fig. 5.3. Breaking prairie sod with a team of horses. Courtesy of the South Dakota Agricultural Heritage Museum.

Fig. 5.4. Bonanza farm in the early 1900s in the Red River Valley showing extensive cultivation for wheat on fertile tallgrass prairie soils. Courtesy of the North Dakota Institute for Regional Studies.

of 1862, incentives from the railroads, and news about rich prairie soils and high crop production.[12] The population of Dakota Territory shot up from 12,887 in 1870 to 428,968 just 20 years later—most were farmers. The speed at which natural ecosystems covering such a large region were converted to farmland was unprecedented in human history.[13] Some prairies were used primarily as pasture or hay land because they were too steep, rocky, wet, or dry to cultivate. The settlers were yet to learn of other impediments to successful farming, especially the occurrence of severe droughts.

In the drier, western Dakotas, homesteading peaked about 10 to 15 years later, between 1900 and 1915, when over 100,000 newcomers ignored warnings about the aridity of the climate. The shorter, thinner cover of the mixed-grass prairie should have been a sign, but optimistic pioneers held on to the belief that droughts and rugged topography could be overcome with determination. One magazine cautioned at the time, "A country where cactus is perfectly at home cannot be farmed the same as where clover grows."[14] The rush for land ran its course during this period and the resilience of a farming system that removed sod and left land bare for much of the year was about to be tested against climatic extremes.

By necessity, the crops of early homesteads were diverse.[15] Wild hay fed the milk cow, beef cattle, and working animals; hay twists (hay ties) were burned for cooking and heating, as described in books by Laura Ingalls Wilder. Specialized hay-burning stoves were available. Later, some farmers replaced wild hay with tame pasture grasses and alfalfa of Eurasian origin. Large vegetable gardens were planted. Grain crops included corn, wheat, oats, barley, and flax. Small grains and some corn were sold, while most corn was mixed with hay to feed livestock during the winter. In the west, more of the farm acreage was allocated to rangeland or hay to feed larger herds of beef cattle. In some places a three-crop rotation was adopted: alfalfa or other forage crops, then corn, and finally small grains. Occurrence of at least some wild hay land on most farms preserved elements of the natural biodiversity in the Dakotas, although overgrazing on small allotments was rampant in the early years. Economic diversification that included chickens and hogs was the trademark of early Dakota farms. Some farmers sold fruit, such as wild plums or currants. Women and children played a large role in diversifying farm income, especially during hard times; turkeys, chickens, ducks, their eggs, and cream were sold for cash. Farm commodities became more profitable when railroads connected the Dakotas to large eastern markets.

Early Obstacles to Farming

The first major setback arose when a drought in 1909 turned corn leaves yellow in August—a harbinger of things to come. Dry weather and hot winds in the summer continued for six years, greatly reducing crop production in the drier parts of both states. Major rivers dried up, and much of the planted seed did not germinate. Desperate settlers ate seed potatoes and young tumbleweeds—an introduced annual weed common in abandoned fields. Some counties lost nearly all farmers and a large percentage of town folks that depended on them for customers. Associated with such droughts were clouds of grasshoppers that blocked the sun (see chapter 2). Relief was provided to some people by the sale of bonds, deferment of payments on homesteads, and gifts of crop seed. The railroads provided free freighting for a short time to bring in relief supplies, including hay, seed grains, and coal. From this drought emerged the attitude still in existence today that the Great Plains is "next-year country," that is, although times are tough this year, farming will be better next year. Many of those who stayed reluctantly asked for federal assistance, establishing a pattern much in evidence in today's agriculture.[16]

Droughts stimulated debates among farmers

about the advantage of raising livestock over grain farming. Sheep and cattle yielded a cash return that drought did not easily destroy. Others worried that a shift back to cattle would lead to further depopulation and dash dreams of building vibrant rural communities. Many farmers compromised by adding more beef cattle or dairy cows to their farming operations and accepting diversified approaches recommended at the time by local universities, such as the planting of alfalfa and other recently introduced forage crops. The desperation induced by drought forced residents to be more flexible in virtually all aspects of life on the Plains, especially in their willingness to adopt new approaches to agriculture in a semiarid land with extreme and unpredictable weather.[17]

The ups and downs of life in the Dakotas continued, with grain prices shooting up during World War I. New settlers arrived as markets improved for farm families and their communities.[18] Breaking the remaining prairie sod was accomplished much faster by steam tractors than by horses or oxen (fig. 5.5). This was followed by a postwar downturn in 1921, when grain prices and land values plummeted. Many banks failed. National and international events had become a force as strong as weather in affecting the livelihood and survival of rural populations. Another rebound occurred when machinery became affordable and agronomists produced new and more drought-resistant crops, but the good times did not last long. The granddaddy of all droughts was just around the corner.

The Dust Bowl in the Great Plains was caused by three main factors: six years of below-average precipitation, from about 1933 to 1939; plowing large

Fig. 5.5. Breaking prairie sod in the early 1900s using a steam-powered tractor, which is much faster than using horses or oxen. Courtesy of the South Dakota Agricultural Heritage Museum.

Fig. 5.6. Dust storm in 1935 in Beadle County, South Dakota. Courtesy of the South Dakota Agricultural Heritage Museum.

areas of prairie sod to grow crops; and overgrazing. Crop failures from drought exposed the soil to strong winds that carried clouds of dust east to the Atlantic Ocean (fig. 5.6). These events and plagues of grasshoppers caused a massive exodus of farmers. Volumes have been written describing the human suffering and environmental damage of the Dust Bowl years. Conditions were less severe in North Dakota, especially the Red River Valley. However, North Dakota history books also report the occurrence of dust storms, grasshopper plagues, crop failures, and farm foreclosures. Throughout the region, highly eroded land and drifts of soil were left in the departing wagon tracks of the environmental refugees (fig. 5.7). Historical records document that the nascent agricultural system failed.[19]

To help farmers and ranchers prevent similar tragedies from happening again, Congress established new programs, such as the Soil Conservation Service and the Taylor Grazing Act. Both promoted better farming and livestock management practices, plus a massive tree-planting program for shelterbelts that was intended to reduce soil erosion (see chapter 6). World War II, which erupted soon after the drought of the 1930s ended, expanded expanded food markets and contributed to another period of farm prosperity.[20]

In general, the shift in how to make a living on the Northern Great Plains was abrupt, from hunter-gather and gardening cultures dependent on solar-powered ecosystem services to one dependent on fossil fuel—primarily to produce grain. The jury is

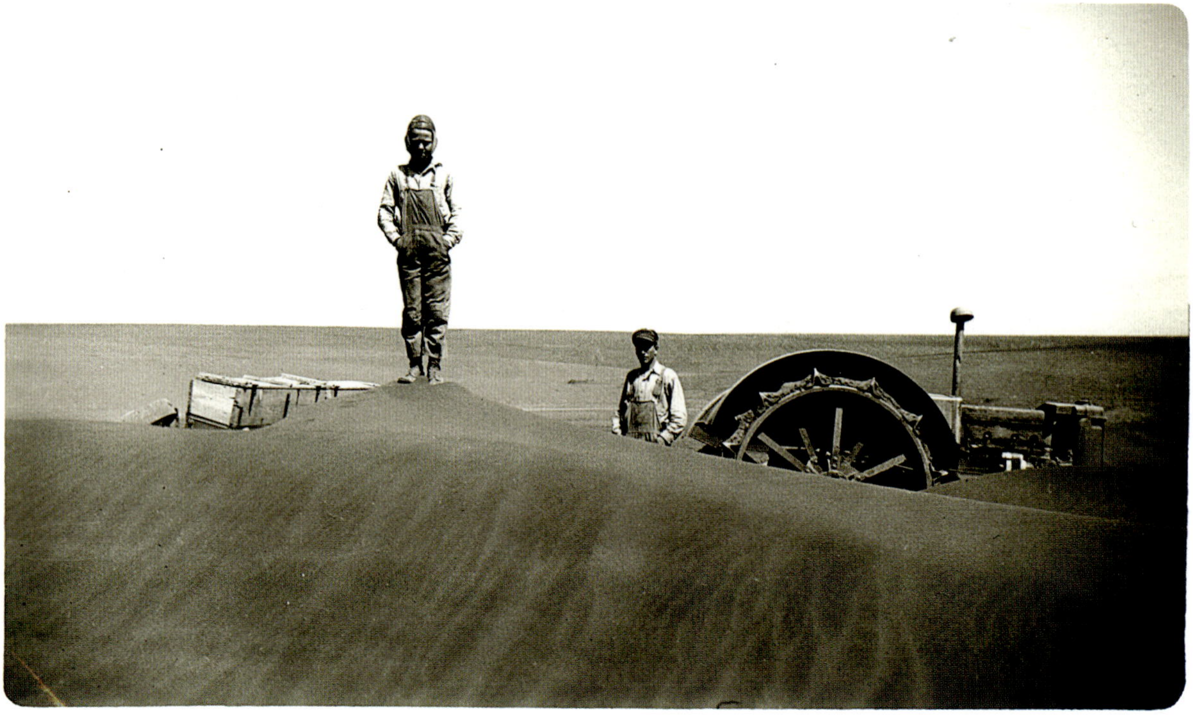

Fig. 5.7. Soil blown by strong winds across bare, dry fields produced drifts of dust near farmsteads, fence lines, and farm equipment, as in this photograph from 1935. Photo courtesy of the South Dakota Agricultural Heritage Museum.

still out on whether the new system is sustainable over centuries of time. Certainly, farming today is a sea change from the days of the prairie pioneers. Many innovations, however, have had undesired consequences, especially in their effects on soils, water quality, and biodiversity.[21]

Trends in Farm Sizes, Crops, and Farmland Cover

Summing the number of farms in the two states shows a decline from about 168,000 in 1935 to less than 60,000 in 2017.[22] During the same period, average farm size increased from about 450 acres to nearly 1,500 acres. The main factors causing these trends were low profit per acre along with increased availability of bank loans, fertilizers, herbicides, pesticides, and labor-saving machinery.[23] Despite these similarities, there are significant differences in agriculture between the two states (fig. 5.8):[24]

- North Dakota ranks second out of all 50 states in the total area of crops planted, second only to Iowa. South Dakota ranks eighth. North Dakota routinely exceeded South Dakota in cropland harvested, in 2017 by about 7 million acres (24 million versus 17 million acres).
- The ratio of grassland to cropland in South Dakota is almost 2:1, while in North Dakota the ratio is the opposite, about 1:2. Rangeland in South Dakota in 2017 was more than double that in North Dakota (22 million and 10 million acres).
- South Dakota produces more livestock (cattle, calves, sheep, and lambs) and hay, ranking seventh and eighth nationally, respectively, while North Dakota is not ranked in the top 10 for these commodities.

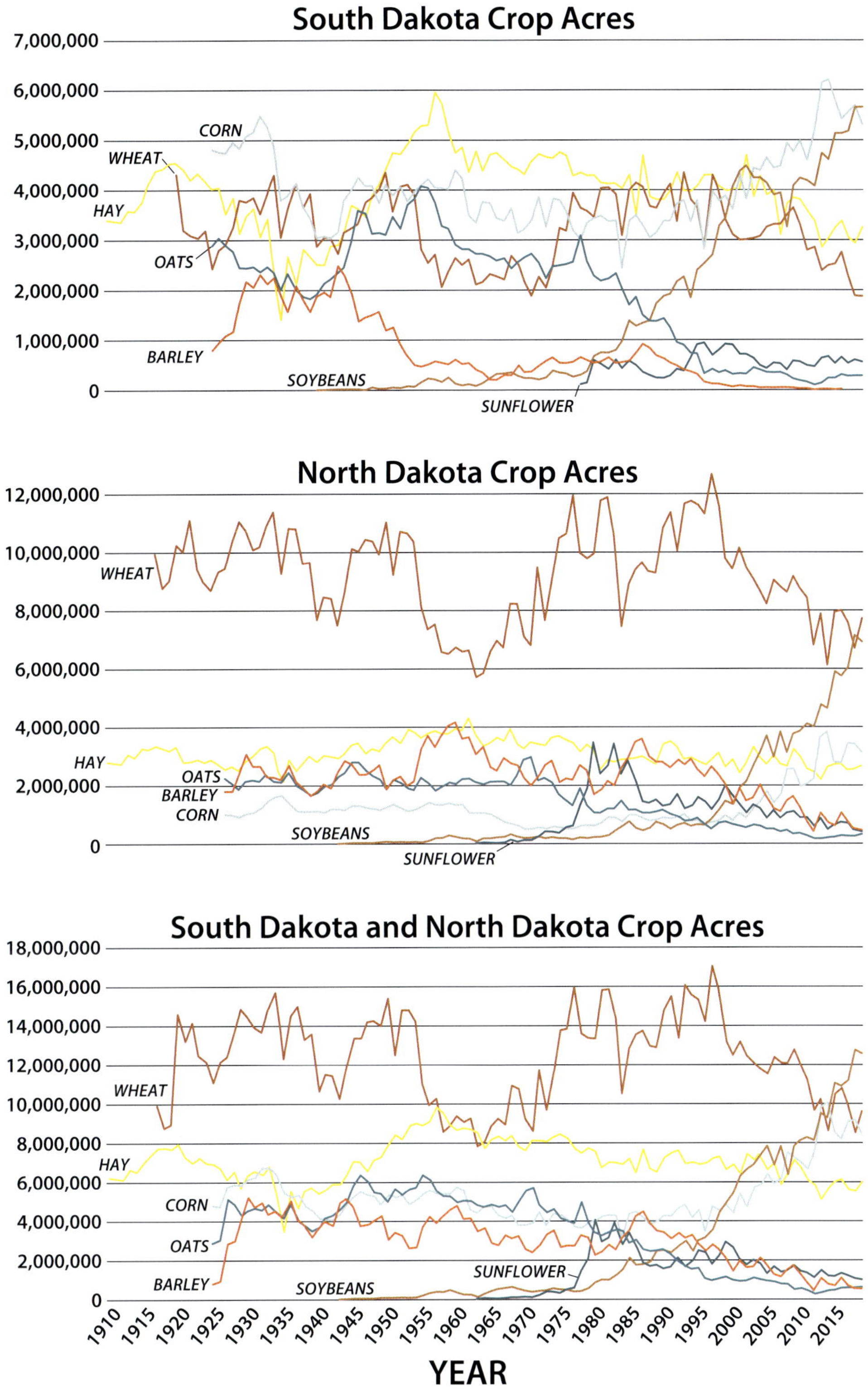

Fig. 5.8. Acres planted or harvested (hay) of main crops over the period of record (1909–2018) for North Dakota, South Dakota, and both states combined. Minor crops such as sugar beets in North Dakota contributed to farm income in certain years. Source: National Agricultural Statistics Service 2017, USDA Census of Agriculture.

- The top crops in South Dakota traditionally have been hay, corn, oats, and wheat, in that order. The most significant shift in the crop mixture has been the decline in small grains: barley dropped out early in the historical record, followed by a similar decline in oats. During the past two decades, acres in wheat and hay declined sharply and corn and soybeans increased. The current mix of crops in South Dakota is strongly dominated by corn and soybeans, with wheat a distant third.
- The mix of crops in North Dakota was overwhelmingly dominated by wheat for many years, but significant shifts have taken place in the past two decades. Wheat acreage has been halved, and soybeans increased eightfold during the same period. In 2016, for the first time in a century, a crop other than wheat was number one in North Dakota—soybeans. Other recent changes are fewer acres in sunflowers and more acres in corn, canola, dry edible peas, and lentils.
- Combing the data from the two states clearly shows reduced crop diversity, domination by corn and soybeans, and a decline in the harvest of most other crops, including hay.

Climate explains much of the agricultural differences between the two states, but geologic history is important as well. More land was glaciated in the north, enabling the formation of younger, more tillable soils over a larger area (see chapter 2).

The ratio between grassland and cropland continues to fluctuate. Early in the 1900s, some cropland on poor soils that was idled or abandoned due to drought changed naturally back to grassland, a process known as *secondary plant succession*. Some of this "go back" land reverted to federal (or public) land known as National Grasslands. More recent land-cover switching has occurred on marginal farmland idled as part of the USDA Soil Bank (1958–1965) and, later, the Conservation Reserve Program (1985–present). These lands were retired from cultivation for periods of 10 to 15 years and were planted to perennial grasses to slow soil erosion, improve soil fertility, increase wildlife habitat, and reduce the chronic oversupply of grain. Easement programs of various agencies and nongovernmental organizations maintain land in grassland cover, either temporarily or perpetually.

But in the twenty-first century, native prairie remnants are being cultivated to meet the demand for corn and soybeans, partially to supply the biofuel (ethanol and bio-oil) industry.[25] Nearly all ethanol refineries in the Dakota portion of the Prairie Pothole Region (PPR) were located in areas where at least 50 percent of the land was planted with corn and soybeans. Corn prices tripled between 2006 and 2012, as did the value of farmland relative to unplowed rangeland or pasture. Grassland conversion in the western Corn Belt was mostly concentrated in the Dakotas east and north of the Missouri River (fig. 5.9). In three years' time (2010–2012) corn and soybean acreage in this region expanded by 27 percent, or about 5,200 square miles, replacing mostly small grain fields and prairie. Some resource managers were concerned about losing still more of the little prairie that remained and the economic and environmental costs of conversion.[26]

Cultivating additional grassland has been the main way to increase cropland area in the Dakotas. However, farmers also learned that excellent crops could be grown at times on wetlands and other low ground when they were drained using ditches and tile (see chapter 11). Despite the fact that ranchers and many farmers with livestock kept wetlands for water and forage, and continue to do so, 40 percent of the original wetlands in the PPR were lost. Wetlands numbered 4.9 million around 1850 in the North Dakota PPR. By the mid-1980s, about half remained (loss of 49 percent). Wetlands in the South Dakota PPR numbered 2.7 million around 1850 and declined to 2.1 million by the mid-1980s, a loss of 22 percent. Wetland losses have continued since the 1980s despite many private and governmental programs to protect them from drainage and efforts to restore them. Losses in both states from the mid-1980s to 2011 amounted to 7.4 percent, or 0.28 percent (15,377 acres) per year. Ninety-five percent of the losses

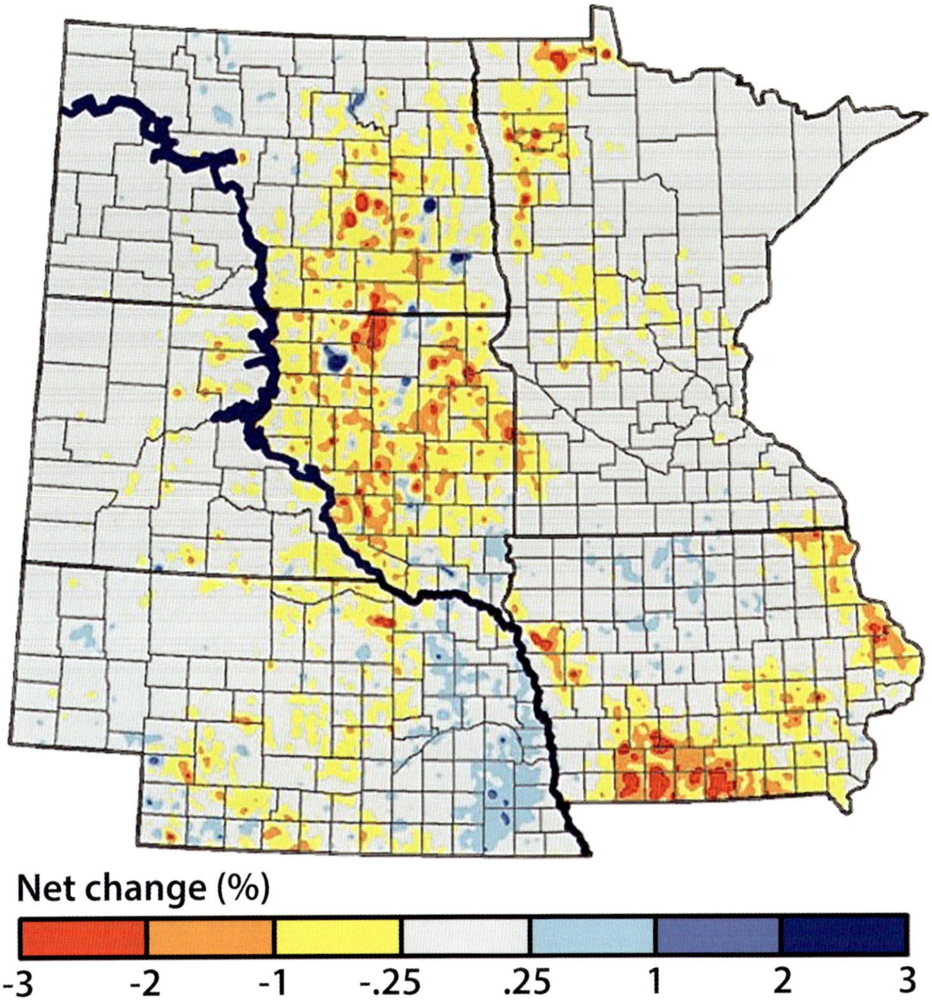

Fig. 5.9. Net rate of change (percentage per year) from grassland to corn or soybeans in the western Corn Belt between 2006 and 2011. Areas colored in shades of red and yellow are losses of grassland; areas in shades of blue are gains in grassland. Grassland area changed little in gray areas. From Wright and Wimberly 2013.

were conversions to agriculture, incentivized by record grain prices from 2008 to 2012 (fig. 5.10). Applications for tiling permits skyrocketed: 1,600 applications were filed in South Dakota in 2009 and 3,500 in 2011.[27]

Consequences of Grassland Cultivation

Motivated by a desire to produce profitable surpluses, early farmers on the Great Plains were destined to plow the vast prairies they encountered. Their fields were small, but the ecosystem was transformed dramatically, with the following consequences:[28]

- Without the prairie root system, soil erosion by wind and water increased; sediment deposition degraded water quality in wetlands, creeks, rivers, lakes, and reservoirs. Soil erosion became an environmental crisis and a threat to national security.
- As the soil was aerated, darkened by tillage, and deprived of aboveground organic matter, the rate of soil organic matter decomposition increased,

Fig. 5.10. Pattern field tiling in the Prairie Pothole Region designed to drain low, wet ground. Tiling can inadvertently drain wetlands if placed too close to them or too deep in the soil. Source: U.S. Fish and Wildlife Service.

providing a short-term spike in nutrients but in the long-term reducing fertility and causing further erosion. With less soil organic matter, the soil was more easily compacted, the rate of water infiltration was reduced, and the crops became more susceptible to drought.

- Wetlands were lost, which reduced the potential for filtering water and processing farm chemicals, maintaining high-quality and accessible groundwater, preventing floods, and providing wildlife habitat.[29]
- Grassland cover essential for the nesting of some ducks and other water birds was removed near wetlands (fig. 5.11).
- Over large areas the habitat for grassland wildlife was destroyed or fragmented, especially with the advent of fence-to-fence farming practices.
- Inorganic fertilizers made up for depleted soil fertility and declining yields from long-term cultivation, but they also led to nitrate pollution in water bodies, including sources of drinking water.
- The bare soil produced by plowing and conventional tilling was favorable for the growth of weeds, requiring frequent cultivation or the application of herbicides to sustain crop yields. Weed control often was extended to adjacent grasslands, killing desirable native plants in the process. Herbicide-resistant weeds evolved.[30]

Fig. 5.11. Removal of grassland cover can have adverse effects on adjacent wetlands and nesting waterfowl. Photo by Dennis J. Larson.

- The cultivation of larger fields of single crops, known as monocultures, facilitated the invasion of damaging pathogens, insects, and other organisms. Pest control became a major challenge and often required the use of pesticides that harmed nontarget species, including insect pollinators and crop pest predators.
- Fences and season-long grazing contributed to range deterioration when too many animals were confined to small areas.
- As bigger machinery became necessary to manage larger farms, farmers plowed through rather than around habitats such as shallow wetlands or subirrigated meadows. Also, some windbreaks were removed.
- Irrigation and cultivating subirrigated land often led to soil salinization.
- Biodiverse prairie land marginal for agriculture was cultivated, quickly depleted of fertility, and enrolled in government-funded set-aside programs, such as the Soil Bank and Conservation Reserve Programs. Extreme examples of grassland conversion, such as from pasture and hay to cropland, used heavy machinery to remove large rocks (fig. 5.12).
- Populations of native grouse (sharp-tailed grouse and prairie chickens) declined as agriculture reduced grassland habitat. This loss was softened by a corresponding increase of the ring-necked pheasant (box 5.1). These species occur together where there is more grassland but still some agriculture. The pheasant has become both a mascot and economic windfall for the Dakotas.

Fig. 5.12. Large granitic rocks deposited by glaciers and pushed to the surface by soil freezing and thawing were often removed by hand and thrown into piles, such as shown in the top photo. Larger boulders can be removed with heavy machinery, thereby enabling the conversion of rangeland to cropland. Bottom photo by Pete Bauman.

These consequences were tolerated for many years because farming and livestock production provided a good living and sustained towns and cities—indeed, they have been and still are the major long-term contributor to the economy of both states. But modern agriculture presents a predicament. The fate of humanity depends on

Box 5.1. Pheasant Capitalism

The ring-necked pheasant is native to China, but South Dakota crowned this cocky alien its official state bird in 1943. What was it about this avian "immigrant" that moved legislators to choose it over homegrown candidates such as the bobolink or the dancing prairie chicken? The answer is easy. The pheasant generated income and provided more recreation, and it was preadapted to living in the Dakotas because of a similar climate to its native China. It took to the agriculturalizing states like a duck to water without being cooped up like its domesticated cousin, the chicken. The pheasant population exploded after first arriving in 1908, justifying the first hunting season in 1919 in South Dakota and 1931 in North Dakota. The bird flourished without needing a handout. Billions of dollars are spent in the United States controlling invasive plants and animals, but except for concerns about mild competition with native grouse, the pheasant had few detractors.

Pheasant numbers have ranged widely, from 1.4 million to 16 million in South Dakota. Their populations tend to be lower in North Dakota, though they are abundant locally. Regulating factors include a favorable balance between grasslands for nesting and cropland with waste grain, along with the presence of wetlands, fencerows, and weedy field edges. Abrupt declines are caused by severe droughts, winter blizzards, weed-free farming, wet springs, pesticides that kill beneficial insects, and overhunting. Farming practices are critical since they influence conditions every year. Unfavorable weather is episodic. Commercial hunting operations often compensate for poor reproduction by stocking millions of pheasants raised in captivity.

The pheasant brings families and friends together, not unlike the national holiday of Thanksgiving. Participants gather at farms on the opening weekend of hunting season to share a meal, get a crop report, and reminisce about past hunting successes. The chance to hunt pheasants in the Dakotas has not escaped sportsmen in other states and countries, as approximately 175,000 nonresident hunters in the Dakotas contribute approximately $350 million annually to the economy. Pheasant capitalism is alive and well, but pheasant prospects have dimmed in recent years as a result of the lowest numbers since surveys began in 1949. The long-held view that agriculture favors the pheasant is being questioned. Has farming removed too much of the grassland needed for nesting? Are fields too "clean" when fencerows are removed, wetlands are drained, and so few weeds and insects survive spraying? Perhaps the pheasant is a "canary in the mine," warning that an imbalance exists. The fact that this introduced bird is celebrated by so many Dakotans for its contributions to their economy, culture, and outdoor sport provides strong incentives to solve this problem.

[a]Flake et al. 2012; Errington and Gewertz 2015.

food production from croplands, but the costs of production and environmental consequences are substantial.[31] Some scientists wonder whether modern agriculture is sustainable, even with generous government subsidies in the form of crop insurance, disaster assistance, and tax exemptions. To address this dilemma, the scientific discipline of agroecology emerged, with the goal of preventing or minimizing further adverse environmental impacts.

Agroecology and Achieving Sustainability

Considering that the native prairie ecosystem evolved over thousands of years, surviving many episodes of drought, fire, and periods of heavy grazing, agroecologists on the Great Plains have arrived at a fundamental principle: the more different a ranch or farm is from the native prairie, the less sustainable it is likely to be unless measures are taken to replace the ecosystem services that enabled prairie ecosystem persistence.[32]

Fig. 5.13. Cow-calf pairs on the Miller Ranch near Fort Rice, North Dakota. Adopting a rest-rotation management program on this big-bluestem pasture increased both grass and beef production while maintaining other ecosystem services. Paddocks were reduced from 40 to 20 acres, and cattle were moved more often than before. Photo by Kenneth Miller, Miller Ranch, Fort Rice, North Dakota.

Fig. 5.14. No-till cropland where soybean seed was planted into stubble and crop residue. Photo by Cody Zilverberg.

Accordingly, a well-managed sheep, cattle, or bison ranch with native grassland is more sustainable than a farm dependent on annual row crops. Rangeland ecosystems are dominated by perennial native plants; there is no need to disrupt the root system and soil structure (fig. 5.13). Similarly, a farm that incorporates cover crops, maintains buffer strips of native species, and uses no-till or minimum-till practices is more sustainable than a farm where these conservation practices have not been adopted (fig. 5.14). Mimicking some or all of the features of natural grasslands promotes farm sustainability, defined here as the ability to sustain production for centuries.

Taking a different perspective on sustainability, the late eminent ecologist Eugene Odum identified contrasts in the flow of energy and nutrients within native ecosystems and croplands.[33] He emphasized the following:

- Native ecosystems are solar powered, whereas fossil fuels and other inputs are required for modern crops and livestock. From the 1950s to the 1990s, the costs of fuel, seed (including genetically modified seed), machinery, fertilizer, pesticides, and human labor rose from about 50 percent of farm income to 75 percent.[34]
- Most nutrients are recycled within native ecosystems; in contrast, croplands require fertilization because soil and nutrients are lost by erosion and

harvesting. Also, the water-holding capacity is diminished as soil organic matter declines.

- Perennial grassland plants are self-maintaining, generally long lived, and tolerant of environmental extremes, in contrast to annual crops that must be replanted at the start of each growing season and cultivated to ensure survival for a few months.

Farmers have been adopting conservation practices for many years. To minimize water erosion, many of them left steep slopes unplowed or reseeded such slopes to perennial grass cover. Windbreaks were planted to reduce wind erosion, and contour farming practices were adopted. Buffer strips of prairie plants along streams help maintain water quality.[35] To increase economic stability, different fields were planted to different crops, and the same crop usually was not planted in the same field in two consecutive years (crop rotation)—two measures that reduced the likelihood of catastrophic losses to pests and pathogens. Nitrogen-fixing plants such as alfalfa, clover, peas, lentils, and soybeans are now planted periodically to reduce the need for inorganic nitrogen fertilizers. Early farmers routinely spread manure from dairy barns on their fields, or allowed their cattle to graze on stubble, thereby restoring some of the nutrients and soil organic matter that was exported by harvesting. Soil microorganisms have come to be appreciated as well. One wheat farmer without livestock commented, "It's not that I don't have any livestock, it's that mine are microscopic."[36] Micromanure is more evenly distributed in the soil than livestock manure. Indeed, as noted in chapter 4, most of the herbivory in grasslands is belowground. Planting late-season cover crops, sometimes referred to as "green manure," is important as well.

After a thorough review of modern agricultural practices, the MacArthur Fellow David Montgomery concluded in 2017 that farm profitability and sustainability could be greatly improved by restoring or maintaining soil health or fertility, which is accomplished by adopting no-till cultivation and complex crop rotations that include cover crops. Such practices are becoming common on Dakota farms, particularly when growing wheat in drier areas. Nationally, roughly half of the farms have used reduced-tillage practices at least once over a four-year period. Such practices include the following:[37]

- Minimum disturbance of the soil, opening the soil only enough to plant seeds
- Keeping the soil covered to minimize erosion, evaporation, nutrient loss, and the establishment of weeds (fig. 5.15)
- Varying the sequence of cash crops and cover crops to minimize the abundance of weeds and other pests
- Intercropping strips of annual crops with forage plants to capture the benefits of livestock grazing, such as manure, and promoting the growth of soil microorganisms that contribute to soil fertility through their micromanure
- Diversifying the farm enterprise to enable livestock grazing on fields with stubble, forage crops, and cover crops
- Encouraging biological nitrogen fixation
- Allowing three years to see benefits, namely, less labor, fewer costly inputs, and increased yield

Unfortunately, as farms became larger and more specialized in the late twentieth century, and more dependent on larger machinery, conservation practices were sometimes set aside. Most farmers no longer have livestock to provide manure, and shelterbelts and even fences often have been removed because they interfered with machinery, cast shade on crops, or occupied valuable farm ground. The habitat for agrarian wildlife became still more fragmented and less diverse. The need to maximize profits sufficient to repay loans and mortgages necessitated more external inputs, such

Fig. 5.15. Cover crop planted in late summer to protect soil prior to initiation of grassland restoration the following spring. Plants include Sudan sorghum, field peas, sunflowers, radishes, red clover, and oats.

as fertilizers, pesticides, herbicides, and fossil fuels, all of which added to the costs of doing business.

Land-grant universities, the U.S. Department of Agriculture, and other federal and state agencies took note of these disturbing trends. New machinery was designed for minimum-till or no-till agriculture, and financial incentives were provided for planting perennial plants on erodible soils and protecting the wetlands that remained. Nongovernmental conservation organizations and some individuals purchased farms for the purpose of maintaining or restoring wildlife habitat and providing opportunities for outdoor recreation, thereby adding diversity to the agrarian landscape as a whole.[38] The value of biological diversity to the public was reflected by passage of the Endangered

Species Act of 1964. As in cities, new technology enabled greater use of wind, solar, and other forms of renewable energy.

Conserving soil, using fertilizers and other chemicals efficiently, and protecting wildlife habitat when possible have all become practical and economically important goals for some farmers on the Great Plains.[39] Many managers strive to maintain the diversity of life in the soil, now considered an important component of soil health. However, drought and floods, the primary causes of crop failure, are beyond their control. The damage to crop plants can only be mitigated, to some degree, by agronomic practices. These include fallowing land to conserve moisture, adopting drought-resistant crops, and breeding drought resistance into conventional crops. Additionally, adopting minimum-tillage or no-tillage systems and leaving crop residue reduces evaporation and increases water infiltration, thereby conserving soil moisture for plant growth. Retaining or restoring wetlands can minimize flood damage (see chapter 11).

Reducing costs is one sure way to increase the likelihood of at least small profits. For example, parasitic wasps kill crop pests, thereby reducing the need for pesticides, but the wasps also require natural habitat to complete their life cycle. And pesticides often kill indiscriminately. Also, many crops require insect pollination, but pollinators have become scarce because they, too, require areas of natural habitat with nectar-producing native and introduced plants. Furthermore, expansion of cropland has reduced natural habitat needed for commercial honey bee apiaries. The loss of pollinators and other beneficial insects is now considered a crisis by many farmers.[40]

Some insects are appreciated for their beauty as well. The best-known example is the eastern North American population of migratory monarch butterflies (fig. 5.16), which includes those in the Dakotas. This population has declined by about 80 percent in the past 20 years, a decline caused by fragmentation of their grassland habitat and "clean farming." All stages of the monarch's life cycle require milkweed plants, from egg laying to the emergence of new adults. A dozen or so species of milkweed historically occupied prairie remnants, croplands, field margins, and rangeland. Prior to the application of glyphosate-based herbicides, this competitive perennial plant was found on virtually all farms. While clean farming increases crop yields, it has greatly reduced the once-common plant that monarchs depend on. About 1.3 billion stems of milkweed were killed by herbicide during the past 20 years. Restoring milkweed is essential to increase monarch populations.[41]

Recognizing the challenges of modern agriculture, some farmers and ranchers have been experimenting with new approaches, often with the support of government subsidies. Many have adopted satisfying innovations that are more profitable than conventional farming practices. Should elements of these experiments become mainstream, agriculture in the Dakotas will have a different look in the future. The following examples illustrate some of what has been learned, first on western rangelands and then on farms.

Innovations for Western Rangelands

The devastation of the Dust Bowl years did not discourage the Mortenson family from establishing a cattle ranch on abandoned farmland west of the Missouri River near Pierre.[42] Clarence Mortenson's stepfather, Ben Young, consolidated about 40 homestead parcels and set about to reclaim the land. Clarence's first job as a 12-year-old, in 1941, was to clean up. Fences were still buried in windblown soil and the area was littered with broken posts and wire cut during the drought to free entangled horses and cattle. While tearing down a dilapidated shanty, Clarence saw his first live deer. Homesteaders desperate for food had virtually eradicated all edible game in the area. Skulls of both deer and pronghorn were common around abandoned buildings. Clarence had sympathy for

Fig 5.16. Monarch butterfly caterpillars (above, Virginia Arboretum CC-ND-2.0) feed on milkweed, while the adults (right) consume the nectar produced by the flowers of numerous plants, including the purple coneflower (echinacea) shown here.

the homesteaders, commenting later: "On a quarter section in this country, no one could've or should've been expected to make a living." Larger homesteads were not allowed until after many small farms had failed.

The condition of the vegetation at the end of the drought was dismal, with every acre overgrazed or farmed and then abandoned to weeds. Cool-season grasses disappeared, and warm-season grasses were severely depleted. The trees and shrubs that had grown in the woody draws (breaks) and along streams all had been cut and burned for cooking and heating or were used for fence posts and corrals. The oldest ash trees currently growing on the ranch date back to about 1940, when many homesteaders left and the weather began to improve.

The Mortenson family chose to raise livestock but converting abandoned fields to productive rangeland took many years of trial and error. The Soil Conservation Service initially recommended planting crested wheatgrass introduced from the steppes of Russia. It established easily, was resistant to drought, and the seed was cheap. Later, the Mortenson family discovered that native species—western wheatgrass, green needlegrass, sideoats grama, and legumes, including lead plant, dwarf indigo, and sensitive briar—were more nutritious for cattle and flourished with the return of normal rainfall. Crested wheatgrass had done its job during the emergency. The sale of native seed improved ranch income in some years.

The heavy weed cover on abandoned fields provided habitat for native grouse as well as pheasants, although pheasants became less common as the weedy fields developed into rangeland. Weight gains made by young steers were excellent, and some parcels of the restored rangeland produced more income per acre than did nearby crop farms. In addition to abandoned farmland, the Mortenson ranch included prairie remnants that gradually recovered with careful livestock management (fig. 5.17).

Another challenge for the Mortensons was Foster Creek, which "scoured out" during every runoff event. In 1942, when looking at the deep gullies on Foster Creek, without a tree in sight, Clarence recalls an old homesteader telling him, "When I came here this creek could be crossed at a trot with team and buggy anywhere; it was tree-lined and grassy-bottomed, had a water hole that never went dry about every mile, and the grass was belly-deep on a team of horses." Fixing this problem required reducing runoff in the uplands by adopting a rest-rotation grazing system, restoring floodplains and shallow aquifers by building dams to slow water and capture sediment,[43] planting native riparian plants, and adopting a comprehensive management approach that includes all parts of the ranch.[44] The goal of restoring wetland grasses and a tree-lined channel to Foster Creek

Fig. 5.17. Mixed-grass prairie in Todd's Draw at the Mortenson Ranch after 70 years of restoration. Oahe Reservoir on the Missouri River is in the distant background. The trees are green ash, hackberry, and, near the creek, cottonwood.

has been partially achieved (figs. 5.18, 5.19, and 5.20). Patches of native cottonwood and willow have established, and a buggy could cross some places at a trot—but you had to hold on.

The family had to earn a profit from their ranch, but they also appreciated the amenities of their restoration work. A summer survey recorded 70 species of birds, many more than during the Dust Bowl. Grassland birds included the lark sparrow, grasshopper sparrow, horned lark, bobolink, western meadowlark, lark bunting, dickcissel, greater prairie chicken, sharp-tailed grouse, long-billed curlew, and the burrowing owl. Woodland birds became more common as well, such as the black-headed and blue grosbeaks, Baltimore oriole, spotted towhee, red-headed woodpecker, yellow- and black-billed cuckoos, and the cedar waxwing (fig. 5.21). Also, large numbers of migratory birds now use the ranch as a resting area. Most significant, all economic and environmental measures for the Mortenson family steadily improved during the long period of restoration and adaptive management.[45]

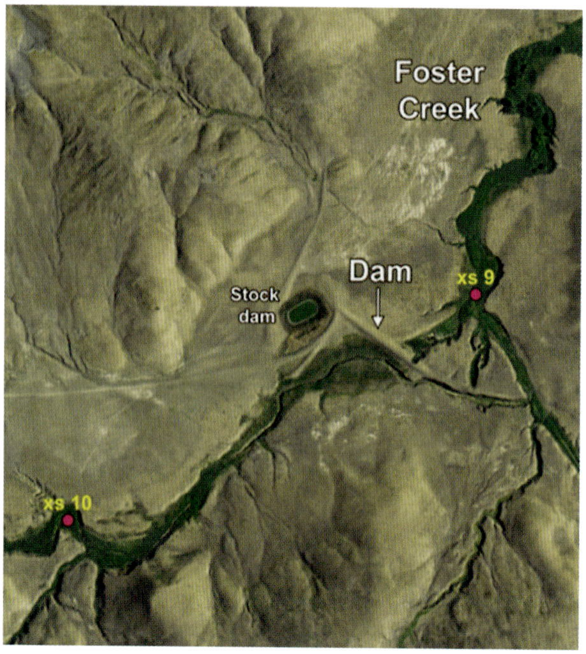

Fig. 5.18. Aerial image of Foster Creek on the Mortenson Ranch that underwent major renovation, including the construction of silt-collecting dams to raise the creek's floodplain. Red dots mark the location of stream cross sections where changes in stream elevation and width were monitored following changes in livestock management.

The Mortenson family also restored woody draws to provide winter pasture and shelter for spring calving. The expansive cottonwood forests where calving occurred historically were destroyed in 1958 by the filling of Oahe Reservoir on the Missouri River. Woodlands expanded when cattle were moved out of the draws during summer to prevent trampling, browsing, and rubbing. Calf survival increased with more shelter from trees and shrubs.

Two additional examples of alternative management in western South Dakota are the Wild Idea Buffalo Company and the Bad River Ranch. Both operations have reaped benefits from replacing cattle with bison (fig. 5.22). The Wild Idea Buffalo Company has two ranches located between the Black Hills and Badlands National Park. The home place is the 28,000-acre Cheyenne River Ranch of Dan and Jill O'Brien that borders the Buffalo Gap National Grasslands and Pine Ridge Reservation. Another 32,000-acre parcel, the Conata Basin Ranch, is within view of Badlands National Park. Their goal is to help preserve the grassland ecosystem while offering consumers 100 percent grass-fed meat as an alternative to feedlot beef. The 141,357-acre Bad River Ranch, located west of Pierre and owned by Ted Turner of Turner Broadcasting Company, strives to achieve ecological and economic sustainability with grass-fed meat and the successful reintroduction of species of concern, such as the swift fox, black-footed ferret, and prairie dog. The ranch sells bison meat in its own restaurants and offers commercial hunting of deer, pronghorn, bison, and turkey. Restored ranches such as these are providing benefits to the owners and the public. Management is less intensive than that of tillage farms and mostly involves the harvesting of cattle and bison. Culling of

Fig. 5.19. Prairie cordgrass grows vigorously on this restored wet meadow along Foster Creek on the Mortenson Ranch, located near the confluence of the Cheyenne and Missouri Rivers.

Fig. 5.20. Early stages of cottonwood-willow woodland recovery along Foster Creek at the Mortenson Ranch. This area was devoid of trees until about 1990. Prairie cordgrass (bright green) also grows on the floodplain.

Baltimore oriole

Common yellowthroat

Cedar waxwing

Blue grosbeak

Fig. 5.21. Woodland birds once absent but now common following restoration of the woody draws on the Mortenson Ranch include the Baltimore oriole, common yellowthroat, blue grosbeak, and cedar waxwing. Photos by Doug Backlund.

bison herds is now done by humans rather than wolves and grizzly bears. While ecologically sustainable, profits depend on skilled management, the demand for meat from grass-fed animals, and, in some cases, the popularity of guided hunting.

Innovations for Farmland

For three decades, the mission of the Dakota Lakes Research Farm in central South Dakota has been to identify "the best methods of stabilizing the

Fig. 5.22. Bison on the Wild Idea Buffalo Ranch near Badlands National Park. Photo by Jill O'Brien.

agricultural economy through promoting agricultural diversity, increasing production efficiency, minimizing negative environmental effects, maintaining soil productivity, and developing techniques to mitigate biological stress effects."[46] This not-for-profit corporation is owned by farmers and dedicated to improving both the environment and profitability by advocating crop rotation, crop diversification, and no-till or zero-till farming. Wind erosion has been greatly reduced, essentially to zero, and the farm is striving to eliminate dependence on fossil fuels by 2025. It is envisioned that biofuel along with wind and solar energy will power farm machinery, and research is under way to increase biological nitrogen-fixation rates in crop fields. One of the largest users of fossil fuels in modern farming is the production of nitrogen fertilizers. There is the potential to improve energy efficiency in farm buildings by using biological foam insulation. Overall, the project emphasizes ecological solutions to problems, a foundational element of agroecology. Dwayne Beck, project leader, writes that "almost all of the agronomic problems we face (seeds, diseases, insects, fertility, etc.) can be traced to problems with ecosystem processes."

Another nonprofit company, EcoSun Prairie Farms Inc., took a different approach, leasing a 475-acre corn and soybean farm near Brookings, South Dakota, and restoring it to native grassland and wetlands (fig. 5.23). This farm had been under traditional tillage for over a century.[47] The goal was to determine how much income could be generated through the marketing of grassland products

Fig. 5.23. Reconstructed, high-diversity prairie at the EcoSun Prairie Farm. Three years prior, this was a 100-acre soybean field. Dominant prairie forbs include purple prairie clover, white prairie clover, Canada milkvetch, prairie coneflower, and purple coneflower. Grasses planted were four cool-season species (western wheatgrass, Canada wildrye, slender wheatgrass, green needlegrass) and three warm-season species (big bluestem, switchgrass, Indiangrass). See box 13.1.

such as forage hay, biofuel feedstock, native prairie seed, and prairie-raised beef. Attempts to restore about a dozen drained wetlands were started. By its very nature, this approach has high potential for sustainability because it regained most tallgrass prairie ecosystem services lost when the grassland was converted to traditional grain production. Not surprisingly, the Prairie Farm's rich soils produced high yields of grass. This seven-year project (2008–2014) demonstrated that a grass farm in the heart of the western Corn Belt can be as profitable as, and at times more profitable than, grain farming.

The key factor in EcoSun's profitability was the low input costs—less than half of typical costs—for fuel, agrochemicals, and equipment compared to conventionally tilled grain farming. The planted native grasses and forbs are perennial, so the expense of annual planting is eliminated, and native plants are more drought tolerant than corn and require very little fertilization. The highest profit per acre came from the seed and hay of prairie cordgrass and sedges in wetlands. Several years after planting, net profit was about $60,000 per year, significantly above the median household income of $49,415 for South Dakota at that time. In that year about 8,000 pounds of seed, 350 tons of hay, and 4,600 pounds of prairie-raised beef were marketed. Bringing grass-fed cattle to market weight on restored grassland and pastures eliminates the cost of shipping cattle out of state as is done now for fattening in confinement, as well as increasing pasture acreage, producing leaner meat locally, and generating more farm income. In addition to crop and beef production, EcoSun's approach reduced soil erosion, increased groundwater recharge, reduced runoff into streams and rivers, augmented soil organic matter, sequestered more carbon, increased native plant diversity, and provided more habitat for at-risk grassland birds (bobolinks, upland sandpipers, and grasshopper sparrows), migrating waterfowl, and chorus and spring peeper frogs, to name a few. Such benefits are not easily valued, although economists are now promoting the discipline of conservation finance, with the primary goal of buying and selling ecosystem services (see chapter 13). The potential for a prairie farm approach would be greater if government-supported cropland insurance available to grain farmers were made available to the farmers of restored grasslands.[48]

Also near Brookings is the Blue Dasher Farm, a 50-acre farm that produces a variety of perennial fruit and nut crops, annual vegetable crops, honey, and small livestock (mainly pigs, goats, and chickens). The animals are raised for meat, eggs, and milk. The land is managed to maintain pollinators and other insects known to be effective for dung recycling and preying on crop pests. In addition to commodity production, the farm staff organizes workshops on agroecology, integrated pest management, risk assessment, soil health, no-till cultivation, and cover cropping.

A well-known example of alternative farming is Brown's Ranch, located on 5,000 acres near Bismarck, North Dakota.[49] The goal of Gabe Brown's family is to improve soil health as a means of reducing or even eliminating fertilizer, fungicide, and pesticide applications. Livestock are fed crops grown on the farm, and the manure is returned to the fields. No-till cultivation reduces soil disturbance, conserves moisture, and builds soil organic matter, thereby sequestering carbon. Soil organic matter has increased from about two percent to over six percent over the last 25 years. Intact prairie soils commonly have 6–8 percent. The Browns sell pastured hogs, grass-finished lambs, and grass-fed beef and eggs (from 1,000 free-range hens) under their own label, Nourished by Nature, with drop-off sites for their products in six North Dakota cities.[50] Their agronomic crops are diverse: corn, spring wheat, oats, barley, peas, rye, hairy vetch, winter triticale, and sunflowers. Direct marketing to customers and reducing input costs by improving soil health are two reasons why the Brown farm is more profitable than some others and likely to be more sustainable.

The journalist Stephanie Anderson's interview with Gabe Brown revealed unconventional thinking about farming in the Dakotas.[51] First, Brown is committed to regenerative or restorative agriculture. Grassland, he says, is the most important element of his operation. For him, one measure of his farming success is the number of prairie grouse and songbirds on his farm. His farming is guided by ecological principles, but he also wants a secure financial future. As Anderson wrote, a cornerstone of his argument is that "conventional farmers work harder for less money, while regenerative farmers work less and earn more. Convincing farmers to accept this fact is the first step in transitioning our nation's agriculture."

One last example is illustrated by the work of The Land Institute in Kansas, founded by Wes Jackson in 1976. The goal has been to use plant breeding to develop the ultimate grain crop—a perennial plant that would not require the cost of tilling and planting every year. Imagine the benefits of a perennial variety of corn or wheat. The soil would rarely be disturbed. Some success has been achieved with the development of Kernza, a variety of intermediate wheatgrass. Creating a perennial variety by crop breeding is a slow process but accomplishing this goal could lead to the first major new crop in over 3,500 years.[52]

These examples illustrate farming approaches that mimic the prairie ecosystem to varying degrees. There are many challenges—on the ground and with regard to marketing and government policy—but they demonstrate that a better balance between economics and environment can be achieved.[53]

David Montgomery, in his book, *Growing a Revolution, Bringing our Soil Back to Life*,[54] described the problem this way. "It boils down to investing in fertility through enlisting trillions of microbial helpers, agriculture's living foundation." He maintains that conventional tillage kills many microorganisms and mines the soil, hence the need for adding expensive agrochemicals and using so much fuel. The focus should be to maintain farm soils filled with beneficial bacteria, mycorrhizal fungi, nematodes, arthropods, and protozoans, all contributing to soil food webs and releasing excrement that provide nutrients for crop plants. Montgomery presents evidence that, within a few years after adopting this approach, farm yield is likely to be the same or higher and profitability will be higher because of reduced input costs. Moreover, fields farmed in this way sequester carbon and hold nutrients and water, thereby contributing to the maintenance of air and water quality. Montgomery thinks of this as *restorative agriculture* and recommends three guidelines to extend the productive life of farms and ranches: minimize soil disturbance (ditch the plow); grow cover crops (cover up); and devise crop rotations that work together as a system (grow diversity). The Food and Agricultural Organization and the World Bank promote these same practices globally, to produce "climate smart" productive crops, to lower greenhouse gas emissions, to capture more carbon in soils, and to bolster resilience to climate change.

Montgomery also identified four reasons such practices have not yet been widely adopted. First, the largest and most influential agricultural agency, the U.S. Department of Agriculture, has lagged behind nongovernmental organizations and private investors who have incentivized alternative agriculture. Second, industry executives and political appointees typically lead government agencies, which perpetuates the status quo. Third, farmers who are losing soil fertility and those who are building it are treated equally for government crop insurance benefits, thereby promoting practices that degrade soils. And fourth, resistance to change exists because so many farmers and agricultural authorities are trained in conventional methods. Montgomery maintains that a more level playing field for conventional and alternative farming could be achieved with changes in bank financing, government subsidization, research funding, and product marketing.[55]

Adapting to Climate Change

Normal variability in temperature and precipitation dramatically affects all aspects of agriculture. In addition to this reality, world economies are bracing for the anticipated effects of climate change on the production of food for a burgeoning world population.[56] Some projections are welcomed in cold climates, such as the warming and lengthening of growing seasons, allowing producers to grow late-maturing crops. For example, warmer weather in North Dakota during the past few decades, along with more precipitation (see chapter 3), probably combined to produce higher yields of corn and soybeans in some areas where wheat and other short-season grains had been grown previously. Corn and soybean acreages have sharply increased in response to the higher yields and expanding international markets. Another expected benefit of warmer winters favors the growth and health of unconfined livestock.

But despite increases in precipitation in the past several decades, especially in the 1990s, there have been episodes of drought. The most severe drought in North Dakota's history occurred from about 2000 to mid-2006.[57] Other dry years were 2012 and 2017. Even if the projected increases in precipitation occur, the future climate is expected to be punctuated by more frequent and longer droughts. The increase in atmospheric carbon dioxide, itself a major plant nutrient, may increase the productivity of certain crops, such as cool-season range grasses, but only when soil nutrients and water are not limiting.

In general, there are more negative than positive outcomes of climate change. Even if precipitation does increase, it may not be sufficient to counterbalance the higher rates of evapotranspiration driven by rising temperatures. The net effect in most years is expected to be lower crop production, the magnitude of which will depend on the degree of imbalance. Increased use of crop irrigation may mitigate the drier climate of the future, but that solution is considered unsustainable over large areas. Only about 2 percent of the harvested cropland in the Dakotas was irrigated in 2017.[58] Another concern is that warmer and more humid conditions are expected to cause more crop diseases, especially for small grains such as wheat. Also, heavier rainstorms (see fig. 3.7) will cause cropland flooding and soil erosion that sometimes will occur during spring planting and fall harvesting. Long-delayed and unsuccessful spring planting throughout the Northern Great Plains in 2019, as a result of record precipitation, is an example of this costly consequence of climate change.[59]

Additional consequences of climate change involve agricultural weeds and insect pests favored by warmer winters and summers. Earlier springs and longer growing seasons will allow some insects to emerge earlier and produce more generations in a single season. Also, higher humidity can cause outbreaks of crop diseases, such as scab (*Fusarium* head blight). Weeds common to the south may move northward into the Dakotas. Rangelands will not be exempt either, as the warmer and drier conditions forecast for the western Dakotas may reduce livestock production and forage quality.[60] Climate change seems to have contributed to the invasion of smooth brome and Kentucky bluegrass into native rangeland (see box 4.3).

Perhaps the most difficult issue facing Dakota agriculture is how to adapt to a more variable and less predictable climate, as if the notoriously variable historical climate were not challenging enough. Short-term strategies include the following:

- Switching to crop types and varieties better adapted to the future climate
- Increasing crop and livestock diversity
- Adopting crop rotation systems that can be evaluated and adjusted annually
- Increasing the proportion of a farm in perennial grassland, including pastures

The long-term strategy for adaptation involves recovering ecosystem services that add resilience to extreme events that are likely to occur. Its centerpiece would be the improvement of soil health—restorative or regenerative agriculture. Fred Kirschenmann, an agricultural futurist and North Dakota farmer, recommends redesigning farms to include more diversity and redundancy, given the likely future of an unstable climate and the end of cheap energy. These changes can start the process of improving soils so croplands can recover more quickly after periods of drought or excess rainfall. David Montgomery wrote: "I find it fitting that *humus* and *human* have the same Latin root, as restoring healthy soil to the world's agricultural lands is one of the best investments we can make in humanity's future. And so as we grapple with the daunting problems of how to feed the world, cool the planet, and stem losses in the natural world, let's not lose sight of a simple truth. Sometimes answers we seek are closer than we might think—right beneath our feet."[61]

Overall, farmers and ranchers have already demonstrated an ability to adapt to new markets and management strategies, sometimes in a rather short period. Restoring ecosystem services whenever possible will help ensure that today's agroecosystems will supply future generations well into the future. Further improvements surely will be made.

Chapter 6 Trees on the Farm

One of the most dramatic changes on the Great Plains during the past century has been the planting of millions of trees in shelterbelts, also known as windbreaks. Immigrants of European descent had come from wooded landscapes, and many were anxious, if not frightened, by the prospect of living on the vast prairie, away from the shade, firewood, and shelter that trees provided. Homes were built near native woodlands whenever possible. Those who arrived later settled on the prairie but soon began planting trees. Shelterbelts in the eastern Dakotas now cover several times more land than native woodlands in some areas.[1]

The success of shelterbelts led ecologists to debate why trees were rare before planting began. Some thought that prairie soils did not have the microbial organisms that tree seedlings required, particularly the right kinds of mycorrhizal fungi, but that idea has been debunked. A better explanation is that small tree seedlings have difficulty competing with well-established prairie plants for water, nutrients, and light. Some are successful, but they survive only until the next fire. Shelterbelt trees usually were transplanted saplings nurtured by weeding and watering, thereby avoiding the difficulties of seedling establishment in a soil fully occupied by prairie plants. Fires were suppressed whenever possible.

Federal and territorial initiatives, along with railroad companies, enticed new settlers to the Great Plains to help provide commodities for sale in eastern markets and help secure the vast lands for a growing nation. The Timber Culture Act of 1873, passed by Congress 11 years after the Homestead Act, granted 160 acres to any homesteader who would plant 40 acres of trees—primarily to protect crops and cropland (figs. 6.1 and 6.2). But as U.S. Forest Service Chief Ferdinand Silcox wrote in 1935, "A larger and more vital value [of trees] . . . is the reinforcement of the people's morale that comes with shade from sun glare, shelter from the ever-prevailing winds, the improved appearance of the countryside, a greater pride in ownership, and a real increase in value of the farmstead—all culminating in a general sense of being at home on the land." Robert Gardner, a scientist, wrote in 2009 that the "vast openness and turbulent climate made trees a practical and psychological necessity." It must have been frustrating for newcomers from the east when many of their newly planted trees died during droughts and prairie fires.

Various incentives were provided for planting trees in the Dakotas, but none was more effective than the Prairie States Forestry Project, started in 1935 during the most severe drought experi-

Fig. 6.1. Parallel shelterbelts, also referred to as windbreaks, are planted because the moderating effects of trees decline downwind. The yellow crop is canola, a variety of rapeseed in the mustard family that is commonly grown in North Dakota to produce cooking oil. Photo by Derek Lowstuter, North Dakota Forest Service.

Fig. 6.2. Approximately 15 years old, these shelterbelts in eastern South Dakota have one or two rows of five species: American plum (1), amur maple (2), Russian olive (1), Rocky Mountain juniper (2), and eastern redcedar (2). On the left is a native riparian woodland (gallery forest) with green ash, boxelder, hackberry, and American elm. Photo courtesy of Nathan Kafer, South Dakota Department of Agriculture.

enced by immigrants of European origin.[2] By that time, some families had worked the land for 50 years or more and were doing well in most years. Gasoline-powered tractors and other machinery had replaced draft animals, larger plows enabled the breaking up of sod over larger areas, and livestock were moved to fenced pastures that were too rocky for cultivation. As the drought worsened, crop yields were greatly reduced, many pastures were overgrazed, and erodible soils were left bare for much of the year. Loose soil and desiccating winds created unimaginable dust storms—veritable "black blizzards" that obscured the sun. Farm families and nearby towns were devastated; dust from the Great Plains drifted eastward to the Atlantic Ocean. Books have been written about the Dust Bowl, which lasted from about 1933 to 1939 and is attributed to prevailing attitudes about conquering nature and a general lack of awareness about what could happen during severe drought if the protective grassland sod is broken over large areas. The historian Donald Wooster wrote in 1979, "The Dust Bowl was the inevitable outcome of a culture that deliberately and self-consciously set about dominating and exploiting the land for all it was worth."

By all accounts, the Dust Bowl years on the Great Plains created a national emergency—one that occurred coincidentally with high unemployment during the Great Depression. Something had to be done. Raphael Zon, of the Forest Service, wrote, "The drought of the last few years clearly brought into focus the need for a coordinated attack on the stern forces of nature and for a planned use of the Plains region if it is to continue as the granary of the United States." This effort, he continued, "must involve land adjustment, control of grazing, diversification of agriculture, water conservation and the building of ponds and reservoirs, shelterbelt and other forms of tree planting, strip cropping, terracing, and the development of new varieties of cereals and soil-building grasses." The Prairie States Forestry Project had the full support of Congress and President Franklin Roosevelt, who envisioned a way of providing work as well as curbing soil erosion. Federal funds were provided to lease private land for planting trees, but landowners soon appreciated the benefits of trees and became cooperators by donating their land.[3]

Also started during this time was the Soil Conservation Service in the U.S. Department of Agriculture (USDA). Later renamed the Natural Resources Conservation Service, the service's goal was to assist farmers and ranchers recover from the drought and adopt practices that would protect them in the future. The Forest Service, also in the USDA, took charge of tree planting, recommending that shelterbelts be located where annual precipitation was 16 inches or more and where the "subsoil moisture is fairly reliable" (fig. 6.3).[4] Enthusiastic silviculturalists envisioned large groves of trees that would create "forestlike" conditions; nurseries were established to produce millions of young trees for transplanting. Not surprisingly, farmers soon became concerned that trees were taking some of their best cropland out of production. Were ten or more rows in each windbreak really necessary? How much would farmers benefit from the additional trees? Encouragement for shelterbelt planting continues to this day, but some landowners thought the Forest Service had become overly zealous in creating woodlands. Consequently, the farmer-friendly Soil Conservation Service became the lead agency in 1942. Still, about 50,000 miles of shelterbelts have been planted in North Dakota since the 1930s, much more than in any other state.[5]

Benefits Provided by Shelterbelts

Shelterbelts provide various benefits for farmers. Wind speeds are reduced for a distance downwind equal to 10 to 30 times the height of the trees, an effect that slows soil erosion, increases crop yield, reduces evapotranspiration during summer, and

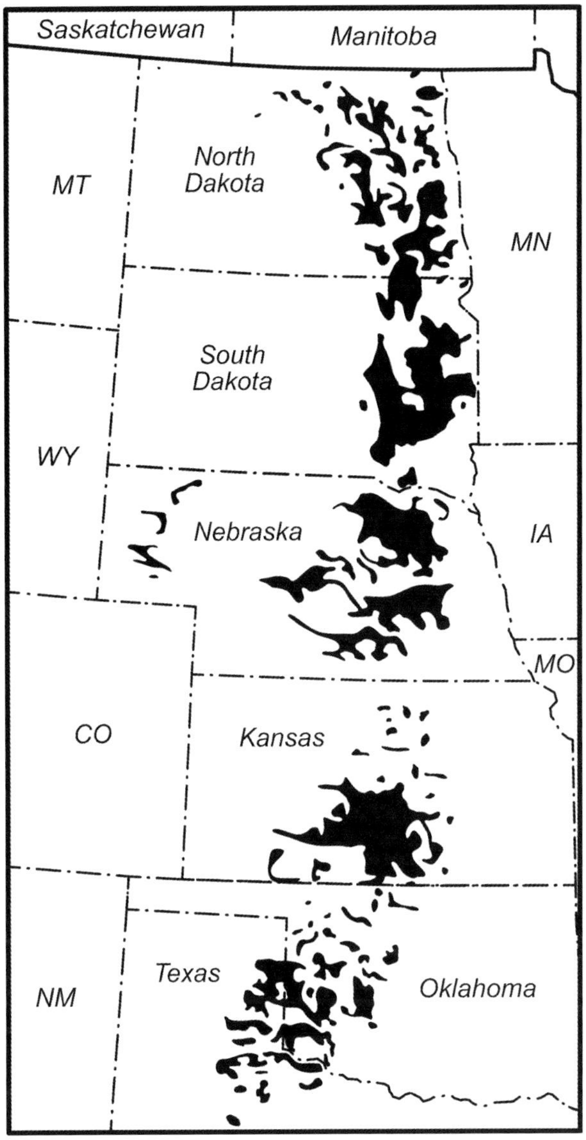

Fig. 6.3. A map illustrating where most shelterbelts were located on the Great Plains in 1954, about 16 years after the Prairie States Forestry Project began. Adapted from Read (1958).

functions like a snow fence during winter (figs. 6.4 and 6.5).[6] Considering the diminishing benefits with distance downwind, parallel shelterbelts at distances of 150 to 300 feet are recommended. Livestock, gardens, and fruit trees are all more productive when shielded from the wind, and groves on two or three sides of buildings reduce heating costs by 25 percent to 40 percent (fig. 6.6). The trees also provide firewood, posts, poles, and habitat for deer and other wildlife.

The effectiveness of shelterbelts depended greatly on the species of trees and shrubs that were planted and the number and spacing of the rows. Up to 17 rows were recommended in some areas, which produced a 165-foot band of trees. Five to seven rows were more common. Once established, the trees and shrubs created enough shade and woodland litter to suppress weed competition and they required very little attention for several decades. However, unlike native ecosystems, shelterbelts require maintenance. Dead trees had to be replaced, but often they were not. Livestock had to be kept out of the trees, as they reduced windbreak effectiveness by opening up the understory. Many farmers managed their trees carefully, remembering the devastation of the Dust Bowl years and the work required to nurture so many small trees in long belts across their land, but newcomers were less motivated. Some farmers planted the number of rows recommended by USDA officials, only to plow up one or more of the rows while planting crops the following spring.[7]

Several considerations were involved in selecting tree and shrub species to plant: availability, ease of transplanting, rate of growth, growth form, and adaptability to hot and sometimes dry summers and cold winters. Early settlers transplanted plains cottonwood because saplings could be found, and they grew rapidly if subsoil moisture was sufficient. In the 1930s the Forest Service recommended evergreen conifers, such as ponderosa pine, Black Hills spruce, blue spruce, eastern redcedar, and Rocky Mountain juniper, because they provided more protection throughout the year and lived longer, but deciduous trees were less expensive and grew more rapidly. Farmers wanted quick results. Thus, the most commonly planted native trees in the Dakotas, in addition to cottonwood, were green ash, hackberry, American elm, boxelder, and peachleaf willow—all found

Fig. 6.4. A shelterbelt of green ash and Siberian elm that is much narrower and more open than recommended for the Prairie States Forestry Project.

Fig. 6.5. Evergreen trees such as blue spruce, Norway spruce, ponderosa pine, Austrian pine, and juniper form dense shelterbelts that provide effective windbreaks and wildlife habitat.

Fig. 6.6. Most farmsteads on the Northern Great Plains are sheltered from the prevailing north and west winds by planting trees. Note the lakes and marshes in this part of the Prairie Pothole Region. Photo by Gray Tappan.

in woodlands associated with river valleys and wooded draws. Eastern redcedar and Rocky Mountain juniper were planted also, but bur oak usually was not selected because it grew slowly. Native shrubs included American plum, chokecherry, hawthorn, juneberry, red-osier dogwood, and silver buffaloberry. Introduced trees and shrubs included Siberian elm, Scotch pine, Russian olive, Tartarian honeysuckle, lilac, and caragana. More conifers have been planted in recent years, sometimes as replacements for dead deciduous trees (fig. 6.7).[8]

Wildlife

The kinds of wildlife found in planted woodlands are different from those in the surrounding prairie and fields. Generally, the abundance of birds and mammals is higher when the shelterbelts are relatively wide and have several species of trees and shrubs. Also beneficial are a few dead trees, known as snags, which are required by cavity-nesting species such as woodpeckers, chickadees, and small owls. Snags also provide perches for mourning doves and raptors. Turkeys and ring-necked pheasants nest on the ground. Mammals include the fox squirrel, eastern cottontail, white-tailed deer, red fox, coyote, long-tailed weasel, and various species of mice, voles, and shrews. The trees sometimes attract large flocks of birds during migration.[9] Some birds, bats, and predatory insects prey on pests in adjacent cropland, suggesting that shelterbelts could reduce the need for pesticides.[10] Combining habitat modification with chemical control is a form of integrated pest management.

Notably, trees also can have a detrimental effect

Fig. 6.7. Snow drifting downwind from a windbreak with sparsely planted evergreen conifers after a blizzard in eastern South Dakota. Such windbreaks can be used to augment soil moisture on croplands and minimize snow drifting on highways. Photos by Jesse Wittnebel, Dragonfly View Drone Service.

on prairie birds—widely recognized as the most threatened group of birds in North America. Their populations have been declining for many years, primarily due to habitat loss and fragmentation. Nearby shelterbelts create an additional problem by providing perches for raptors that prey on these birds, which are easily spotted in the prairie patches that remain. If too small, prairie remnants become *ecological traps*. Prairie birds that instinctively avoid trees include the upland sandpiper, Henslow's sparrow, greater prairie chicken, bobolink, savannah sparrow, sedge wren, dickcissel, and clay-colored sparrow.[11] Prairie birds also are susceptible to nest predation, mostly by small mammals and birds that consume eggs or fledglings, and there is evidence that nest (brood) parasitism by brown-headed cowbirds is enhanced by the presence of shelterbelts. The cowbirds lay their eggs in the nests of other birds, and usually their chicks hatch first, commanding more of the food brought to the nest by adults of the host species and thereby reducing reproductive success of the species that built the nest. For these reasons, and because many grassland birds are threatened, conservation biologists sometimes recommend cutting trees on or near tracts of prairie that still exist or are being restored.[12] Grassland restoration projects intended to benefit native grassland birds should not include trees.

Generally, the benefits that shelterbelts provide for wildlife depend on farm location and management goals. If surrounded by croplands rather than prairie, the conservation of prairie species is not an option and planted woodlands enhance the variety of species that can be seen and enjoyed; but if trees are on or near prairie remnants or

Fig. 6.8. The seeds of eastern redcedar commonly are dispersed by birds into nearby grasslands, often adjacent to a rock where the bird has defecated. Rainwater drains off the rock, providing a better environment for tree seedlings than would be available away from the rock. Planting eastern redcedar in windbreaks has accelerated unwanted invasion of adjacent grasslands, especially in South Dakota. See Meneguzzo et al. (2018).

reconstructed prairie, the diversity of wildlife could be diminished. To be confident about this prediction for a specific locality, other variables must be considered, such as the size of the prairie remnant. Predators find their prey much more easily in small patches of habitat. Birds that are adapted to open field and fence-line habitat, such as meadowlarks, are generally abundant and not likely to be severely affected by trees.

Complicating the potential adverse effects of trees in prairie and CRP land is the fact that some tree species are invasive; that is, they are becoming established in places where they were not planted and are not wanted. The best-known example in South Dakota is eastern redcedar. A juniper despite its name, this evergreen conifer was originally common in the southeastern part of the state and eastward, but it is now more widespread because of fire suppression, the intentional planting of the tree in windbreaks, and the effectiveness of seed dispersal by birds (fig. 6.8). Seedlings often become established near rocks and along fences where birds perch. Two other introduced invasive plants with bird-dispersed seed are buckthorn, a shrub that is abundant in some shelterbelts and woody draws, and Russian olive in some riparian woodlands—especially in the western Dakotas. Removing invasive trees can be expensive.

Shelterbelts on the Modern Farm

Agriculture has changed dramatically since the Prairie States Forestry Project began almost

Fig. 6.9. The trees in a windbreak adjacent to this field of soybeans and corn have been cut and stacked. The wood might be burned in place, when that can be done safely, but it also can be converted to biochar. The value of biochar and incentives for restoring old windbreaks are described on the websites of the North Dakota Forest Service and the South Dakota Division of Resource Conservation and Forestry.

90 years ago. Farms are now much larger, and many farmers irrigate their crops and apply more fertilizers, herbicides, and pesticides. Farm management in general has become a business as much as a lifestyle. Precision agriculture is the goal, where all aspects of a farming operation are considered quantitatively for short- and long-term costs and benefits as farmers attempt to maximize profits in a way that is sustainable for future generations.[13] Shelterbelts must be justified along with other choices. Not surprisingly, some have removed shelterbelts and fences that are in the way of machinery.[14] Wind erosion is still a concern, but other potential solutions are sometimes thought to substitute for trees, such as no-till agriculture and strip farming (see chapter 5). Significantly, while some shelterbelts have been removed, new ones have been planted in more strategic locations. The first farmers to break prairie sod in the 1870s and 1880s could not have imagined how farming would change, but one practice is the same: having a windbreak around one's home is appreciated today as it was then.

Over two-thirds of the shelterbelts that remain are in poor to fair condition (fig. 6.9).[15] States now provide grants to assist with restoring shelterbelts, the Natural Resources Conservation Service provides replacement trees at essentially no cost, and the USDA has a National Agroforestry Center. But some farmers have other priorities. Indeed, the task of shelterbelt maintenance was formidable when many trees died at about the same time, as they often did. Most were of the same age and comprised just a few species that would succumb to the same pest, disease, or drought. Also, trees were sometimes killed by herbicide drift caused by unanticipated wind gusts. Livestock grazing, often unintentional, opened up shelterbelts so they became less effective. Questions arose about how much benefit the trees provided, considering that they sometimes interfered with the turning of machinery or occupied land that could be cultivated. Straight shelterbelts often did not mesh well with the circular patterns of pivot irrigation.

Still, various studies suggest that rows of trees should be part of the equation when calculating farm profitability.[16] Crop and livestock production—and the efficiency of using water and fuel—can be higher in the shelter of trees. One study concluded that windbreaks could have a positive or neutral effect on corn production if a shelterbelt persists for 30 years and if the cornfield has a protected leeward zone of 12 times the height of the trees.[17] If farmers can be convinced that shelterbelts lead to higher profits for the farm as a whole, then they have an additional incentive for planting trees. Such is the nature of precision farming.

Planted trees also can help farmers with other challenges, for example, reducing costs associated with feedlots and buildings for rearing stock. Planting trees upwind is known to increase livestock weight gain, and planting them downwind creates turbulence that increases the dispersion of the hydrogen sulfide odor that neighbors find objectionable. Also, tree foliage filters odorous particles and chemicals.[18]

The benefits provided by trees are examples of ecosystem services. There are costs associated with cultivating trees, but once established they function for many years and are "powered" by the sun. Very little burning of fossil fuels is required. Economists now work with farmers to determine ways ecosystem services can be sold to provide an additional revenue stream. A good example is the USDA's Conservation Reserve Program, for which the buyer can be thought of as the public at large. Willing farmers are paid to provide benefits for society as a whole, namely, soil conservation, wildlife habitat, and improved air and water quality—benefits achieved by planting grasses and other perennial plants rather than annual crops in fields with highly erodible soil. The income previously generated from the sale of traditional crops is replaced by CRP payments, which for some farmers is a good deal. In the future farmers may receive payments for the services provided by trees. For example, county road departments could subsidize living snow fences in strategic locations, thereby reducing the costs of snowplowing. Outdoor enthusiasts might pay for the privilege of enjoying the plants and animals found in shelterbelts, including hunting and the gathering of fruit and firewood. Properly managed, such uses could provide additional incentives for investing in shelterbelts.

Moreover, in an age when many are concerned about climate change, the planting of trees is commonly recommended as a means of reducing the adverse effects of emitting still more carbon dioxide from new industrial developments. Carbon immobilized in wood does not contribute to climate change until the wood decays or is burned, often a century or more later. Corporations seeking permits to increase emissions in some states are required to fund efforts that mitigate adverse environmental impacts, which could include tree planting for carbon sequestration, initially in wood and eventually in biochar and soil organic matter (biochar is a form of charcoal used to enhance soil fertility and sequester carbon). Also, tree planting in appropriate places could partially replace degraded or lost wildlife habitat. As the market for ecosystem services develops further, shelterbelts may become a larger part of a farm's revenue.

Trees also could help landowners adapt to climate change. For example, a study in Saskatchewan found that crops sheltered by trees produced more than nonsheltered crops, and that windbreaks could compensate for the adverse effects of climate change in all but the most extreme scenarios.[19] Any investments in the potential benefits of trees on the farm must consider that trees are not as drought tolerant as prairie plants. A successful strategy requires that farmers select trees that seem well adapted to climatic conditions anticipated in the future.

Thinking broadly about modern agriculture and trees on the farm, it is likely that calculating carbon budgets for a farm will become routine. Indeed, the USDA has developed the Whole Farm and Ranch Carbon and Greenhouse Gas Accounting System.[20] The ultimate goal is for farms to become carbon neutral, which means that the amount of carbon sequestered on a farm is equal to or greater than the amount emitted to the atmosphere. Achieving carbon neutrality would require more fuel-efficient machinery, greater reliance on alternative energy sources for heating and cooling buildings, and adoption of agricultural practices that minimize carbon emissions and maximize carbon sequestration. Planting trees could help achieve carbon neutrality, not only because trees sequester carbon but also because windbreaks around buildings lower carbon emissions by reducing the amount of energy consumed for heating and cooling. Also, some farmers will find it economically beneficial to restore their cropland to something similar to native prairies, such as CRP lands, which sequester more carbon in living roots and humus than do annual croplands.[21] Using remote sensing and readily available environmental and socioeconomic data, such calculations will become more common. Properly located and strategically designed and managed, shelterbelts will continue to provide benefits.

Chapter 7 Upland Deciduous Forests, Woodlands, and Parklands

Forests and woodlands in the Dakotas were scarce when French explorers first met Mandan Indians nearly 300 years ago. Cottonwood and willow were scattered along the floodplains of some creeks and rivers, as described in chapters 9 and 10, but they would have been abundant only along the larger rivers, especially on the wide floodplain of the Missouri River. Diaries indicate that other kinds of hardwoods were found in some ravines and along the shores of some lakes—green ash, American elm, boxelder, basswood, hackberry, and bur oak (table 7.1). Groves of aspen were common in relatively moist locations, such as on Turtle Mountain, the Killdeer Mountains, and in the Black Hills. Evergreen conifers formed extensive forests only in the Black Hills. Woodlands of all kinds are still scarce in the Dakotas, but today there are more trees than ever before, primarily because of planted shelterbelts.

When discussing ecosystems with trees, various terms have been used depending on the number and stature of the trees. Vast prairies with widely separated trees are known as *savannas*, but the term is not widely used in the Dakotas. What might be considered an oak savanna is more commonly known as an *oak opening*—a place where a few open-grown oaks intermingle with grassland plants. The presence of large horizontal branches indicates that the tree has grown where light was available from all sides of the tree. There was no benefit in growing straight up. The term *woodland* refers to places where trees are dominant rather than grasses or shrubs, but where the trees are relatively short, less dense, and less widespread than in a *forest*. Distinguishing a woodland from a forest is a matter of personal preference, but for most ecologists, forests cover a larger area and create a noticeably shady environment in the understory, one that only some kinds of shrubs and herbaceous plants can tolerate. The trees tend to be tall and straight, indicating an environment where light limits plant growth. The term *gallery forest* refers to the narrow, intermittent forests and woodlands found on the floodplains and breaks of some rivers and creeks. *Riparian woodlands* are found specifically on floodplains. Whatever terms are used, the presence of trees provides habitat for birds and other wildlife not usually seen elsewhere.[1] Upland hardwood forests in the eastern Dakotas, those with minimal livestock grazing, are likely to have popular spring flowers such as trillium, bellwort, bloodroot, and wild ginger.

Upland Hardwood Forests

Nowhere in the Dakotas are native hardwood forests so extensive as on Turtle Mountain in north-

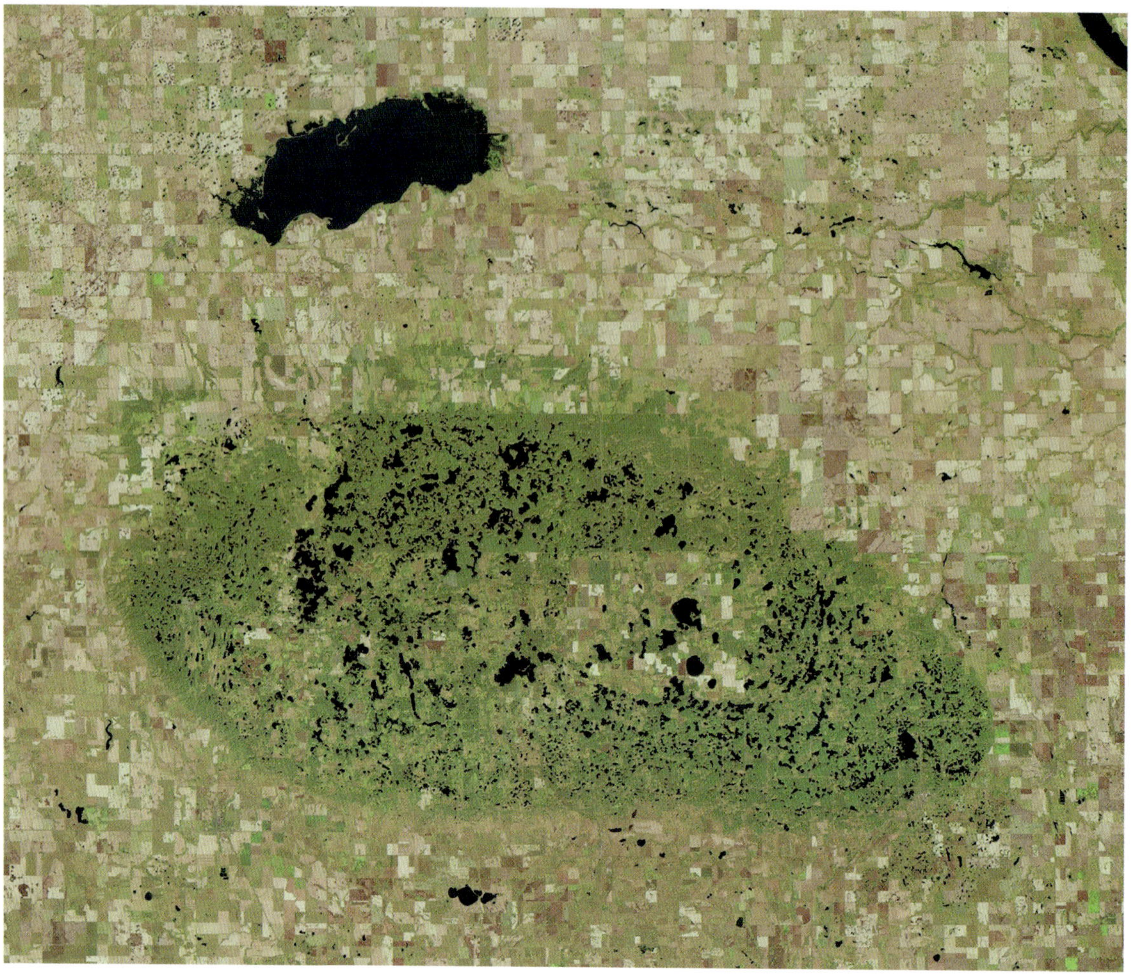

Fig. 7.1. A 2020 satellite image of Turtle Mountain, a forested highland in north-central North Dakota that straddles the border with Manitoba. The surrounding prairie was converted to cropland, mostly fields of wheat, canola, and barley. More of the forests were cleared for agriculture in North Dakota than in Manitoba, which led to an international borderline that is visible from space. See fig. 1.2 for location. Source: Landsat 8, U.S. Geological Survey, EROS Center.

central North Dakota (fig. 7.1). This oval-shaped range of low hills on the border with Manitoba extends about 40 miles from east to west and 20 miles from north to south. The highest point, Boundary Butte, is 2,541 feet above sea level and about 1,000 feet above the surrounding croplands. Geologists describe the mountain as a glacial moraine that overtopped one or more buttes, formed in much the same way as the Missouri Coteau to the southwest and the Prairie Coteau to the southeast (see chapter 2).[2] Agriculture is possible on Turtle Mountain, where fine-textured loamy soils have developed on relatively flat terrain, but most of the soils are coarse textured with gravel and rocks. The lakes, ponds, marshes, meadows, and forests create a northwoods environment that is much appreciated for hunting, fishing, camping, and water sports. Wildlife managers strive to maintain habitat for ruffed grouse, white-tailed deer, moose, and numerous other species; the

Fig. 7.2. Views of deciduous forest on Turtle Mountain. Lakes and wetlands are common on this elevated glacial moraine in the Prairie Pothole Region. Clones of aspen are common, as are forests of bur oak, green ash, boxelder, and American elm. Aerial photo taken in 2014 by Aaron Bergdahl, North Dakota Forest Service.

gray wolf and American marten have been seen. Numerous lakeside cabins and resorts are found in the area, along with Turtle Mountain Provincial Park, Lake Metigoshe State Park, Twisted Oaks Recreation Area, the Turtle Mountain Chippewa Indian Reservation, Turtle Mountain State Forest, and the International Peace Garden.

With a relatively cool mean annual temperature and a mean annual precipitation of 16 inches (see figs. 3.1 and 3.3), the first European explorers on Turtle Mountain saw widespread woodlands dominated by aspen.[3] Also found on north-facing slopes and in ravines were American elm, green ash, bur oak, balsam poplar, and boxelder (fig. 7.2). Paper birch was common near lakeshores. Understory shrubs included beaked hazelnut, red-osier dogwood, serviceberry, highbush cranberry, chokecherry, American plum, and wild currant. Native grasslands were found on relatively dry, south-facing slopes, often with scattered, open-grown bur oaks and patches of western snowberry.

Surprisingly, there are no coniferous forests on Turtle Mountain except in plantations. Black spruce, white spruce, and tamarack were there centuries ago, as indicated by pollen in lake sediments, and these trees are found only 60 miles to the northeast, but none were mentioned by explorers of European origin.[4] The lack of conifers probably is due to frequent fires, which favor hardwood trees because, unlike conifers, they are capable of root sprouting after the aboveground part of the tree is burned. Notably, some of the bur oak and American elm on islands are older than those away from lakes, up to 570 and 280 years (in 1948), respectively, suggesting that the islands burned less frequently. Most of the Turtle Mountain forests had burned or were harvested by the mid-1900s, but recovery was rapid. Given enough time, conifers probably would have become reestablished naturally. Thousands of white spruce and Scots pine saplings grew well after being transplanted in the first half of the twentieth century. Other transplanted conifers were ponderosa pine, red pine, jack pine, white pine, Austrian pine, balsam fir, white cedar, Siberian larch, and Colorado blue spruce.

Fire suppression continues to have a significant effect on Turtle Mountain forests. With fewer fires, trees invade grasslands and tree density increases nearly everywhere. Fires had been frequent, with most of Turtle Mountain burning in 1897 and 1903. Much of Lake Metigoshe State Park was burned in 1886 and 1921. Photos taken in 1873 show bare hilltops that are now heavily wooded; and tall, straight, forest-grown trees now surround open-grown oaks that matured in grasslands. Long-term observations on unburned islands—which now have no aspen—suggest that, while aspen is the primary pioneer species that invades after a fire and other disturbances, woodlands of green ash, boxelder, bur oak, and American elm will eventually develop where fires are suppressed. Indeed, many aspen forests are now considered overmature, that is, the trees are growing slowly and many are dying, creating gaps in the canopy that favor the invasion of other species—a natural process known as succession and a problem where there is a desire to maintain aspen as the dominant tree.[5] Fire suppression also has led to a more uniform forest cover, with fewer patches and less diversity in the ages of the patches than those that most likely existed about 150 years ago. Timber harvesting is used in some places to reverse these trends, especially where prescribed fires are not an option.

Because of effective fire suppression and the accumulation of fuels, the forests of Turtle Mountain have become even more flammable than before. Climate change is likely to heighten the risk of fires due to more frequent periods of drought. Vigorous stands of new aspen will develop by root sprouting after these fires—unless the climate becomes too dry.

One of the most intriguing and ecologically significant patterns on Turtle Mountain is the sharp contrast in land cover along the border between the United States and Canada. As is apparent in figure 7.1, much of the forests on the North Dakota side have been converted to croplands, partially due to subsidies that have encouraged cultivation. Another possible factor is that Manitoba is more inclined to protect one of its largest tracts of hardwood forest. Clearly, federal and state or provincial policies have caused this striking pattern, as no climate or soil differences would conform so precisely to a straight-line political boundary.

The ecological effects of these contrasting land management policies have not been well studied, but several possibilities are plausible. First, birds, mammals, and other wildlife that benefit from extensive forest cover will be more common in Manitoba, whereas those that benefit from forest edges will be more common on the North Dakota side, where the forests have been fragmented (less so on Chippewa land). An example would be white-tailed deer, which are known to be favored by farmlands and which are heavy browsers on young trees. Second, lakes with forested watersheds are likely to have higher water quality than lakes surrounded by more erodible croplands. Considering that most hardwood forests and woodlands in the Dakotas are small, covering only

Table 7.1. Characteristic trees and shrubs of deciduous hardwood forests, woodlands, and river breaks that grow naturally in different parts of North Dakota and South Dakota

Common name[a]	Latin name	W	NND	END	NESD	SESD
TREES						
Ash, green	*Fraxinus pennsylvanica*	X	X	X	X	X
Aspen	*Populus tremuloides*	x	X	X	x	–
Basswood (linden)	*Tilia americana*	–	–	X	x	X
Birch, paper [R]	*Betula papyrifera*	x	x	X	x	–
Boxelder	*Acer negundo*	x	x	X	x	X
Coffeetree, Kentucky	*Gymnocladus dioica*	–	–	–	–	X
Cottonwood, plains [R]	*Populus deltoides*	x	x	X	x	X
Elm, American [R]	*Ulmus americana*	x	x	X	x	X
Elm, slippery	*Ulmus rubra*	–	–	X	x	X
Hackberry	*Celtis occidentalis*	–	–	X	x	X
Honeylocust[R]	*Gleditsia triacanthos*	–	–	–	–	X
Hophornbeam (ironwood)	*Ostrya virginiana*	x	–	X	x	X
Maple, silver [R]	*Acer saccharinum*	–	–	–	–	X
Maple, sugar [SH]	*Acer saccharum*	–	–	–	x	–
Oak, bur	*Quercus macrocarpa*	x	x	X	x	X
Poplar, balsam	*Populus balsamifera*	–	x	–	–	–
Walnut, black	*Juglans nigra*	–	–	–	–	X
Willow, peachleaf [R]	*Salix amygdaloides*	x	x	X	x	X
SHRUBS						
Buffaloberry, silver	*Shepherdia argentea*	x	X	–	–	–
Chokecherry	*Prunus virginiana*	x	X	X	x	X
Cinquefoil, shrubby [R]	*Dasiphora fruticosa*	x	X	X	x	–
Cranberry, highbush [R]	*Viburnum opulus*	–	X	–	–	–
Creeper, Virginia (woodbine)	*Parthenocissus quinquefolia*	x	x	x	x	X
Currant, wild	*Ribes* spp.	x	X	X	x	X
Dogwood, red-osier [R]	*Cornus sericea*	x	X	X	x	X
Grape, riverbank [R]	*Vitus riparia*	x	X	X	x	X
Hawthorn	*Crataegus* spp.	x	X	X	x	X
Hazelnut, beaked	*Corylus cornuta*	x	X	–	–	–
Hazelnut, American	*Corylus americana*	x	X	X	x	X
Ivy, poison	*Toxicodendron rydbergii*	x	X	X	x	X

(continued)

Table 7.1. (*continued*)

Common name[a]	Latin name	W	NND	END	NESD	SESD
Plum, American	*Prunus americana*	x	X	X	x	X
Rose, Wood's	*Rosa woodsii*	x	X	X	x	X
Serviceberry (juneberry)	*Amelanchier alnifolia*	x	X	X	x	X
Silverberry (wolfwillow) [R]	*Elaeagnus commutata*	x	X	X	–	–
Snowberry, western (buckbrush)	*Symphoricarpos occidentalis*	x	X	X	x	X
Sumac, smooth	*Rhus glabra*	–	X	X	x	X
Willow, sandbar[R]	*Salix exigua*	x	X	X	x	X

[a]W = western half of both states, NND = northern North Dakota, END = eastern North Dakota, NESD = northeastern South Dakota, SESD = southeastern South Dakota (see tables 8.1 and 12.1 also). Abundance: X = common, x = occasional; a dash indicates that a plant is absent or not common. Specific habitat requirements of some species are indicated with superscripts: R = riparian or wetter environments, SH = Sica Hollow and nearby river breaks only.

a few hundred acres at most, Manitoba's large tract of hardwood forest on Turtle Mountain is important for maintaining the biological diversity of the Northern Great Plains.[6]

Killdeer Mountains

Other large tracts of hardwood forest are found in the Killdeer Mountains in western North Dakota. Much smaller than Turtle Mountain, the "Killdeers" are an unglaciated ridge that is eight miles long, two to three miles wide, and rise about 900 feet above the surrounding plains. The ridge extends from South Killdeer Butte to North Killdeer Butte and was carved from the Missouri Plateau by the Little Missouri and Knife Rivers (see chapters 2 and 12).[7] The maximum elevation is 3,314 feet, which is 192 feet lower than White Butte—the highest point in North Dakota. Vernon Bailey wrote in 1926, in a paper on the natural history of North Dakota: "Probably no area in North Dakota is better suited for game refuges and parks than the Killdeer Mountains. . . . On pleasant Sundays 50 to 100 automobile parties even now visit the mountains for picnics in the cool shade, for drafts of pure, cold water, the sight of strange flowers, plants, trees, birds, and mammals, rugged climbs, and a glorious view over wide country." Today the mosaic of woodlands, shrublands, and grasslands is used largely for livestock grazing and hunting, with small amounts of logging. Elk, mule deer, white-tailed deer, and cougar are common, and bear and moose have been seen. Homes are found in the foothills, with public access available primarily on state land managed for wildlife on the west side of the ridge (Killdeer Wildlife Management Area). The badlands of the Little Missouri River valley are nearby, about 18 miles to the west.

An aerial view of the Killdeer Mountains shows numerous patches of bur oak, aspen, and paper birch dispersed among drier slopes and ridges dominated by mixed-grass prairie (figs. 7.3 and 7.4). Less common trees include green ash, American elm, and Rocky Mountain juniper. Tree density and species diversity is higher on north-facing slopes and ravine bottoms, and aspen trees often fringe the edge of oak woodlands.[8] Common shrubs include beaked hazelnut, serviceberry, silver buffaloberry, chokecherry, and American plum. Fire suppression has enabled higher tree density and the spread of trees and shrubs into grasslands.

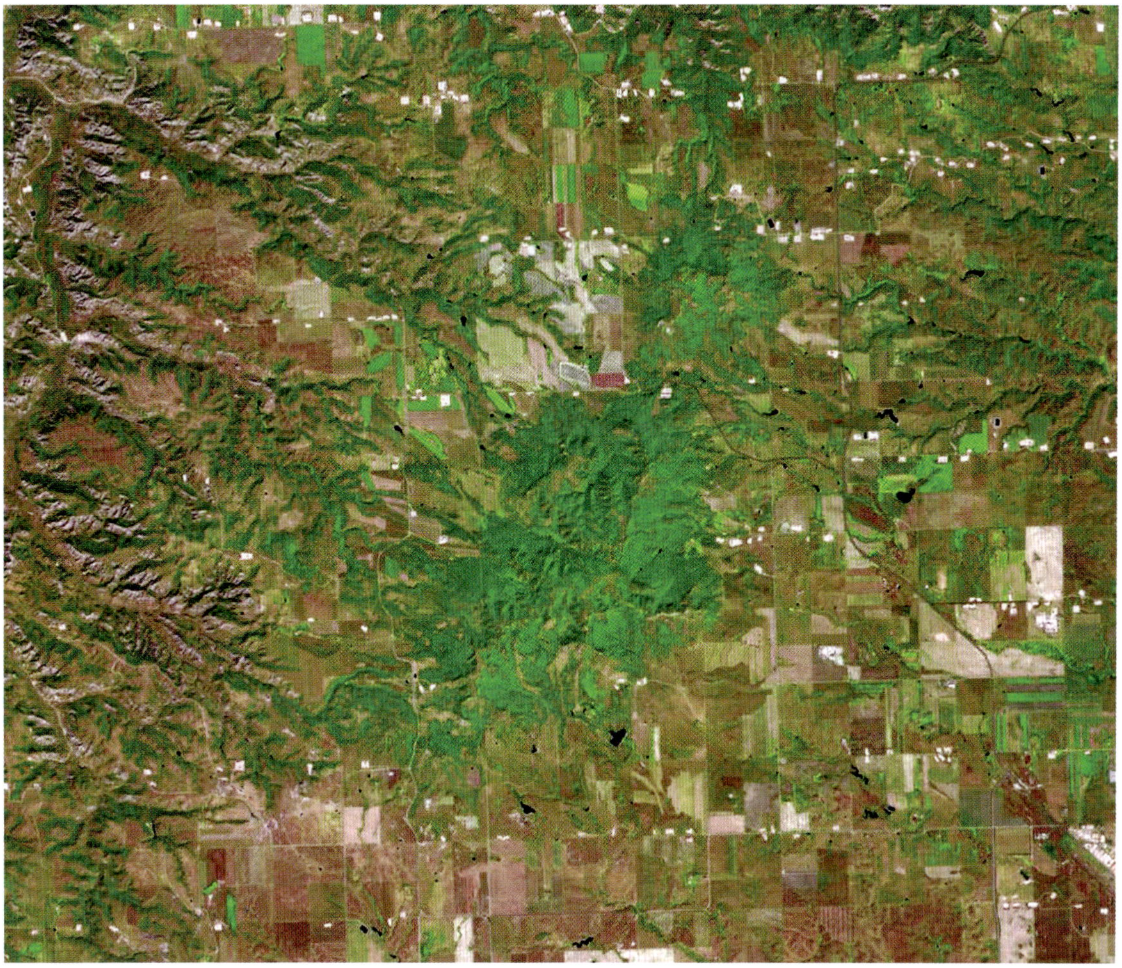

Fig. 7.3. Satellite image of the Killdeer Mountains in western North Dakota that illustrates a mosaic of cropland, grasslands and shrublands on drier sites and dense hardwood forest at higher elevations and on north slopes, where moisture is sufficient for tree growth (green). The breaks of the Little Missouri River are visible on the left. Source: Sentinel 2 (August 2020), U.S. Geological Survey, EROS Center.

River Break Woodlands

Ravines and steep valleys were a mixed blessing to Native Americans and European immigrants alike. These *breaks* in the plains slowed their travel, but trees found there also provided firewood, shade on hot summer days, shelter from the wind, and a nearby source of water—good places to camp. Villages, farmsteads, towns, and cities would be built in such places. Farther west, where the climate is drier, some of the river breaks became too steep or barren for settlements; some were labeled *badlands* (see chapter 12).

Generally, river breaks slope down rather abruptly from the surrounding plains. Small drainages coalesce into larger valleys until a major river is reached. Walking down well-vegetated slopes, one wonders how the small, often dry creeks could have formed such large fissures in the landscape. Geologists remind us, however, that present-day landscapes are the result of thousands of years of water erosion, and that most of the

Fig. 7.4. Four representative photos of woodlands in the Killdeer Mountains, taken in May 2018. Top, bur oak, aspen, and green ash dominate the forests on a north slope; chokecherry is common on the forest edge. Bottom, bur oak in the foreground, with a clump of chokecherry in the grassland on the right; western snowberry, which increases with fire suppression, is abundant in the foreground and distant ravine. On the horizon are two dome-shaped aspen groves. On the next page, top, a stunted aspen grove with sprouts invading the adjacent mixed-grass prairie, due to fire suppression; the taller grass is little bluestem, and aspen invasion and mulch accumulation in grasslands results from fire suppression. Bottom, the tall, gray-green shrub on the right is silver buffaloberry, which is common on the edge of woodlands in western North Dakota, less so in South Dakota; western snowberry is invading the grassland, and the trees are bur oak (left) and aspen (dark green).

erosion occurred when rivers were much larger, fed by melting glaciers. Also, some of the ravines were shaped when the stabilizing plant cover was less, such as during heavy rains after an extended multiyear drought.

The Vegetation Mosaic of River Breaks

The slopes of some creek and river valleys are gentle with a dense plant cover that is similar to the surrounding grasslands; elsewhere, steep embankments and escarpments have other kinds of plants. In such places, north-facing slopes with less direct sunlight provide a cooler, wetter environment than on other slopes or the flatlands above. An important part of the microenvironment is snow drifting near the top, along ridges that are leeward to the prevailing winds. Downslope from these drifts, plant density and growth are higher because the soil is deeper and water is more available. Such differences also affect the abundance of some animals.

The effects of rugged topography on plant life was noted by early explorers who described how patches of various kinds of shrubs and trees would be found in the breaks, and how the tops of trees were sometimes at or near the level of the surrounding windswept upland—barely visible except at the valley edge (fig. 7.5).[9] Three factors explain the presence of woody species in ravines: (1) drifting snow, higher humidity, and shelter from wind, which enabled the survival of delicate shrub and tree seedlings; (2) less frequent fires in the uneven ravine topography; and (3) springs found in some ravines, providing a wetter environment. Candace Savage wrote in 2004 that prairie woodlands "tuck themselves into the prairie wherever the lay of the land improves the supply of water."[10]

Fig. 7.5. Woody draws along the Bad River in south-central South Dakota dominated by green ash, wild plum, chokecherry, hawthorn, and an occasional plains cottonwood.

Some of the river break forests cover a hundred acres or more. The best examples in South Dakota are in Sica (pronounced "she-cha") Hollow State Park, on the northeastern edge of the Prairie Coteau; along the Big Sioux River in Newton Hills and Good Earth State Parks; and in the Big Sioux Recreation Area. They also are common in the breaks of the Missouri River, especially to the south. In North Dakota the best examples are at Cross Ranch State Park along the Missouri River; along portions of the Souris and James Rivers, including the Arrowwood National Wildlife Refuge; in Fort Ransom State Park and near Mirror Pool along the Sheyenne River; and where the Pembina River cuts through the Pembina Escarpment (figs. 7.6, 7.7, 7.8, and 7.9). These isolated woodlands, often linear and referred to as *gallery forests*, are small examples of the eastern deciduous forest that is more widespread in Minnesota, where the annual precipitation is higher. Common trees are green ash, American elm, hackberry, and occasionally basswood and aspen (see table 7.1). Some forest-grown trees are quite large; for example, along the Red River in the 1880s, one explorer wrote about "basswood over two and one-half feet in diameter, tall and straight."[11]

Further to the west and coinciding with the distribution of mixed-grass prairie, the river breaks have shorter woodlands that cover less land area. Green ash is the most common tree, but bur oak, hackberry, American elm, boxelder, eastern redcedar, and Rocky Mountain juniper are found there also.[12] Often the ravines have shrub thickets but no trees at all. Meriwether Lewis observed about the Dakotas in 1805, in a letter to his mother, "This country on both sides of the river, except some of its bottomlands . . . is one continued open plain, in which no timber is to be seen except a few . . . clumps of trees, which from their moist situation, or the steep declivities of the hills, are sheltered from the effects of fire."[13]

Bur oak grows in many river breaks all across the Dakotas, and it becomes quite large in the southeast, such as in the Big Sioux Recreation Area near Sioux Falls, the Newton Hills near Canton, and on islands in large lakes.[14] Before prairie fires were suppressed, ridge-top prairies with scattered oaks most likely burned every three to ten years, maintaining oak openings (fig. 7.10). The large oaks had thick bark that enabled many of them to survive fires that burned only around the base of their trunks. The presence of open-grown trees surrounded by younger forest-grown trees that are straight indicates that the forest developed after a period of fire suppression.

With fire suppression, the oaks, eastern redcedar, and other woody plants grow more densely, causing declines in the abundance of prairie species.[15] Redcedar was previously found mostly along the Missouri River and eastward, but it has expanded westward because of seed dispersal by birds and the planting of redcedar windbreaks. In some places it is an undesired invasive species. The distribution of bur oak is sometimes a puzzle, as often it is absent from what seems like suitable habitat. Acorn dispersal largely depends on squirrels and blue jays, but climate and other limiting factors surely are part of the explanation.[16]

Lakeside Woodlands

The shores of lakes in the eastern Dakotas sometimes have woodlands if the land slopes abruptly downward to the lake. The reasons for the scarcity of trees elsewhere around lakes could be frequent prairie fires that burned up to the shore so frequently that trees could not become established. Evidence for this is the presence of trees on the leeward side of some lakes and on some islands, where the probability of fire is much lower (fig. 7.11). Perhaps the best example of an island with upland hardwood forests is Grahams Island State Park, located in Devils Lake. Similar forests occur on islands in Waubay Lake. The common trees found on lakeshores and islands are the same as in river breaks, although riparian species such

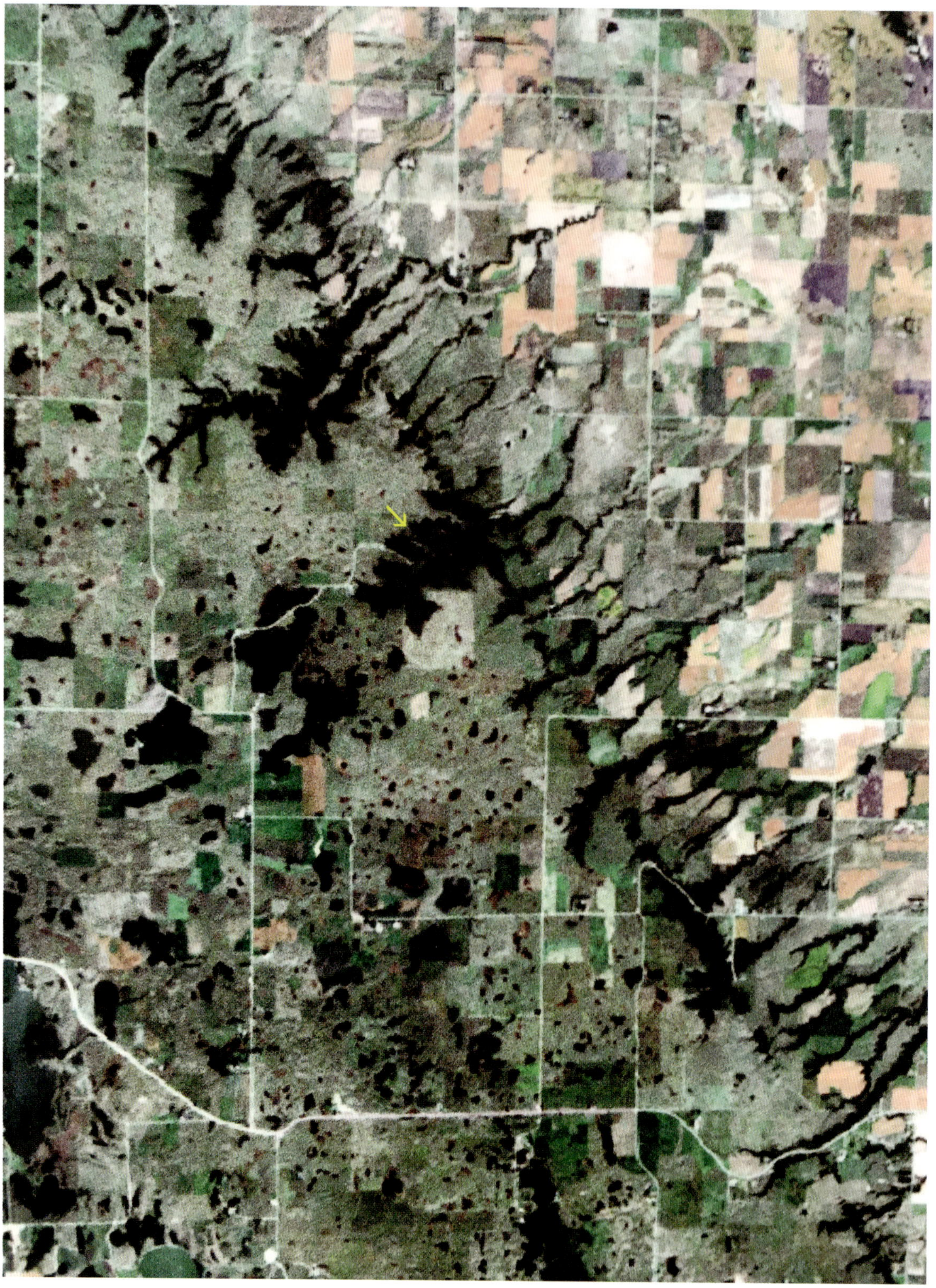

Fig. 7.6. Satellite image of the northeastern part of the Prairie Coteau, west of Sisseton in northeastern South Dakota (see fig. 1.2 for location). The dark patches are river break woodlands and woody draws along creeks flowing eastward. Note the abundant lakes and wetlands on the highland, where mixed-grass prairie and pastures (lighter green) are common. Croplands occur on the Coteau but are more conspicuous on flatlands to the northeast, which formerly were dominated by tallgrass prairie. An arrow identifies Sica Hollow State Park. Source: U.S. Geological Survey, EROS Center.

Fig. 7.7. River break woodland dominated by green ash, bur oak, American elm, and hackberry on the northeast edge of the Prairie Coteau. Photo taken from the top of Nicollet Tower (west of Sisseton on State Highway 10), the namesake of Joseph Nicollet. Severson and Sieg (2006) review woodland observations of Nicollet and other early explorers in the Dakotas.

Fig. 7.8. Contrasting deciduous woodlands in October: Above, where the Pembina River passes through the Pembina Escarpment in northeastern North Dakota, with bur oak, aspen, green ash, and chokecherry (photo by Dave Bruner); and on the right, along ephemeral creeks on the much-drier east slope of Slim Buttes in northwestern South Dakota, with green ash, chokecherry, and silver buffaloberry in ravines and ponderosa pine on the ridge top. Such woodlands are referred to as *gallery forest* when trees are abundant, such as along the Pembina River, and as *gallery woodlands* or *woody draws* where trees are sparse, such as near Slim Buttes.

Fig. 7.9. The forest of Newton Hills State Park is found in breaks along the Big Sioux River in southeastern South Dakota. The most common trees are green ash, basswood, hackberry, American elm, ironwood, and bur oak. Walnut is found in the area also. Common understory plants include Virginia waterleaf, Virginia creeper, bedstraw, bloodroot, sweet cicely, Canada wood-nettle, white snakeroot, and garlic mustard. Some herbs, known as spring ephemerals, grow only in early spring and include trillium, bloodroot, Dutchman's breeches, bellwort, and jack-in-the-pulpit.

Fig. 7.10. An oak opening in Newton Hills State Park where open-grown oak trees are found in grasslands that are maintained by frequent prescribed fires. Without prairie fires, small trees begin to invade after a few years. Older bur oak trees have thick bark that is not easily ignited, which enables them to survive grass fires. The dark-green conifer on the right is a tall eastern redcedar, an undesired invader.

Fig. 7.11. The shorelines of most lakes in the Prairie Coteau and elsewhere in the Prairie Pothole Region have only a few trees, except in places where fires have not occurred for many years, such as on the island and other places protected by water visible on this August 2020 aerial photograph of four glacial lakes west of Altamont. Relatively high atmospheric humidity and lack of fire on the island would facilitate tree seedling establishment. Source: National Aerial Photography Program, U.S. Geological Survey, EROS Center.

as cottonwood and peachleaf willow often grow at water's edge. Immigrants often homesteaded near lakes because, like river breaks, they provided a nearby source of water, wood, and food. Indeed, another explanation for the scarcity of trees around some lakes could be tree-cutting homesteaders; the oldest trees are sometimes found on less accessible islands.

The land adjacent to some lakes is now highly valued real estate. Many of the native trees have been cleared to provide space for homes, lawns, flower gardens, and evergreens such as spruce, juniper, and pine. This comfortable, manicured environment often includes the application of fertilizer and herbicides that can have adverse effects on the lake ecosystem. Leaking septic systems also can be a problem. Such concerns usually are taken seriously, as the owners know that the value of their property is dependent on the quality of lake water. Still, reliable monitoring often is not done, and identifying the cause of water-quality problems is difficult and controversial. Tracts of native woodlands and prairie along lakeshores assist in maintaining water quality and slowing shoreline erosion—two examples of much-appreciated ecosystem services.

Aspen Groves

Trembling aspen is the most widely distributed tree in North America, growing from coast to coast and from Alaska and northern Canada down through the Rocky Mountains to northern Mexico. Obviously, there is great genetic diversity within the species. Where aspen occurs in the Dakotas, it is an indicator of a relatively cool, moist environment. Usually it occurs with other hardwoods in the forests of the Black Hills, of the Killdeer Mountains, and on Turtle Mountain, and sometimes in river breaks such as in Sica Hollow State Park and along the Sheyenne and Little Missouri Rivers. In northwestern Minnesota, southern Manitoba, and nearby northern North Dakota, aspen forms conspicuous clones dispersed among croplands, shrub thickets, prairie, and wetlands (fig. 7.12)—an area widely known as *aspen parkland* that extends across Canada in the transition zone between grasslands and boreal coniferous forests.[17] Common shrubs include serviceberry, hazelnut, wild rose, western snowberry, chokecherry, highbush cranberry, and silverberry where the soil is well-drained. Red-osier dogwood and various willows are common where the soil is wetter.

The availability of water, a variety of habitats, and other factors explain the great diversity of wildlife found in aspen parklands: moose, white-tailed deer, elk, gray wolf, coyote, red fox, black bear, weasel, northern pocket gopher, beaver, snowshoe hare, sharp-tailed grouse, ruffed grouse, sandhill crane, black-billed magpie, Baltimore oriole, great horned owl, and various kinds of woodpecker—not to mention all the invertebrates (roundworms, snails, segmented worms, centipedes, mites, spiders, mosquitoes) and the animals of adjacent grasslands and wetlands. Some investigators have concluded that the native diversity and density of animals in aspen parklands is greater than in most other habitats in North America.[18]

Like all ecosystems, the parklands have changed over the years. Scientists report that the soils supporting aspen groves often are typical of grassland ecosystems, which suggests that trees are relatively new arrivals.[19] Early explorers in the 1800s did see aspen, but the groves at that time apparently were restricted to relatively wet depressions that were not burned during prairie fires. Now, fires are suppressed whenever possible and the groves have expanded. Once established, the enlarged groves persist because fires do not kill the roots, which are capable of sprouting.[20] As noted previously, conifers in the area are not capable of this kind of vegetative, asexual reproduction. Dome-shaped aspen groves, with older, taller shoots in the center and younger shoots on the edge, are formed by root sprouting around the perimeter of a single clone. An abundance of seed is produced, but the seeds are very small and soon lose their viability. Rarely do seedlings survive. Notably, each clone has developed from a single seedling, one that did become established—most likely several centuries ago, perhaps during a wet period that followed some kind of disturbance that reduced prairie plant competition, possibly the digging of pocket gophers and badgers.

Fire, drought, late-spring freezes, large herbivores, and various diseases and insects can disrupt

Fig. 7.12. Aspen parklands commonly have groves of aspen in depressions surrounded by croplands, such as in this hayfield. Such parklands are found in north-central and northeastern North Dakota, northwestern Minnesota, and southern Manitoba, Saskatchewan, and Alberta. Similar aspen groves are found on sand dunes and sand sheets, such as on the Souris and Sheyenne River deltas (see chapter 12).

the growth and expansion of aspen groves. Fires, whether started by lightning or humans, occur every three to 15 years, as in the adjacent prairie, often during dry years. Water-stressed trees, or trees wounded by browsing and trampling, whether by elk, deer, or livestock, are also susceptible to diseases that kill the aspen shoots, such as black canker (*Ceratocystis fimbriata*).[21] Notably, the forest tent caterpillar (*Malacosoma disstria*) and the large aspen tortrix (*Choristoneura conflictana*) periodically defoliate aspen groves, which can slow photosynthesis and cause severe dieback. However, healthy trees often are not killed and may even produce new leaves during the same growing season.

Throughout much of its North American range, aspen groves have been observed to die within a few years, a puzzling development known as *sudden aspen decline*. The cause appears to be a combination of insect defoliation, drought, and early spring freeze-thaw events that damage buds and roots. Some aspen roots may survive to produce new growth. In Canada the weakened groves may become dominated by white spruce and balsam fir, at least until the next fire, after which the aspen clone likely will be restored.

Ecologists sometimes refer to aspen parklands and oak savannas as ecological tension zones, a place where the relative abundance of woodlands and grasslands is dependent on subtle changes in the climate. Trees and shrubs are favored during periods with higher precipitation and fewer fires. In contrast, grasslands are favored during droughts, when many of the trees are burned or die from water stress. Where aspen and bur oak occur together, the oak is on drier sites—an indication that bur oak may be better adapted than aspen for anticipated climate changes.

Biological Diversity, Invasive Species, and Climate Change

Immigrants of European descent quickly exploited the relatively scarce forest and woodland resources found in the Dakotas. Many of the trees and larger shrubs were cut for construction and fuel, and livestock grazing of understory plants often was heavy. Today there is less demand for the wood, and many of the trees have regrown, both by root sprouting and by seedling establishment. Excessive livestock grazing still occurs in some areas, but less so now than in the mid-1900s. Most ranchers provide widely dispersed sources of water to distribute livestock over a larger area, and some now use woodlands only for wintering cattle and calving.[22]

Appreciating the shelter and other benefits provided by woodlands, farmers and ranchers soon sought to moderate the prairie environment by creating woodlands around their buildings (see chapter 6). The abundance of shelterbelts in the Great Plains today is a testament to their success. The rapid growth of these trees and shrubs suggests that fires, now usually suppressed, may have been a more important factor restricting the presence of trees than aridity, at least in the eastern Dakotas. However, the saplings that were transplanted for shelter usually benefited from hoeing or plowing that removed the competition of prairie plants. A notable exception is aspen, which invades grasslands by root sprouting. Ranchers in northern North Dakota and the prairie provinces of Canada are acutely aware of how aspen expansion reduces forage availability for their livestock, prompting them to control aspen spread with prescribed fire, herbicides, and livestock grazing.[23]

Prairie woodlands now cover less than 3 percent of the Dakotas (see table 1.1). Though occupying a small area, they provide woodland-prairie edge habitat that greatly augments the biological diversity of the region. Plants growing under the trees are different, and forest-dwelling animals such as squirrels, sharp-tailed grouse, woodpeckers, and various cavity nesters coexist with grassland species. Numerous birds and small mammals benefit from the thousands of invertebrates per square meter in both grasslands and woodlands, and these "insectivores" are in turn the prey of the coyotes, foxes, various raptors and snakes, and other predators. Wildlife in such areas can be extraordinarily diverse and abundant. For these reasons, there is considerable interest in protecting the woodlands that remain.[24]

Trees and shrubs in some areas may now occupy more land than in the early 1800s, but many native woodlands have been cleared and the understory is often heavily grazed by domestic animals, which can damage the trees. Also, large numbers of white-tailed deer eat the bark and twigs of saplings, especially aspen, further restricting tree growth in some areas. And diseases infect many trees, such as the introduced Dutch elm disease, which has killed a large proportion of the American elm in the Dakotas.[25] More recently, the emerald ash borer has become a threat. This beetle has killed millions of green ash in eastern North America and has been reported in southern Manitoba and southeastern South Dakota (fig. 7.13). In addition to diseases and insect pests, invasive plants such as eastern redcedar, buckthorn, and garlic mustard (figs. 7.14, 7.15, and 7.16) are viewed as threats to native woodlands.

Fig. 7.13. The emerald ash borer, measuring about a half inch long, is an Asian beetle first seen in Michigan in 2002 that has since killed more than 50 million ash trees east of the Dakotas. The insect feeds on four species of ash in North America, including green ash—the most common native tree on the Northern Great Plains. Trees are killed by the beetle's larvae, which feed on the inner bark. The insect deposits eggs on the bark surface which are then spread over long distances on harvested wood. Campers are urged to use firewood obtained at their destination. In the Dakotas, the insect had been documented only in Sioux Falls by 2020. Photo by Debbie Miller, USDA Forest Service, www.bugwood.org.

Fig. 7.14. Eastern redcedar, the dark-green conifer in the foreground and in the distance along a Missouri River reservoir in South Dakota, is a small tree that is native in southeastern South Dakota, southward over much of the eastern Great Plains and eastward to the Atlantic Coast. It has been planted for windbreaks and wildlife habitat. Birds disperse the seeds, thereby facilitating the invasion of grasslands if fires are suppressed. Like Rocky Mountain juniper, eastern redcedar does not sprout and is easily killed by cutting or prescribed fires. The two species can live for over 500 years and they commonly hybridize. Male and female cones are produced on different trees. Only male trees should be planted where redcedar invasion is considered a potential problem.

Fig. 7.15. Common buckthorn is an invasive Eurasian shrub or small tree that is shade tolerant and outcompetes many native plants in the understory of some woodlands. The fruits and foliage, if eaten, are toxic to people and at least some other mammals, but birds eat the mature fruits and disperse the seed in their droppings. The shrub is an alternate host for some diseases on cereal crops and is an important overwintering host for the soybean aphid. Once established, buckthorn is difficult to control because of sprouting. Photo by Leslie J. Mehrhoff, University of Connecticut, www.bugwood.org.

Fig. 7.16. Garlic mustard is an aggressive Eurasian herbaceous plant that could invade river break woodlands and meadows in the Dakotas. Currently it is found along the Big Sioux River in southeastern South Dakota (Ley 2012). Its expansion north- and westward along rivers is expected unless an effective control measure is discovered. Photo by David Cappaert, www.bugwood.org.

Confounding all of these effects is climate change. As ecologist Norman Henderson and his associates concluded in 2002, woodland species on the otherwise-grassy Great Plains are at the limits of their environmental tolerances and are vulnerable because the places where they live are often small and isolated. As a result, the changes that occur because of rising temperatures and severe drought—as is forecast to occur more often, along with wildfires and insect epidemics—may be "sudden and dramatic." Trees were killed even by the relatively short droughts of the 1930s.[26] Proactive approaches to woodland management may be required, which could include suppressing fires, which previously were deemed desirable, a counterintuitive decision that might be made if woodlands become even more rare than before. Another recommendation might be the introduction of drought-tolerant varieties of trees and shrubs to replace those that are more vulnerable. The numerous shelterbelts and windbreaks planted on the Great Plains provide a means of judging which species are likely to be tolerant of the climatic changes now under way (see chapter 6).[27] It remains to be seen whether this level of intrusive conservation management will become widespread, but a changing climate affects everything—the nature and frequency of disturbances, the spread of invasive species and pathogens, the success of seedling establishment, the synchrony of pollinator abundance with host plant flowering, food availability for sensitive species of wildlife, and more.

In summary, prairie woodlands of all kinds cover a small percentage of the Dakotas and are widely dispersed (see chapter 1), but the entire prairie ecosystem is enriched and more diverse because of them. These woodlands have a history of expansion and contraction in response to a naturally changing climate since the most recent glacial retreat. They have now been affected by a century or more of fire suppression and livestock grazing, yet they contribute significantly to the economy of the plains—often via ecosystem services. Highly valued for the biological diversity and watershed protection they contribute to the region, woodlands are likely to be early indicators of human-driven climate change.

Chapter 8 Ponderosa Pine, Pine Ridge, and the Black Hills

Extending from Wyoming through Nebraska and into South Dakota is a series of escarpments and rocky slopes with ponderosa pine (figs. 8.1, 8.2, and 8.3). Known collectively as Pine Ridge, it is the longest outcrop of sedimentary rocks on the Missouri Plateau. The presence of trees on ridges seems unusual because the rocky soils appear to be too dry, and often they are. Many ridges do not have trees of any kind. However, water infiltrates more quickly to greater depths in the rocky or coarse-textured soils that have developed, compared to the fine-textured soils of surrounding grasslands, thereby creating the relatively moist environment trees require. Thus, on the semiarid western Great Plains, fine-textured soils are drier than coarse-textured soils—the inverse texture effect.[1]

Lower flammability is another factor that enables trees and some other plants to survive on escarpments. Candace Savage, in her 2004 book on North American prairies, suggested that the best explanation for "scarp woodlands" is "their top-o'-the-world location." She writes, "Where better to find refuge than atop a natural firebreak, an outcropping of safety in a world of flame?" Forest ecologists determined that, several hundred years ago, these woodlands had low-intensity surface fires that would kill many of the small ponderosa pines but usually not the larger trees, which have a thick, protective bark and high branches. Fire suppression has enabled denser tree growth, the expansion of pines into adjacent grasslands, and an increase in the potential for catastrophic wildfires—an issue of great concern in the area.[2]

Some wonder why ponderosa pine, a magnificent tree, does not grow naturally in the eastern Dakotas where rainfall is higher. Part of the explanation is that pine seedlings do not compete well with tall grasses, forbs, and other kinds of trees. Also, lacking the ability to sprout from charred stumps, conifer seedlings do not survive where prairie fires occur every few years. The first explorers found hardwood trees growing on the woodlands of Turtle Mountain, on the Killdeer Mountains, and on the wooded draws of the Prairie Coteau, but no pine, spruce, or fir. The successful planting of conifers in such areas is possible only if fires are suppressed and young saplings are planted instead of seeds.

The Pine Ridge landscape is sometimes referred to as *parkland* or *savanna*, with widely dispersed trees in prairie, and it has attracted people on the Great Plains for thousands of years. Many animals depend on these ridges for nesting, shelter, and food, which makes them ideal hunting grounds, and today people often build homes in such places—sheltered from the wind and with

Fig. 8.1. Ponderosa pine savannas and woodlands on a portion of the Pine Ridge landscape. Pine Ridge extends from Wyoming into Nebraska and South Dakota, marking the northern edge of the southern Great Plains. The various ridges are formed from sandstones, claystones, and siltstones, all Oligocene in age and made erosion resistant by periodic deposits of volcanic ash originating west of the Dakotas.

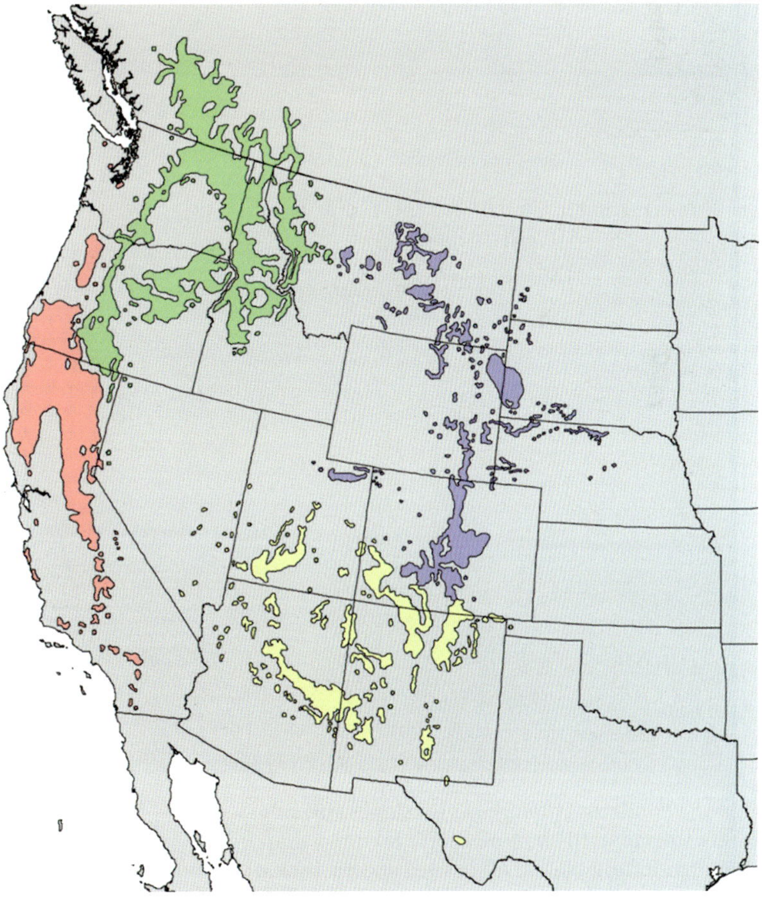

Fig. 8.2. Map showing the distribution of the four subspecies of ponderosa pine in North America, based on Elbert Little's atlas. Subspecies are commonly designated for plants that have large distribution ranges that encompass different environmental conditions. The map reveals the locations of Pine Ridge, Black Hills, Slim Buttes, Cave Hills, the badlands of the Little Missouri and White Rivers, and other buttes and ridges. In the Dakotas, subspecies *scopulorum* is found above approximately 2,500 feet elevation. Source: U.S. Department of Agriculture.

Fig. 8.3. Ponderosa pine grows on slopes and ridges where soils tend to be more coarse textured than on the surrounding plains. Associated species include little bluestem, sideoats grama, and, in the foreground, western snowberry. The flowering plant is soapweed yucca.

panoramic views. Away from the escarpments, the mixed-grass prairie provides excellent forage for livestock, as it did for bison, and often it is plowed for croplands, especially if irrigation water is available. Large mammals found in the vicinity are elk, mule deer, white-tailed deer, and pronghorn. Wild turkeys and sharp-tailed grouse are common, and fishing is popular in some tree-lined creeks draining north to the White River. Ponderosa pine reaches ages over 250 years old. Other plants include Rocky Mountain juniper, green ash, snowberry, chokecherry, American plum, little bluestem, and sideoats grama. Trees in nearby creek bottoms and ravines include plains cottonwood, hackberry, boxelder, and American elm (table 8.1).[3]

The Black Hills

Nowhere on the Great Plains is ponderosa pine so abundant as in the Black Hills. It dominates 90 percent of the area (fig. 8.4),[4] growing with Rocky Mountain juniper on dry ridges in the foothills and with bur oak, aspen, white spruce, and occasionally paper birch in moist locations. Spruce is found in small groves where the temperature is relatively cool, such as at high elevations and on north slopes or in ravines (fig. 8.5). Native Americans have lived in the area for thousands of years, an island in the plains that provided opportunities and resources not readily available elsewhere.[5] Some anthropologists suggest they spent most of their time in the foothills, a transition zone that provided access to the resources of both mountains and plains.[6]

European immigrants depended on beaver trapping and gold mining initially, and then timber harvesting, livestock grazing, and tourism.[7] Homesteaders flocked to the area in the late nineteenth century, especially along creeks where water was readily available. Today about 11 percent of the land is privately owned, mostly in small parcels. The remainder is public land—82 percent in the Black Hills National Forest, which includes the Black Elk Wilderness and the Bear Lodge Mountains. Smaller tracts of public lands are Wind Cave National Park, Mount Rushmore National Memorial, Jewel Cave and Devils Tower National Monuments, and Custer and Bear Butte State Parks.

One of the first to write about the natural history of the Black Hills was Lt. Col. George Armstrong Custer of the Seventh Cavalry. In 1874 he led an expedition from Fort Abraham Lincoln near Bismarck to reconnoiter a route to the Hills, explore their interior, and validate rumors that gold was found there. On July 25, he wrote: "Every step of our march that day was amid flowers of the most exquisite colors and perfume. So luxuriant in growth were they that men plucked them without dismounting from the saddle. . . . It was a strange sight . . . the men with beautiful bouquets in their hands, while the head-gear of the horses was decorated with wreaths of flowers fit to crown a queen of May. Deeming it a most fitting appellation, I named this Floral Valley." Another member of his expedition observed: "Everybody was making bouquets. . . . Some said they would give a hundred dollars just to have their wives see the floral richness for even one hour."[8] A journalist traveling with Custer wrote: "None of us ever saw so dense and so extended a growth of pine."[9]

The Custer expedition also would have seen wolves, grizzly bears, cougars, and raptors—and numerous scavengers feeding on carcasses. Wildlife diversity in the Black Hills is still much appreciated, although the nearest grizzlies are now in the Greater Yellowstone Ecosystem, 250 miles to the west. Occasional wolves venture into North Dakota from time to time, but they are rare. Bird watchers flock to the Hills to see species not easily found elsewhere on the Great Plains, such as the northern goshawk, Lewis's woodpecker, red crossbill, and American dipper. The enabling legislation passed by Congress for Wind Cave National Park specified protection of the cave, but reestablishing

Table 8.1. Characteristic trees and shrubs of woodlands in the Black Hills, on Pine Ridge, and elsewhere in the western Dakotas[a]

		Pine Ridges		Black Hills		
Common name[b]	Latin name	ND	SD	Foothills	Montane	Riparian
TREES						
Ash, green	*Fraxinus pennsylvanica*	X	x	X	–	X
Aspen	*Populus tremuloides*	X	x	X	X	X
Birch, paper	*Betula papyrifera*	X	–	X	X	X
Birch, water	*Betula occidentalis*					
Boxelder	*Acer negundo*	x	x	X	–	X
Cottonwood, narrowleaf	*Populus angustifolia*	–	–	–	–	X
Cottonwood, plains	*Populus deltoides*	–	–	–	–	X
Elm, American [M]	*Ulmus americana*	x	–	–	–	X
Elm, slippery	*Ulmus rubra*	–	–	X	–	–
Hackberry	*Celtis occidentalis*	–	x	–	–	X
Hawthorn	*Crataegus* spp.	–	–	X	–	X
Ironwood (hophornbeam)	*Ostrya virginiana*	–	–	X	X	X
Juniper, Rocky Mountain	*Juniperus scopulorum*	X	X	X	–	X
Oak, bur	*Quercus macrocarpa*	x	x	X	–	X
Pine, limber	*Pinus flexilis*	rare	–	–	Rare	–
Pine, lodgepole	*Pinus contorta*	–	–	–	Rare	–
Pine, ponderosa	*Pinus ponderosa*	x	X	X	X	-
Spruce, white (Black Hills spruce)	*Picea glauca*	–	–	X	X	X
Willow, peachleaf	*Salix amygdaloides*	–	–	–	–	X
SHRUBS						
Barberry, creeping	*Mahonia repens*	–	x	X	X	–
Bearberry (kinnikinic)	*Arctostaphylos uva-ursi*	–	–	X	X	–
Buffaloberry, russet	*Shepherdia canadensis*	–	–	–	X	–
Buffaloberry, silver	*Shepherdia argentea*	x	x	–	–	–
Chokecherry	*Prunus virginiana*	x	x	X	X	X
Currant	*Ribes* spp.	–	x	X	X	X
Dogwood, red-osier	*Cornus sericea*	–	–	–	–	X
Hazelnut, beaked	*Corylus cornuta*	x	–	X	X	X
Huckleberry, dwarf (grouse whortleberry)	*Vaccinium scoparium*	–	–	–	X	–

Common name[b]	Latin name	Pine Ridges ND	Pine Ridges SD	Black Hills Foothills	Black Hills Montane	Black Hills Riparian
Ivy, poison	*Toxicodendron rydbergii*	–	–	X	–	X
Juniper, common	*Juniperus communis*	–	x	–	X	–
Juniper, horizontal	*Juniperus horizontalis*	x	x	–	–	–
Mahogany, mountain	*Cercocarpus montanus*	–	x	x	–	–
Ninebark, mountain	*Physocarpus monogynus*	–	–	X	–	–
Plum, American	*Prunus americana*	x	x	–	–	X
Raspberry	*Rubus idaeus*	–	–	X	–	–
Rose, prickly wild	*Rosa acicularis*	–	–	–	X	–
Rose, Wood's	*Rosa woodsii*	x	–	X	–	X
Sagebrush, big	*Artemisia tridentata*	x	x	X	–	–
Sagebrush, silver	*Artemisia cana*	x	x	–	–	X
Serviceberry (juneberry)	*Amelanchier alnifolia*	x	x	X	X	X
Snowberry, common	*Symphoricarpos albus*	–	–	X	X	–
Snowberry, western	*S. occidentalis*	x	x	X	–	X
Spiraea, white	*Spiraea betulifolia*	–	–	–	X	–
Sumac, skunkbush	*Rhus trilobata*	x	x	X	–	X
Willow, Bebb	*Salix bebbiana*	–	–	–	–	X
Willow, sandbar	*Salix exigua*	–	–	–	–	X

[a]For more detail see Girard et al. (1989) and Hansen and Hoffman (1988) for southwestern North Dakota (ND) and northwestern South Dakota (SD), respectively, and Hoffman and Alexander (1987) and Marriott and Faber-Langendoen (2000a, 2000b) for the Black Hills region.

[b]For plant identification, see Larson and Johnson (1999) and Barkley (1986); X = common, x = occasional. A dash indicates that the plant is absent or not common.

bison, elk, and pronghorn soon became another part of the park's mission. These charismatic animals, along with mule deer and white-tailed deer, are common in the park and in adjacent Custer State Park, the largest state park in the nation and the place where bison stampedes have been filmed for Hollywood movies. The cougar is now the top predator for large mammals, as are hunters, who harvest approximately 700 elk and 4,500 deer each fall.[10] Without predation, the herbivore populations would degrade their habitat. Also, with so many roads in the Black Hills, wildlife-vehicle collisions would be more frequent.

Plant Life

About 50 years after Custer expressed such appreciation for the wildflowers in Floral Valley, botanists wrote about the unusual assemblage of plants in the Hills.[11] As might be expected, Rocky Mountain species are common, such as ponderosa pine, Oregon grape, Rocky Mountain juniper, common juniper, narrowleaf cottonwood, grouse whortleberry, and heartleaf arnica. Their presence is best explained by close proximity to the Rockies and the fact that Rocky Mountain plants were more widespread about 10,000 years ago, when the

Fig. 8.4. Ponderosa pine forest in the Black Hills. Without frequent surface fires, young pine trees become established, making the forest more dense and increasing the probability of a crown fire. The herbaceous plants in this area are various grasses and forbs, including Kentucky bluegrass, poverty oatgrass, bedstraw, wild bergamot, and western snowberry (buckbrush). Common juniper, a shrub, is found at higher elevations.

Fig. 8.5. Groves of white spruce, also known as Black Hills spruce, are common in cooler, wetter locations at high elevations, on north slopes, and along some valley bottoms with cool-air drainage. In locations with high humidity the tree branches sometimes have a pendulant lichen known as old man's beard, not shown in this photo.

climate of the Great Plains was cooler and more humid.[12] More puzzling is the presence of plants that typically are found much farther away in the deciduous forests of eastern North America, such as American elm, boxelder, bur oak, hackberry, hophornbeam, Virginia creeper, and bloodroot, or those more common in the boreal forests of Canada, such as bunchberry dogwood, Canada scurvyberry, paper birch, and white spruce. At least some of the eastern species have survived along relatively moist west-to-east-flowing tributaries of the Missouri River.[13] Boreal species migrated southward as the continental glaciers advanced. The Black Hills were not glaciated, but boreal plants survived south of the ice where the climate was favorable—and still is today in some places. Conservation biologists place great value on disjunct populations because they contribute significantly to regional biodiversity.

Ponderosa pine now dominates Black Hills forests, but streams and floodplains at low elevations often are lined with meadows, plains cottonwood, and various willows. Green ash, boxelder, hackberry, American elm, and bur oak occupy nearby slopes.[14] Woody draws or ravines have thickets of chokecherry, American plum, hawthorn, western

Fig. 8.6. Bur oak forms woodlands at lower elevations in the Black Hills and Bear Lodge Mountains, and sometimes grows in the understory of ponderosa pine. Elevation 4,250 feet.

snowberry, and red-osier dogwood. Bur oak occurs most often on the relatively moist uplands in the northern parts of the Hills and is commonly associated with hophornbeam, green ash, American elm, and western snowberry (fig. 8.6). On Mowry shale, the oaks grow in open woodlands with few understory plants. Sometimes the oak is a shrub under ponderosa pine, and because of its ability to sprout, the oak can become the dominant tree if the pines are killed by fire or harvesting.[15]

Aspen occurs throughout the Black Hills on relatively moist uplands, such as where drifting snow accumulates or near the bottom of slopes. It grows occasionally with paper birch or bur oak and commonly forms a narrow band at the edges of some forests (fig. 8.7).[16] Associated species include hazelnut, bracken fern, and wild sarsaparilla. Aspen woodlands are sometimes overtopped by taller white spruce or ponderosa pine, but sprouting aspen and birch regain dominance more quickly following disturbances.

Nearby grasslands are mostly mixed-grass prairie, although tallgrass species such as big bluestem, Indian ricegrass, and prairie sandreed are common on sandy soils.[17] Streamside meadows have prairie cordgrass, tufted hairgrass, wild iris, and a variety of wetland sedges (fig. 8.8). Where the soils are saline and frequently moist, inland

Fig. 8.7. Aspen commonly grow on the edges of ponderosa pine forests and in meadows where the soil is not too dry or too wet. Though deciduous, aspen has chlorophyll in the bark that is capable of photosynthesis. Thus, this tree can be thought of as a deciduous evergreen. All the aspen "trees" visible in this photo probably are part of one clone, that is, all the stems have sprouted from the roots of one plant.

Fig. 8.8. Lt. Col. George A. Custer and his troops camped in this meadow in 1874, near the confluence of Silver Creek with Castle Creek about one mile north of Deerfield Reservoir on U.S. Forest Service Road 110. The meadow appears much the same today as it did then, according to a comparison of this photo with photos taken by the expedition photographer William H. Illingworth. However, introduced species such as Kentucky bluegrass, timothy, and smooth brome are now common, along with green needlegrass, stiff goldenrod, and numerous other native plants. The surrounding ponderosa pine forests are now more dense. Illingworth's photos of this meadow and others can be viewed in Progulske (1974) and Grafe and Horsted (2002). Elevation approximately 6,000 feet.

saltgrass is common. Several large grasslands, including Gillette and Reynolds prairies, still exist, but most have been partially or completely converted to introduced hay grasses, mainly timothy and smooth brome. Native prairie in these areas, known as Black Hills montane grasslands, are identifiable by the presence of three grasses—prairie dropseed, Richardson needlegrass, and timber oatgrass.[18] Shrublands are common, typically on dry escarpments in the foothills.[19] Skunkbush sumac and Rocky Mountain juniper are widespread; snowberry and russet buffaloberry are common in the north, and mountain mahogany occurs in a small part of the south. Big sagebrush is not common but can be found on relatively deep soils where snowdrifts develop.

Geologic History and the Landscape Mosaic

The Black Hills provide a window for understanding the development of mountain ranges and how variation in soils and topography affect the ecology of an area. As described in chapter 2, the Hills originated with a regional uplift about 50 million years ago.[20] Hard as it is to imagine, water erosion has washed thousands of feet of sedimentary material off the top of the Hills and out onto the Great Plains, exposing the more erosion-resistant granitic core (fig. 8.9). Erosion occurred largely toward the east, the direction in which the two major rivers flow—the Belle Fourche River around the north end and the Cheyenne River around the south end. Devils Tower, an unusual geologic feature, was exposed by this erosion (fig. 8.10).

The domelike uplift and subsequent exposure of Black Hills bedrock led to the formation of five more or less concentric geomorphic regions with distinctive landscape patterns.[21] On the perimeter is the Hogback Rim, composed of Mowry shale, Lakota sandstone, Fall River sandstone, Minnewasta limestone, and other erosion-resistant sedimentary strata that were tilted upward as the Hills developed. These rocks form a sharply defined ridge to the north, east, and south and have woodlands similar to those on Pine Ridge to the southeast. On the interior side of the Hogback Rim is Red Valley, a name that stems from the red shales of the Spearfish Formation that give a distinctive red color to the fine-textured soils.[22] Also known as the *racetrack* because of its oval shape around most of the Hills, the valley was formed by the erosion of softer shales, siltstones, and sandstones underlying the more resistant rocks of the Hogback Rim. Grasslands predominated in the valley when European immigrants arrived, as they do today where the land has not been cultivated. Much of Wind Cave National Park is in the Red Valley, as are the towns of Hot Springs, Sundance, Spearfish, and the western part of Rapid City. The elevation of the valley is 3,000 to 3,600 feet.

Further to the interior, above the Red Valley, are the Minnelusa Foothills, which are composed of harder, erosion-resistant sedimentary strata. The vegetation is mostly ponderosa pine forest, with grasslands at lower elevations. The hills north of Newcastle are part of this region. Still higher and farther to the west, Madison Limestone forms a comparatively flat area known as the Limestone Plateau (6,200 to 7,200 feet elevation). The vegetation of the plateau is primarily ponderosa pine forest and savanna, commonly intermingled with bur oak woodlands in the north and groves of white spruce on north slopes or in canyons. The underlying limestone is permeable to water, which has enabled the formation of numerous caverns, including Wind Cave and Jewel Cave. Some creeks disappear into the bedrock before emerging again, such as Spearfish and Boulder Creeks in the north. Spearfish Creek formed the most spectacular canyon in the Hills, cutting through 400 feet of Paleozoic sandstones.

A fifth geomorphic region is the Central Area where Precambrian granitic rocks have eroded into pinnacles (easily viewed along the Needles Highway, fig. 8.11). The elevation is mostly 5,000–6,000 feet. Again, the vegetation is predominately ponderosa pine forest, with occasional groves of

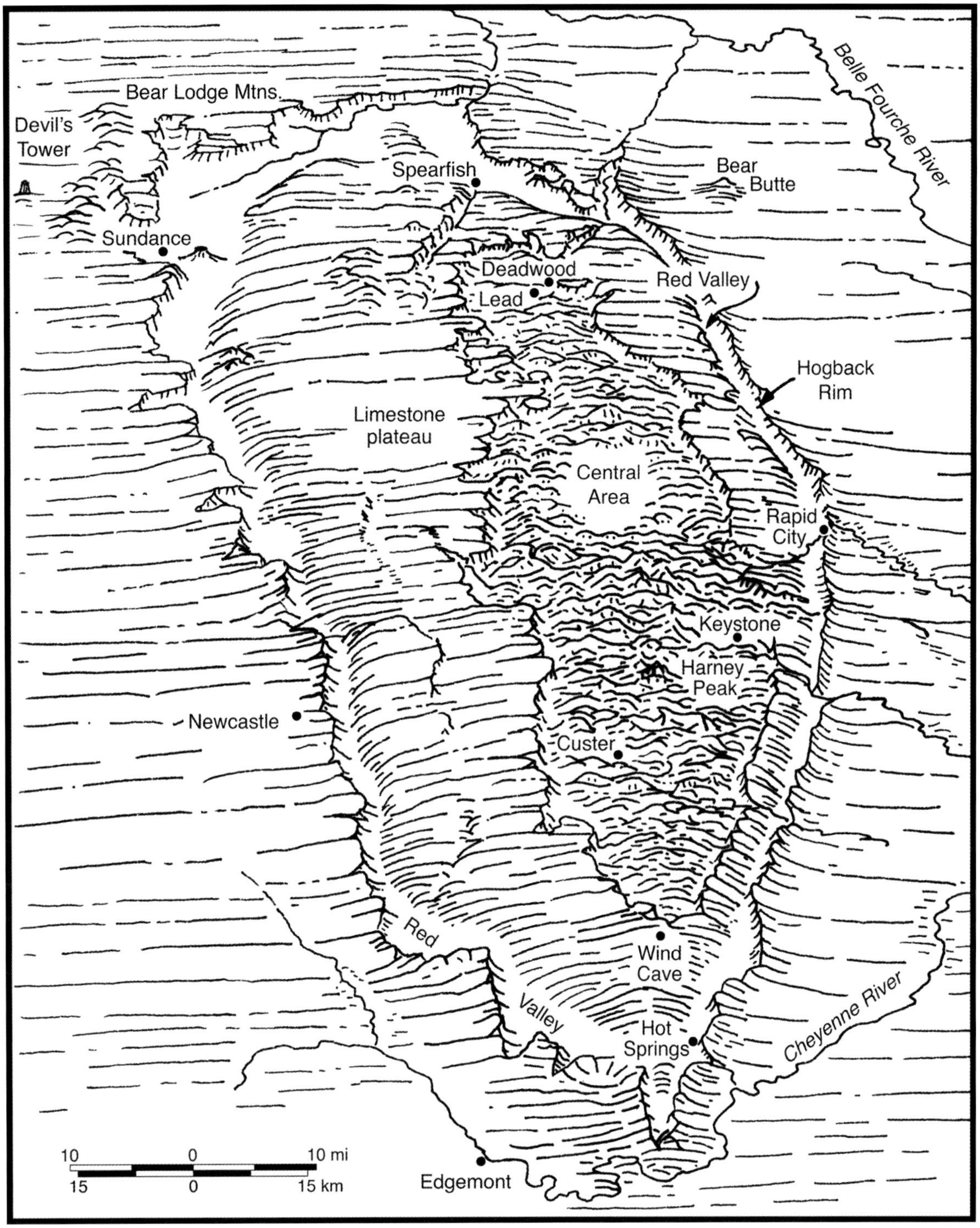

Fig. 8.9. Geomorphic features of the Black Hills. The Central Area is where granitic rocks have been exposed by erosion of the sedimentary rocks of the Limestone Plateau, such as along the Needles Highway. From Knight et al. 2014, with permission from Yale University Press.

Fig. 8.10. Devils Tower, flanked by ponderosa pine woodlands, savanna, and mixed-grass prairie, is one of 12 igneous intrusions in the Black Hills region. Others include Inyan Kara Mountain, Little Missouri Buttes, and Warren Peaks in Wyoming; and Bear Butte, Black Butte, Crow Peak, Custer Peak, and Terry Peak in South Dakota. Elevation at the top of Devils Tower is 5,117 feet. Flowering plants in the foreground are upright prairie coneflower and pale purple coneflower. Plants found on top of Devil's Tower include skunkbush sumac, big sagebrush, fringed sagewort, pricklypear cactus, selaginella, bluebunch wheatgrass, junegrass, and blue grama (H. Marriott, personal communication). Photo by Hollis Marriott.

Fig. 8.11. The granitic core of the Black Hills is easily observed along the Needles Highway in Custer State Park. The forests are dominated by ponderosa pine, with groves of aspen and white spruce in relatively moist areas. Elevation 5,000 to 6,000 feet. Photo by Justin Meissen.

white spruce and aspen. The region is devoid of natural lakes, but several reservoirs occur in the area, such as Pactola, Sheridan, and Sylvan.

The smaller Bear Lodge Mountains, located in Wyoming to the northwest of the Black Hills, have a similar geologic history but less-well-defined geomorphic regions. Warren Peak is the highest point—6,655 feet—about 600 feet lower than Black Elk Peak (formerly Harney Peak) in the Hills. Except for the absence of white spruce and a few other plants, the vegetation of the Bear Lodge Mountains is similar to that of the Black Hills.

Forest Management

Surprisingly, ponderosa pine probably was not present in the region until about 4,000 to 6,000 years ago. This conclusion is based on two lines of evidence: First, fragments of ponderosa pine in packrat middens throughout the western United States have been found only in southwestern North America at the beginning of the Holocene, about 10,000 years ago. Second, the first pine fragments found in the Dakotas, in the Black Hills, were in middens carbon-dated to 3,850 years old. Exactly when ponderosa arrived is not known, but pine fragments in older middens have not yet been found.[23] Additional evidence for a postglacial, mid-Holocene arrival is provided by what is now known about the climatic preferences of ponderosa pine, namely, relatively warm with more precipitation in the summer than in the winter. The climate in the Dakotas probably would have been too cold while glaciers were not far away, near the Missouri River, but pine expanded as the climate warmed and most of the boreal species slowly disappeared—surely a curious development that Native Americans talked about through the ages.

Located off the beaten track of travelers on the Missouri River to the north and the Oregon Trail to the south, the Black Hills were one of the last frontiers for the first Europeans. Most came for beaver pelts and gold. The Treaty of Fort Laramie in 1868 gave ownership of the Hills and all country west of the Missouri River in South Dakota to bands of the Lakota, Dakota, and Arapaho Nations—the Great Sioux Reservation.[24] But that didn't stop European trappers and prospectors. Their aggressive trespassing and the aftermath of the Battle of Little Bighorn in 1876 led to the U.S. government nullifying the Fort Laramie treaty in 1877, only nine years after it was signed. Tribes were relegated to five smaller reservations and the Black Hills became part of the public domain; much of the land was available for sale or homesteading.

Well before the 1860s, trappers had nearly exterminated the beaver throughout the Black Hills and much of North America—arguably the first major ecological impact of Europeans.[25] That exploitation ended when the demand for beaver pelts faded during the 1830s. Beaver populations in the Hills and elsewhere have never fully recovered. Their dams shaped the riparian landscapes that are now preferred for homes, farms, resorts, and towns. With fewer beaver, some riparian landscapes changed dramatically. Streamside habitats also were altered by placer mining for gold after Custer verified its presence in 1874. A year later an estimated 15,000 miners were in the Hills.[26] Panning for gold soon was replaced with dredging and sluicing, which disrupted hundreds of miles of streambed. Livestock were driven to the Hills for meat and dairy products, with many of the animals spending most of their time grazing in the riparian zone near water. As is true throughout the west, then and today, streamside habitats were affected more than any other ecosystem.

The first mobile sawmill was powered by steam and was brought to the Hills in 1876 during the Gold Rush.[27] For many years there would be great demand for the timbers and lumber required for mining, railroad ties, buildings, and the firewood that everyone used—in the Hills and on the surrounding plains.[28] By the 1890s, whole mountainsides had been cut without regard for watershed protection. Forest fires were frequent, and bark beetles were killing large numbers of trees. Dispatched to evaluate the situation in 1897, Henry Graves wrote that the Hills were "densely timbered, but the forest is broken in many places by parks and mountain prairies, and enormous tracts have been entirely denuded by forest fires."[29] Fire suppression was his recommendation. Graves also observed a "reckless waste of timber which has been going on for years." In the same year, Charles Sprague Sargent wrote that the Black Hills "has suffered seriously from fire and the illegal cutting of timber." He recommended the forests "should be protected and made permanently productive." The Black Hills Forest Reserve was established a few months later and became the Black Hills National Forest in 1905.[30] The goal was to harvest timber in a way that would not damage the forest or the watershed. Also passed at this time was the Forest Management Act of 1897, which established rules for determining which trees could be cut and sold. The first federal timber sale in the Black Hills—and anywhere in the nation—was in 1899. Unlike before, now only the timber could be sold, not the land. As the historian John Freeman wrote in 2015, there was a "fundamental change from selling public lands to encourage settlement and earn income for the federal government to keeping public lands in public hands for the public good and still earn income for the government."

Changes in national forest management were sparked by developments in the Black Hills and continued with President Theodore Roosevelt's appointment of Gifford Pinchot as chief of the Forest Service. Both saw the value of sustainable timber-harvesting practices; they also urged fire suppression and the protection of watersheds, wildlife habitat, and the environment in general. At the same time, prominent South Dakotans promoted the Black Hills as a tourist destination, encouraging the construction of scenic highways and resorts, and the establishment of state and national parks. Bison and elk were reintroduced. Many people came to the Hills to fish, hunt, and see animals not found on the surrounding plains. Aesthetics became a valuable natural resource. Concerns were expressed about too much timber harvesting even though tree regeneration was so dense that thinning was usually necessary to promote the commercial goal of rapid tree growth (figs. 8.12 and 8.13). Midwesterners appreciated the big pines that many had not seen before; travel to the Hills became relatively easy as automobiles and roads improved. Today there are more roads than in any other national forest—four to five miles of roads per square mile. John Freeman observed that the Black Hills National Forest "borders on thousands of backyards"—the result of so many people having homesteaded or purchased a piece of land with big trees, a moderate climate, and creeks with clear water.

Concerns about excessive timber harvesting and livestock grazing on public lands eventually led Congress to pass the Multiple Use–Sustained Yield Act of 1960. This law mandated that national forests be managed for outdoor recreation, water, and wildlife as well as for timber and livestock. Nine years later, in 1969, passage of the National Environmental Policy Act (NEPA) required more careful planning and public involvement. The National Forest Management Act of 1976 placed further restrictions on forest management. Archaeologists, biologists, hydrologists, and landscape architects were hired by the Forest Service to assist in developing timber-harvesting methods that minimized adverse effects on other forest values.[31] For example, selective harvesting usually is now preferred over clearcutting, by which all trees are cut, and some standing dead trees (snags) and downed logs are left for the benefit of wildlife and the maintenance of soil fertility and biological diversity.[32] Pine forests often are thinned to provide more light, water, and nutrients for the growth of plants such as aspen, grasses, forbs, and shrubs—not solely for the growth of the surviving pines.[33] Maximizing tree growth is no longer the only objective. Low-intensity surface fires are sometimes ignited as a management tool, and livestock grazing is regulated more than before.[34]

Fig. 8.12. This thinned forest of ponderosa pine will soon become very dense and flammable if a fire or mechanical thinning does not kill some of the tree saplings that have become established.

Fig. 8.13. Older ponderosa pine trees with thick bark often survive surface fires even though the heat of smoldering pine needles scar the base of the tree. Scars form when the thin, delicate cambium separating the bark and sapwood is scorched. Multiple scars can be created on the same tree and the resulting annual rings around each scar can be used to determine the year a fire occurred. Collectively the scars provide an estimate for the fire-return interval where that tree is located.

The Black Hills National Forest was the first in the national forest system to complete a NEPA-mandated management plan.[35] Since its adoption in 1983, the plan has been amended and revised to incorporate new findings and priorities. Still, management decisions are often controversial.[36]

Fig. 8.14. Timber harvesting and slash removal can be used to create a less flammable ponderosa pine forest, such as here in the southern Black Hills. Where feasible, prescribed surface fires are used to control invading saplings.

Fire Suppression, Tree Density, and Streamflow

During the past century, the most far-reaching effect of land management in the Black Hills and Pine Ridge has been fire suppression, greatly increasing ponderosa pine density and aerial extent. Previously, surface fires occurred every 10–35 years, depending on topographic position, elevation, and slope exposure.[37] Usually the fires were low intensity but sufficiently hot to kill most of the small pines. Some saplings would survive because of dense clusters of needles that protected the terminal buds of the saplings, especially if the flames moved rapidly on a windy day. Larger trees usually would not be killed because the bark was not highly flammable and was thick enough to protect the cambium, a thin cylinder of cells that produces new sapwood to the interior and new phloem and bark to the exterior. Some trees were scarred by the scorching of a portion of the camblum. Because of these varied effects, plus the combustion of much of the fuel, frequent surface fires maintained an open forest in many areas (figs. 8.13 and 8.14).[38] Early on, forest fires were relatively easy to put out, but suppression for long periods led to the accumulation of large amounts of fuel that enabled stand-replacing crown fires that could not be stopped—the wildfires that worried everyone.

Low-intensity surface fires may have been characteristic of most pre-European ponderosa pine forests, but crown fires occurred as well.[39] For example, in 1880 the geologists Walter Jenney and Henry Newton wrote, "The Black Hills has been subjected in the past to extensive forest fires, which have destroyed the timber over considerable area. Around Custer Peak and along the limestone divide, in the central portion of the Hills, on the headwaters of the Box Elder and Rapid Creeks, scarcely a living tree is to be seen for miles. . . . Some portions of the parks and valleys, now destitute of trees, show by the presence of charred trees and decaying stumps that they were once covered by forest, but generally the pine springs up again as soon as it is burnt off, though sometimes it is succeeded for a time by thickets of small aspens."[40] Similarly, Lieutenant Colonel Richard Dodge wrote in 1876: "Throughout the Hills the number of trees which bear the marks of the thunderbolt is very remarkable. . . . The woods are frequently set on fire. . . . There are many broad belts of country covered with tall straight [dead] trunks of what was only a short time before a splendid forest of trees." Photographs taken in the 1870s show an open forest of live trees intermingled with many dead standing trees.[41] Notably, bark beetles rather than fire may have killed many of those trees.

Thus, even before effective fire suppression, extensive crown fires still occurred in some areas. Ecologists Douglas Shinneman and William Baker studied fire history in the Black Hills and concluded that frequent surface fires maintained open savannas on drier sites, such as in the southern Hills, but some forests at higher elevations and in the northern Hills had less frequent, more dramatic stand-replacing crown fires. They concluded that much of the forests had a mixed fire-intensity regime, with both low- and high-intensity fires. They challenged the commonly held view that mechanical thinning and prescribed surface fires are the way to conserve natural conditions wherever ponderosa pine forests occur in the Hills.[42]

Contributing to the debate with tree-ring evidence, paleoecologist Peter Brown and his associates concluded that present-day forests are now more homogeneous across the landscape and have fewer large trees, a conclusion that supports the work of William Baker and his colleagues.[43] Brown also found that ponderosa pine seedlings, with or without fire, were more likely to become established during wet periods, with the most extensive recruitment associated with an extended wet period from the late 1700s to early 1800s.[44] From this he concluded that large numbers of trees often become established because of a sequence of years with favorable climatic conditions, not solely because of a catastrophic fire or the initiation of fire suppression. Studying small tracts of ponderosa pine in southwestern North Dakota, Loren Potter and Duane Green wrote in 1964 that a new cohort of seedlings "gets started periodically when a good cone crop corresponds with several years of favorable moisture."

But with fire suppression, tree density increases in already forested areas and trees invade adjacent grasslands, causing a concomitant reduction in the amount of forage in both places for livestock, deer, and elk.[45] Many herbaceous plants cannot survive in the understory environment, whether due to insufficient light, thick pine-needle accumulation, or competition with trees for water and nutrients.[46] Increased tree density and forest expansion also can reduce streamflow, as evapotranspiration is higher in forests and woodlands than in grasslands.[47] Watershed managers know that streamflow increases after tree cutting or burning. Also, before the near extirpation of beaver, their dams enabled a higher level of water storage in stream banks, more plant growth in riparian meadows, and possibly streamflow later into the summer and with fewer flash floods.[48] Management for one resource always affects others; there are ripple effects, some would say a cascade of effects.

The value of periodic surface fires in many places is now widely recognized—reduced fuels, a lower probability of hard-to-control crown fires, more streamflow, more forage production, and, with more native, fire-adapted grasses and forbs, fewer invasive plants (fig. 8.14).[49] Prescribed fires frequently are proposed on both private and public lands but are not easily implemented because some forests have become so dense.[50] Another challenge is the large number of homes on private inholdings on the National Forest. Prescribed fires almost always burn as planned, and usually they can be put out at will, but there are still risks in the wildland-urban interface. Thus, fire is used less often than some managers would like, and when they are ignited—only when weather conditions are favorable—the slow-moving fires usually burn a small portion of the land that would benefit from burning.[51]

The largest known fire in the Black Hills was the Jasper Fire, which burned 83,508 acres. Ignited on August 24, 2000, in the southern Black Hills west of Custer, in and around Jewel Cave National Monument, it was of mixed severity. Some of the forest sustained raging crown fires that killed all the trees, on some days burning a hundred acres a minute; other areas had low-intensity surface fires that killed less than a quarter of the trees. A patchy mosaic was created.[52] New pine seedlings could be found a few years after the fire, but fire intensity was severe enough to burn most of the seeds over large areas. Consequently, 20 years later it appears as though large tracts of forest have been converted to prairie (box 8.1, figs. 8.15–8.18). Some understory plants capable of sprouting became more common, such as snowberry, yarrow, sideoats grama, needle-and-thread grass, Canadian horseweed, and cudweed sagewort. Other native plants became less abundant, such as common juniper and ninebark. Introduced plants benefited locally, such as bull thistle, Canada thistle, leafy spurge, and Russian and spotted knapweed. The potential for invasive plants further complicates decisions about where and when prescribed burning and timber harvesting are advisable. Limiting the severity of fires and minimizing forest floor disturbances can alleviate these issues.[53]

Mountain Pine Beetles and Flammability

As in Rocky Mountain forests to the west, an epidemic of the mountain pine beetle was of great concern in the Black Hills during the period 1996–2016, with much speculation about increased fire risk caused by so many dead trees and what land managers might have done to avoid it (figs. 8.19 and 8.20).[54] Such forests are sometimes referred to as tinderboxes, but research has shown that forest

Box 8.1. Explaining the Absence of Trees

Ponderosa pine is so abundant in the Black Hills that ecologists are curious about why this tree is absent from some places. The explanation is easy where the soils are often wet, such as on floodplains or wet meadows: pine seedlings do not tolerate anaerobic soils. But there also are large areas in the Hills without trees where the soil is better drained and aerated. Two puzzling examples are Reynolds and Gillette Prairies, both of which are well over a square mile in size (fig. 8.19). One possible explanation is that the soils are unique in some way that favored the establishment of prairie plants, perhaps caused by an abrupt change in the geologic substrate. An alternative explanation is that a severe fire burned essentially all of the ponderosa pine seed over large areas, or the seedlings did not survive after a fire, and by the time wind-dispersed pine seed reached most of the burned area, grassland plants had formed a dense sod that excluded pine seedlings. Thus, prairie now prevails where a forest might have grown.[a]

The lack of tree establishment over large areas after the Jasper Fire in 2000 supports the severe fire hypothesis. Burning over 130 square miles, this is the largest known fire recorded for the Black Hills. Twenty years later, hardly any young trees can be found over thousands of acres. When the downed wood decomposes, this area could resemble a prairie. Where new trees do occur, nearly all of them are clustered around trees that survived the fire (figs. 8.20, 8.21, 8.22). Ponderosa has wind-dispersed seed, but apparently not enough seeds reached the interior of the burn—or the seedlings succumbed to drought or competition. The Forest Service has successfully planted ponderosa pine saplings where reforestation seems important and most feasible. Aspen is now growing again in the burned area where it grew before, but aspen clones survived by root sprouting. Pine lacks that adaptation.

Thus, it is reasonable to hypothesize that a severe fire could convert a densely forested area to a prairie under certain circumstances. Managers strive to assure natural forest regeneration, but climate, weather, and wind are beyond their control. With the suppression of fires in these puzzling prairies, pines do invade when climatic conditions are favorable. Apparently, the soil is not a limiting factor.

[a]Dodge 1876; Gartner and Thompson 1972; Froiland 1990; Keyser et al. 2008; Stevens-Rumann et al. 2018; Ziegler et al. 2017; Coop et al. 2020.

Fig. 8.15. Reynolds Prairie is a large montane mixed-grass prairie in the north-central Black Hills. Surrounded by forest, the essential absence of trees in the prairie is a puzzle (box 8.1). Common native plants in grasslands such as this include western wheatgrass, blue grama, little bluestem, and prairie smoke. Introduced species include Kentucky bluegrass and smooth brome, especially in moist drainage bottoms (see table 4.1 and Marriott 2012).

Fig. 8.16. There is very little natural reforestation over large areas 20 years after the Jasper Fire in the southern Black Hills (in 2000). Perhaps another large prairie is forming, such as Reynolds Prairie shown in fig. 8.15 (see box 8.1). One-year-old containerized seedlings are now planted in the spring, but success depends on sufficient rainfall during the first month or two. The Jasper Fire was followed by seven years of drought.

Fig. 8.17. This aspen grove developed from root sprouting after the Jasper Fire in 2000. Ponderosa pine becomes reestablished naturally only if small seedlings can survive the first year or two after seed germination. The scattered young pines on the hillside will soon produce more seeds, some of which will produce new trees. An uneven-aged forest is likely if climatic conditions are favorable. If unfavorable, a savanna may develop.

Fig. 8.18. Some mature ponderosa pine survived the Jasper Fire and provided seed for new trees nearby, as illustrated in this photo.

flammability is determined by drought and wind more than by the proportions of trees that are alive or dead. Supporting this conclusion, the Jasper Fire occurred before the most recent mountain pine beetle epidemic had killed large areas of trees in the area.[55]

Numerous species of beetles have coexisted with pines for millions of years,[56] and so far all that affect pine in the Black Hills are native. Their life history is fascinating. Adult mountain pine beetles emerge from the bark in July and August and usually fly toward large trees (more than 6–8 inches in diameter). The beetles bore holes through the bark and are soon cutting egg-laying galleries in the inner bark and some of the adjacent sapwood. Healthy or vigorous trees produce enough resin and other chemicals that the beetles are stopped before any damage to the tree occurs; small globs of resin can be found on the bark, often with a dead beetle. In contrast, less vigorous trees—whether due to age, water stress, or insufficient nutrients—are unable to produce enough resin or chemical defenses.[57] Sawdust around the holes and at the base of the tree indicates successful beetle invasion (fig. 8.21). By instinctively selecting larger trees, the beetles are more likely

Fig. 8.19. Mountain pine beetles killed the reddish-brown trees during the previous year. Millions of ponderosa pine have been killed by the beetles during the past 20 years. Without periodic fires, ponderosa pine seedlings will invade meadows and grasslands. The saplings in this photo may survive fires if the flames are pushed rapidly by wind.

Fig. 8.20. A type of bark beetle, the mountain pine beetle (*Dendroctonous ponderosae*) and other species in this genus are native to North America and have coexisted with their hosts for thousands of years. The adults are about 0.2 inch long. Some bark beetles do not kill the tree, but mountain pine beetles will if the tree cannot produce sufficient resin to defend itself. Photo with wings extended by Dion Manastryski, from the collection of Lorraine Maclauchlan; the other photo courtesy of the U.S. Forest Service.

Fig. 8.21. Numerous spots of resin on the bark, some with sawdust, indicate that a tree has been successfully invaded by mountain pine beetles. Each spot is caused by one beetle burrowing through the bark. The beetles introduce the spores of a blue-stain fungus that grows into the sapwood, thereby obstructing sap flow and killing the tree.

to infest trees that have more food for their larvae and are old enough that resin production is relatively low. In addition to visual clues, the beetles are attracted by odors emitted by stressed or weakened trees.[58] Where deemed practical or advisable, prescribed fires or mechanical thinning can help maintain the vigor of larger trees, thereby reducing the likelihood of an epidemic, but once under way an epidemic is nearly impossible to control.[59]

Curiously, after the first female beetles burrow successfully through the bark, they emit an aromatic chemical known as an *aggregating pheromone,* which attracts other beetles to the same tree—an adaptation for survival of the species.[60] Moreover, once a certain number of beetles have been attracted to the tree, the successful beetles emit a *disaggregating pheromone,* which repels other beetles, presumably to help avoid beetle overpopulation on that tree. Such responses to specific stimuli by both the beetles and the tree are a reflection of the long history of coevolution of the insect and its host.[61]

Adult beetles and their larvae consume the inner bark all around the tree within a few months, thereby stopping the flow of carbohydrates from the leaves to the roots. This *girdling* reduces the effectiveness of the roots and surely contributes to the death of the tree. However, trees subjected only to girdling can live for an additional four to five years. In sharp contrast, beetle-infested trees typically die within a year because the beetles introduce blue-stain fungi.[62] The fungal filaments, known as hyphae, plug the pores of the sapwood, thereby restricting the flow of water to the leaves. Thus, the trees die from water stress. The fungus creates bluish streaks in the wood, hence its name, but the wood is still useful for construction. The novel coloration is preferred by some homeowners.[63] Aside from the fungi, the beetles introduce mites, nematodes, and bacteria.

In addition to vulnerable trees, periods of relatively warm weather are required for a beetle epidemic. Beetle larvae are so small and delicate that it is surprising any of them survive freezing winter temperatures, but some do. This is possible because the larvae produce antifreeze-like chemicals in their bodies. They also eliminate substances that could be ice-nucleating agents. No single lethal temperature threshold applies in all places and at all times, as great genetic and physiological variations occur among different beetle species, but the larvae are most vulnerable to cold in late fall and early winter, before the cold-hardening process is completed, or in late winter and early spring when warmer temperatures have stimulated the breakdown of cold-hardening compounds. Ecologist

Teresa Chapman and colleagues found that recent outbreaks of mountain pine beetle in northern Colorado and southern Wyoming lodgepole pine forests closely followed a period of warming temperatures, which included several years of reduced precipitation.[64] Large numbers of trees were vulnerable to beetle attack because they were stressed by drought and were of sizes preferred by the beetles. The same probably occurred in the Black Hills.

Beetle outbreaks have now subsided in the Rockies and the Black Hills, apparently because so many of the vulnerable trees were killed. There is no evidence that cold weather curbed the epidemic.[65] By selectively killing larger trees, beetles had reduced total tree growth for several years after the peak of the outbreak. However, the smaller, unaffected pines often grow more rapidly because there are fewer large trees competing for water, nutrients, and light. Small aspen and oak also grow more rapidly, as do shrubs and herbs, thereby enabling the recovery of total forest plant growth within a few years.[66] Looking at the effects of bark beetles and other native insects on ecosystems over long time frames, some ecologists have suggested that plant-feeding insects actually help maintain high levels of total plant growth by killing older, slow-growing trees and freeing resources for understory plants, whether herbaceous or woody.[67]

Some observers think of forests with large numbers of dead trees as being unhealthy, but such forests existed in the Black Hills long before the advent of forest management (fig. 8.22).[68] Early explorers commented on the abundance of dead standing trees, known as snags. The paradox of identifying such natural forests as unhealthy, and a growing appreciation of the benefits that dead trees provide, suggests that healthy forests can have dead trees. The benefits of snags include habitat for cavity-nesting animals, perches for raptors, shade for young tree seedlings, and, just as important, a source of coarse downed wood when they fall. Downed wood provides valuable microenvironments for maintaining biological diversity and contributes to soil development. Dead trees are a problem only where the goal is rapid

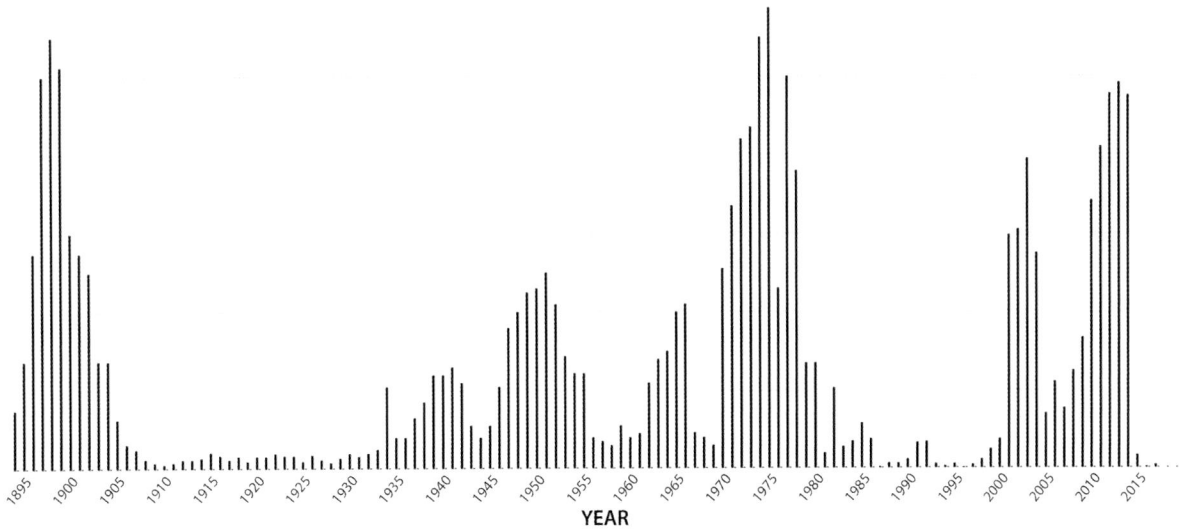

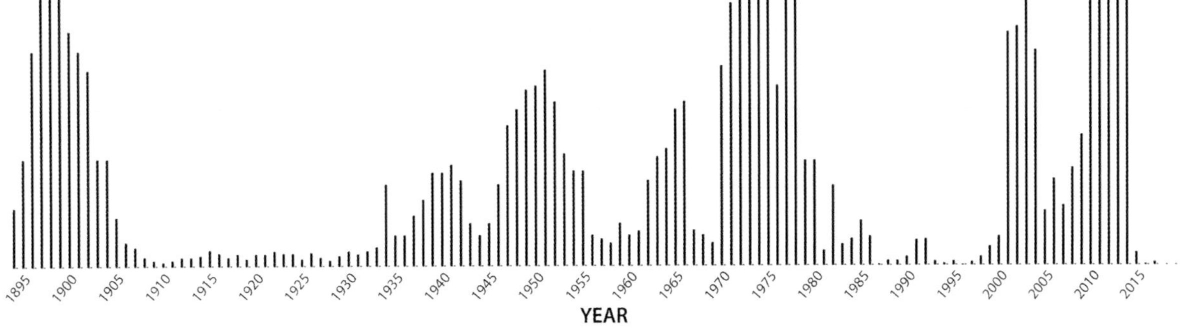

Fig. 8.22. The number of trees affected by mountain pine beetles in the Black Hills has fluctuated greatly during the past 125 years. Source: U.S. Forest Service, courtesy of Kurt Allen.

wood production rather than multiple-use management, or as a hazard around campgrounds and along roads. The Forest Service and timber industry have collaborated in removing many of the beetle-killed trees and excess downed wood, often too slowly for those who think this forest health restoration work is important. Others view this work as salvage logging with the potential of habitat degradation. How many dead trees and how much dead wood should be retained?[69] The debate becomes more intense when still more road construction is proposed. Governors, state legislators, congressional representatives, and county commissioners become involved—a measure of how much the forests of the Black Hills are appreciated and valued.

Presently, more people recognize that suppressing all fires will lead to problems, and also that the prevalent insects and diseases are native species—part of the natural biodiversity. To manage a forest with the goal of eliminating insect epidemics, or so that large fires do not occur, could create a kind of forest that never existed before and many people would not like. How to resolve such issues calls for collaboration and compromise, which managers hope will create landscapes that recover quickly from disturbances, or, in other words, that are ecologically resilient. Current approaches include forest thinning and slash removal to reduce fuel load, increase the vigor of remaining trees, and perhaps make them more resistant to beetle invasion—an indirect approach. The original structure of some ponderosa pine forests is thereby restored to some degree, but it often lacks the large trees of prelogging times. Prescribed fires are sometimes used when weather conditions suggest that the risk of a wildfire is minimal and there are no mountain homes or businesses nearby, which is unusual. Livestock grazing is encouraged with the reasoning that it will reduce the abundance of *flash fuels* (grass), although heavy grazing is needed to accomplish this. In some places the spread of aspen and other desired species is promoted by cutting the pines, which adds diversity to the landscape that could slow the movement of pine beetles. Managers are vigilant in anticipating and controlling invasive plants to the extent they can. Accomplishing all of this over an area large enough to be effective in a timely manner, as described in the Environmental Impact Statement for the U.S. Forest Service's Black Hills Resilient Landscapes Project, is challenging and controversial.[70]

Invasive Plants

Thus far, the invasive weed problems in ponderosa pine forests are minor and usually are in road ditches, campgrounds, and other places where the native plants have been disturbed. Notable exceptions are in foothill woodlands and on pine ridges where two species of annual brome grasses—downy brome (also known as cheatgrass) and Japanese brome—can become abundant. They reduce native plant diversity and make the landscape fire-prone earlier in the summer than usual, simply because they grow quickly in the spring and then die and become flammable by early July. Thus, if annual grasses or other invasive plants become more abundant because of excess disturbance during tree thinning, the benefits of that thinning are reduced.

Invasive plants are a bigger problem in grasslands and shrublands (table 8.2). Smooth brome, timothy, and Kentucky bluegrass have become the dominants in some grasslands and meadows (see box 4.3). Planted initially for the purpose of providing livestock forage, these introduced perennial grasses provide many of the soil-stabilizing and habitat benefits of the native species, but they suppress most native grasses and forbs. To replace them with native species may require disruptive, costly plowing and replanting. Canada thistle also occurs in dense patches throughout the Hills. Although these patches usually do not cover large areas, this spiny plant usually is not eaten by either domestic livestock or wildlife. Occasional eruptions of sweetclover, as happened in 2019, leave

Table 8.2. Some introduced invasive plants in the vicinity of the Black Hills, Pine Ridge, and the western Dakotas[a]

TREES AND SHRUBS	
Buckthorn	*Rhamnus cathartica*
Redcedar, eastern	*Juniperus virginiana*
Russian olive	*Elaeagnus angustifolia*
Saltcedar	*Tamarix* spp.
GRASSES	
Bluegrass, Kentucky	*Poa pratensis*
Brome, Japanese	*Bromus japonicus*
Brome, smooth	*Bromus inermis*
Cheatgrass (downy brome)	*Bromus tectorum*
Timothy grass	*Phleum pratense*
FORBS	
Baby's breath	*Gypsophila* spp.
Bindweed, field	*Convolvulus arvensis*
Burdock, common	*Arctium minus*
Chickory	*Cichorium intybus*
Cinquefoil, sulfur	*Potentilla recta*
Daisy, oxeye	*Leucanthemum vulgare*
Henbane, black	*Hyoscyamus niger*
Horehound, white	*Marrubium vulgare*
Houndstongue	*Cynoglossum officinale*
Knapweed, diffuse	*Centaurea diffusa*
Knapweed, Russian	*Acroptilon repens*
Knapweed, spotted	*Centaurea stoebe*
Mullein, common	*Verbascum thapsus*
Mustard, garlic	*Alliaria petiolate*
Sowthistle, perennial	*Sonchus arvensis*
Spurge, leafy	*Euphorbia esula*
St. John's-wort	*Hypericum perforatum*
Sweetclover, yellow	*Melilotus officinalis*
Tansy, common	*Tanacetum vulgare*
Thistle, bull	*Cirsium vulgare*
Thistle, Canada	*Cirsium arvense*
Thistle, musk	*Carduus nutans*
Thistle, scotch	*Onopordum acanthium*
Toadflax, Dalmatian	*Linaria dalmatica*
Toadflax, yellow	*Linaria vulgaris*
Wormwood, absinth	*Artemisia absinthium*

[a]A complete list, including aquatic plants, is updated periodically by the Black Hills Invasive Plant Partnership. Some species are of higher priority for control than others, including those that are still uncommon but relatively easy to contain or eliminate.

stands of dead stalks that can irritate the eyes of grazing animals. This legume does add nitrogen to the soil, but that may benefit other invasive plants more than native species.

Riparian zones are particularly susceptible to invasive plants because of their abundant moisture. Russian olive, widely planted in shelterbelts in the surrounding plains, is problematic along streams, particularly in the southern Hills. In the northern Hills, the common tansy and meadow sage have escaped from gardens and become locally abundant along some streams and adjacent meadows.

In any setting, whether forest, grassland, or riparian zone, managers must balance the benefits of using herbicides to control problematic species against their expense and the risk of killing desired plants or affecting water sources and animals. Biological control—introducing insects that controlled a plant in its native range on another continent—often works well, but finding effective biocontrol agents and making sure they will not harm native species is time consuming and expensive. And effective biocontrols do not exist for many invasive species. For invasive grasses, correctly timed prescribed fires and targeted livestock grazing can be effective, although much work needs to be done to determine the specifics of this approach. Quick eradication of invasive plants when they first arrive is more feasible than eliminating them after they become established.[71]

Climate Change

All natural resource managers must now consider climate change. Mean annual temperature has risen by about 2.2° F since 1900, and the best-available models predict an increase of 5° to 12° F by the end of the twenty-first century (see chapter 3).[72] Average precipitation is projected to remain the same or increase slightly, but warmer temperatures will lead to higher rates of evapotranspiration and longer growing seasons, thereby increasing the frequency and intensity of droughts (see chapter 3). The timing and intensity of precipitation events will also change, with less snow, earlier snowmelt, more rain, and more intense thunderstorms.

Climate change is likely to affect the abundance of ponderosa pine, as its regeneration is correlated with spring weather. The U.S. Forest Service scientist Gerald Rehfeldt and his associates published a study in 2006 that used a model relating the current climate of areas where ponderosa pine now exists to predict the effects of climate change on ponderosa pine in the Black Hills.[73] Their model suggested that ponderosa pine forests will be much less dense and less widespread in the Black Hills by 2060, and that the species will no longer be reproducing by 2090 in some places where it now seems to thrive—assuming that unmitigated warming continues. A more recent study suggested that the establishment rate of ponderosa pine seedlings will increase in the next 30–40 years but will decline after that.[74] Such studies emphasize that, while mature trees can tolerate temperature fluctuations and drought stress, new seedling establishment is likely to be less frequent as older trees die from beetle epidemics, fire, wind, or timber harvesting.

The ultimate fate of ponderosa pine in the changing climate will depend not only on the direct effects of climate on tree establishment and survival but also on how the future climate affects fire patterns. An average of 123 fires per year occurred in the Black Hills from 1970 to 2016, an average that has not changed much in recent decades. However, in the past 20 years the fires have been more difficult to control and have averaged ten times larger (from less than 1,000 acres to nearly 10,000 acres). Most of the larger fires occurred in the drier, southern part of the Hills.[75] As predicted, the establishment of new trees in some burned areas has been slow, raising concerns about the resiliency of the forests now and in the future (see box 8.1). Similarly, although aspen is the most widely distributed tree in North America, some think that much of the Black Hills is marginal habitat for this species and that it also could become less common.[76] Along with spruce, it may persist only in relatively cool, moist creek bottoms.

The magnitude of such changes is startling and can be debated, but the trends are clear. In general, forests dominated by ponderosa pine are likely to be more open and less widespread less than a century from now. As trees die from various inevitable disturbances, it would be ideal if landscapes could be managed to maintain the populations of sensitive species, but will that be possible with a warmer climate and altered precipitation patterns? Other plants may become more prevalent if the climate becomes favorable for them, species such as western snowberry, russet buffaloberry, Rocky Mountain juniper, and various prairie plants. It is difficult to predict how the interactions between these new species and those already present will play out. As in the past, changes are under way that, when understood, will provide a basis for developing new management guidelines.

In summary, fire suppression, habitat fragmentation, road and home building, fencing, and the effects of invasive species are well-known management issues throughout the Black Hills, Pine Ridge country, and other places where ponderosa pine grows. And now there is the challenge of rapid climate change and the prospect of a very different landscape. Two guiding principles are fundamental: conserve the diversity of native plants and animals and maintain the productive capacity of soils, both of which have been developing for millennia. One aspect of the Black Hills and pine-dominated ridges that will persist is an environment suitable for plants and animals not usually found on the surrounding grasslands and shrublands.

Chapter 9 Rivers and Riparian Ecosystems

People have historically settled near rivers, where abundant water, wood, fertile soils, and fish and game could be found. The Mandan, Hidatsa, and Arikara tribes established the first known gardens along the Missouri River, about 1,000 years ago, and much later, capital cities were located along the Missouri, first Yankton—the capital of Dakota Territory—and then Pierre and Bismarck. The largest river towns, Sioux Falls and Fargo, were founded on the Big Sioux River and the Red River of the North, respectively. Traditional uses are still important, but rivers also provide hydroelectric power and recreational boating and fishing.

Rivers and their adjacent riparian zones are expressions of their watersheds. Geologic history, flow regime, channel slope, and sediments determine the kinds of plants and animals that will be found and whether the rivers are meandering, straight, or braided. For example, trout thrive in Black Hills streams with clear water and gravel bottoms, while catfish prefer turbid rivers with muddy bottoms. An important theme of this chapter is the importance of high and low flow rates in maintaining biological diversity. Some species require periodic floods, such as cottonwood and willow. The modification of natural flows by dams, water diversions, excessive runoff and erosion, and inputs of nutrients and various biocides have impaired most rivers. Reservoirs provide recreational benefits, but fish production declines and sediments accumulate as they age. Restoration options are considered at the end of this chapter, along with the implications of climate change.

Rivers in the Dakotas can be divided into two general groups: ancient rivers that generally flow from west to east on unglaciated terrain, such as the Heart, Cannonball, Moreau, Bad, Little Missouri, Grand, Cheyenne, and White, and youthful rivers that formed more recently as continental glaciers melted and generally flow north, south, or in circuitous paths reflecting the irregularity of glacial deposits (see fig. 1.10). The James, Vermillion, and Big Sioux flow southward into the Missouri River and the Souris, Pembina, and Sheyenne flow eastward into the north-flowing Red River of the North. The Missouri River separates the two types of rivers, following a course fixed by the position of continental glaciers.

The Missouri is by far the largest river flowing through the Dakotas, with a mean daily flow rate that is six times greater than the Red River, the next largest (table 9.1). The longest river completely contained within the two states is the James River (710 river miles).[1] The 15 largest rivers usually flow all year every year and are classified

Table 9.1. Hydrologic characteristics of the 16 largest rivers in North Dakota and South Dakota, listed in order of mean daily flow (cubic feet per second)

River name	Stream gage location	River length (miles)[a]	Drainage area (square miles)	Minimum daily flow (cfs)	Mean daily flow (cfs)	Period of record	Maximum daily flow (cfs) and date
Missouri	Sioux City, IA	2,341	314,600	3,000	29,641	1929–2015	441,000 April 14, 1952
Red	Drayton, ND	550	34,800	110	4,929	1936–2015	124,000 April 24, 1997
James	Yankton, SD	710	20,962	1	1,500	1982–2015	29,200 March 28, 2011
Big Sioux	Akron, IA	419	7,879	4	1,404	1928–2015	108,000 June 18, 2014
Cheyenne	Plainview, SD	527	21,351	0	824	1920–2015	73,200 June 7, 2008
White	Oacoma, SD	580	9,915	0	590	1929–2015	51,900 March 30, 1952
Little Missouri	Waterford City, ND	560	8,310	0	551	1935–2015	110,000 March 25, 1947
Belle Fourche	Elm Springs, SD	290	7,004	0	388	1927–2015	47,500 June 6, 2008
Vermillion	Vermillion, SD	199	2,254	4	382	1984–2015	21,400 June 23, 1984
Souris	Westhope, ND	435	16,900	0	344	1930–2015	30,400 July 5, 2011
Sheyenne	West Fargo, ND	591	8,870	1	293	1903–2015	4,830 April 29, 2011
Moreau	Whitehorse, SD	200	4,889	0	279	1954–2015	34,200 March 20, 2011
Grand	Little Eagle, SD	227	5,316	0	272	1959–2015	32,400 March 22, 2009
Heart	Mandan, ND	180	3,310	0	272	1924–2015	30,500 April 19, 1950
Cannonball	Breien, ND	295	4,100	0	252	1906–2015	94,800 April 19, 1950
Bad	Pierre, SD	161	3,147	0	171	1905–2015	70,000 July, 1905

[a] River lengths were determined from medium resolution U.S. Geological Survey National Hydrography Datasets and other published sources. All river sources were started at line segments farthest upstream bearing the river's name. Except for the Vermillion River, river length does not include forks.

as perennial. However, with the exception of the Missouri, all have dried up at one or more of their gaging stations during the period of record. The lowest recorded flows were during the drought of the 1930s.

Riverine ecosystems appear as arteries in a landscape, with water flowing from small rivulets downslope into creeks until it reaches small rivers that flow into larger rivers. Tributaries add water to the stream, and some are fed by groundwater seepage. A stream's drainage basin includes numerous interconnecting channels organized hierarchically by stream order.[2] Streams that have no easily identified tributaries are known as first-order streams; those that have only first-order streams as tributaries are second order. This sequencing is continued until the last stream (usually a river) enters the ocean (fig. 9.1).

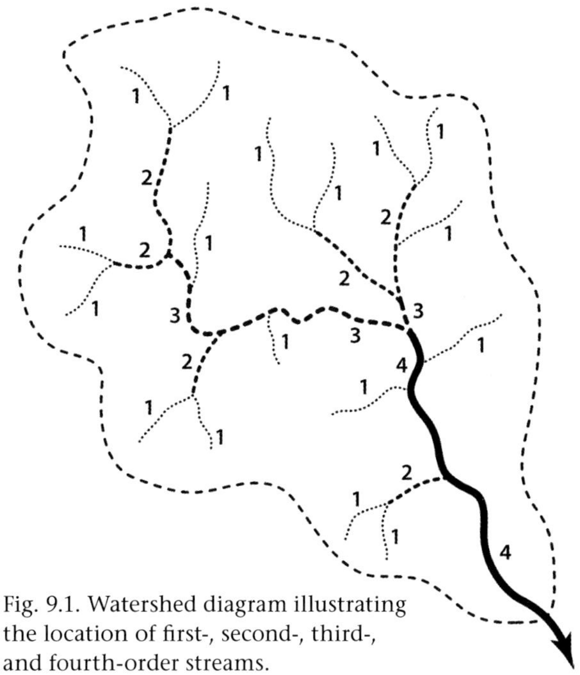

Fig. 9.1. Watershed diagram illustrating the location of first-, second-, third-, and fourth-order streams.

Each river has a unique drainage basin in terms of land area, geologic history, climate, soils, vegetation, and land use—factors that determine the amount and quality of water draining from the watershed at different times of the year. But water is not the only substance transferred downstream. Soil particles and minerals are eroded from adjacent uplands, especially during high runoff events, and floating debris includes driftwood, leaves, seeds, humus, and an abundance of other organic materials—living and dead. Research has enabled predictions about how specific disturbances and land management practices affect aquatic ecosystems. Each waterway is different. The Missouri River is the most altered because of extensive damming and large reservoirs. Seventy percent of its length in the Dakotas is now reservoir, as discussed in the next chapter.

Rivers and their tributaries are highly valued, but human activities have affected them adversely nearly everywhere. The runoff and effluent from cities and farms pollute the water with excess nutrients and pesticides; erosion from roads and croplands adds turbidity to the water and silt that covers stream bottoms. Livestock graze floodplain plants, sometimes more than is recommended, while introduced plants outcompete native plants on the floodplain and streamflow has been altered by dams and water diversions.

There are tens of thousands of miles of rivers and streams in the Dakotas. Recent surveys in both states found that less than 30 percent could be used safely for swimming, fishing, and boating.[3] Also, many once-common aquatic species have become rare. Degradation was attributed to siltation, nutrient enrichment, riparian habitat alterations, and high levels of fecal coliform bacteria and methylmercury in some places. Implicated causes included agricultural runoff, livestock grazing, and inadequate septic systems. Remediation is possible and often recommended.

Ancient Floodplains

One of the most remarkable features of rivers on glaciated terrain is how wide their floodplains appear to be. How could the small rivers that we see today carve such wide valleys? There are

Fig. 9.2. The Red River of the North meanders northward across the flat land formed by Glacial Lake Agassiz. Commonly referred to as the Red River Valley, the actual river valley is narrow and lined with woodlands (see fig. 9.9). The native tallgrass prairie has been almost entirely converted to cropland, principally soybeans, corn, wheat, and sugar beets. Photo by Wark Photography Inc.

several possible explanations. For the Red, Sheyenne, and James, large portions of what appears to be their present-day floodplain is actually the bed of glacial lakes that dried up thousands of years ago (see chapter 2). The best-known example is the Red River of the North, which flows across the bottom of Glacial Lake Agassiz—an unusually flat surface that is sometimes referred to as the *Red River Valley Lake Plain* (fig. 9.2)—not the true valley formed by the Red River itself, which is shallow and only a few hundred feet wide.[4] The last one-quarter of the Sheyenne River also flows across the Red's lake plain. Similarly, portions of the James River in northern South Dakota flow across the flat sediments of Glacial Lake Dakota (see fig. 2.8).

Other rivers in the eastern Dakotas, not flowing on glacial lakebeds, have incised and widened valleys. They originated when morainal dams on glacial lakes failed, creating very large, highly erosive rivers. The best example is the Sheyenne River which today is only about 50 feet wide where it joins the Red. Astonishingly, about 10,000 years ago when Glacial Lake Souris drained, the Sheyenne River was large enough to carve the valley that exists today—several miles wide and 200 feet deep in some places (see fig. 2.9).[5] Similarly, the now-small Pembina River was once large enough to form a valley that is two miles across. All rivers are erosive, but those on glaciated terrain were much more so while the glaciers were melting.

The Big Sioux River, which winds its way through the most populous part of South Dakota,

Fig. 9.3. The Big Sioux River originates on the Prairie Coteau and flows south through undulating ground moraines and croplands. At Dell Rapids, Palisades State Park, and Sioux Falls, the river has cut through Precambrian Sioux quartzite. The riparian gallery forest is dominated by green ash, boxelder, hackberry, and bur oak. Less common trees are cottonwood and peachleaf willow. Photo by Greg Latza, PeopleScapes Photography.

is quite different (fig. 9.3).[6] It originated as glacial ice melted on the Prairie Coteau, now the highest land in eastern South Dakota (about 200 feet lower than the highest point on Turtle Mountain). The gradient of the Big Sioux averages about 2.4 feet per mile, compared to 2 feet per mile for the Sheyenne, 1.6 feet per mile for the James (less than 1.1 feet per mile on the Lake Dakota plain), and 0.5 foot per mile for the Red. Notably, the Big Sioux cut through Sioux quartzite where this bedrock was near the surface at Dell Rapids, Palisades State Park near Garretson, and in Sioux Falls (see fig. 2.2).

Types of Rivers

All rivers are separated into two major groups by their ability to adjust their shape and slope.[7] Bedrock-controlled rivers are confined between outcrops of rock. This type of river in the Dakotas is typical in the Black Hills (especially steeper, upstream reaches) and along the Big Sioux River in the vicinity of the palisades. In contrast, alluvial rivers adjust their dimensions, shape, pattern, and gradient as conditions change, often rather quickly after major floods or land-cover changes (agriculture, deforestation, urbanization). Alluvial rivers transport significant amounts of sand, silt, and clay picked up from their bed and banks.

Alluvial rivers are a complex group that has been subdivided into two broad types: meandering and braided (fig. 9.4).[8] The major factor separating them is the type of sediment transported or the mode of its transport, as bedload or suspended load. Bedload is that part of the sediment (coarse material such as sand and gravel) that is pushed along the channel bottom by gravity and the force

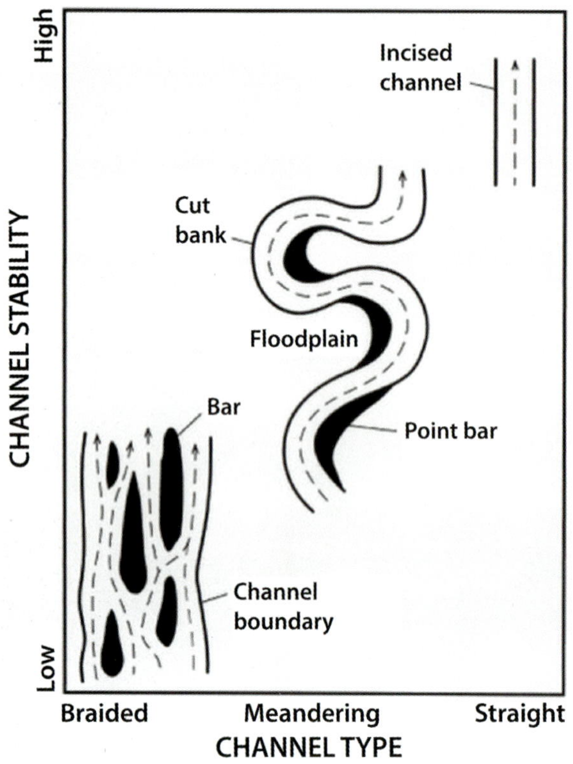

Fig. 9.4. Different combinations of sediment size, valley slope, and streamflow produce three channel types: straight, meandering, and braided. Meandering streams are most common in the Dakotas. Arrows indicate the direction of flow. From Knight et al. (2014), with permission from Yale University Press.

of the water; suspended load is silt and clay held in suspension by the forces of flowing water. Suspended sediments make the water murky. Streamflow velocity, affected by channel slope, affects the sediment composition of both types.

Most Dakota rivers are of the single-channel, meandering type that twist and turn because of relatively gentle slope and moderate velocities, except during floods. The sediment transported is roughly balanced between bed and suspended loads. Meandering rivers gradually erode the outer "cut" banks of channel bends, the major source of sediment for the formation of point bars downstream on the opposite side of the river (figs. 9.5). This swapping of material between banks leads to a relatively constant channel width but a reshaping of the floodplain. In addition to creating new point bars, floods also fill in old channels and raise floodplain surfaces. Occasionally tight loops in the channel are cut off, forming an oxbow pond or lake (fig. 9.6). Curiously, a section of the James River near Sand Lake Wildlife Refuge seems to violate the principle that water always flows downhill: the stream gradient north of Huron drops only 0.2 feet per mile, and heavy rains can cause one of its tributaries, the Mud River, to push water upstream for a few miles.[9]

Braided rivers have a steeper slope, higher velocity, and a sediment load dominated by coarse material such as sand and gravel. These rivers have

Fig. 9.5. Cut bank on the outside of a bend on the Garrison Reach of the Missouri River in North Dakota. The mature cottonwoods are about 75 years old. Shorter understory trees are green ash, boxelder, and occasionally bur oak.

Fig. 9.6. Multichannel segment (reach) of the Heart River delta entering Lake Patterson in western North Dakota. Flow is from left to right. The uppermost channel will soon form an "oxbow" pond by cutting through the narrow neck of a tight bend. Such cuts isolate channel sections and shorten river length. Photo by Greg Latza, PeopleScapes Photography.

multiple channels that often surround vegetated islands (see fig. 9.6). The best example in the region is the Platte River in Nebraska. The Souris and Sheyenne have segments, known as *reaches*, that are braided where they flow through glacial outwash sands. Many rivers exhibit more than one form; braided upstream and meandering downstream.

Floods

Some of the groundwater flowing through the Dakotas comes to the surface in intermittent creeks that merge downstream into generally slow, turbid, meandering streams. Peak flows typically occur in the spring when snow is melting, rainfall is higher, and soils are wet and possibly frozen. Flow often stops during drought years, with only a series of pools persisting. The longest low-flow period recorded in the eastern Dakotas was during the 1930–1940 drought, when at times the Red River was completely dry.[10] But floodplains attest to the fact that floods sometimes can be severe, relocating the channel and forming new islands and oxbows. Floods are required for maintaining riparian ecosystems, but they can be catastrophic for people who live there.

The 1997 flood on the Red River captured the attention of the entire nation, displacing tens of thousands of people and causing $3.5 billion in damage—the worst flood on the Red in recorded

Fig. 9.7. Aerial image of the 1997 flood on the Red River near its confluence with the Sheyenne River. The riparian woodlands along the river channel are evident above the floodwaters. Riparian vegetation is tolerant of short-term flooding. Photo by Donald P. Schwert, Department of Geosciences, North Dakota State University.

history (fig. 9.7).[11] Previous floods had prompted the construction of levees, diversion canals, and deeper channels to protect homes and businesses on floodplains, but they proved inadequate in most places.[12] With each flood, weather forecasters strive to become more proficient at predicting floodwaters far enough in advance to minimize damage. Floods along the nearby Sheyenne River have been less damaging, some would say because of a 28-meter-high dam and reservoir, Lake Ashtabula, north of Jamestown. Less flood damage along the Sheyenne also might be attributed to the fact that relatively few people live there.

The cause of the extraordinary 1997 Red River flood has been attributed generally to the following:

- High precipitation during the previous fall
- An usually cold winter with heavy snow accumulation
- A cool spring followed by heavy rainfall during the thaw
- Drainage of wetlands that contributes additional flow to the river

All rivers on the Northern Great Plains would flood under these circumstances, but what proportion of a watershed must be subjected to these conditions? What are the effects of additional variables that are known to influence streamflow, such as draining wetlands and converting more of the watershed from prairie or woodland to crop production. Such land management practices can be

Fig. 9.8. The green riparian ecosystem along the Cheyenne River in western South Dakota, dominated by plains cottonwood, contrasts sharply with the senescent mixed-grass prairie. The dominant tree on the floodplain is plains cottonwood. This river originates in Wyoming and flows eastward across the Missouri Plateau to the Missouri River. The Sheyenne River is in eastern North Dakota and flows into the Red River.

as important or more so than weather patterns.[13] Also, for the Red River, ice jams form downstream to the north, where spring comes later, and contribute significantly to flooding. There are numerous variables involved with forecasting and mitigating the hazards associated with living on floodplains. For each prediction of high water there is this question: will existing flood abatement measures be enough?

Characteristics of Riparian Ecosystems

The noun *riparia* is used by scientists to describe the biotic communities living on the shores of lakes and the floodplains of streams and rivers.[14] Riparia form distinct corridors in prairie landscapes because of the taller and often greener vegetation (fig. 9.8). Plant roots and stems impede the flow of water and increase bank resistance to erosion, which can slow the rate of meandering. When bank erosion does occur, downed tree trunks and other woody debris stuck in the riverbed initiate the formation of sandbars and possibly vegetated islands. By slowing streamflow, riparian vegetation captures sediment and builds higher surfaces, incrementally burying the trunks of trees during floods. This process of sediment deposition and channel meandering can raise the floodplain by 10 feet or more, depending on the river and its sediment load. Beaver, also, can convert flowing streams into pond and meadow habitat that store massive amounts of sediment. In short, riparian vegetation is strongly determined by interacting physical and biological forces.

Riparian zones are defined by the presence of a creek or river along with plants that require more water than those on higher ground. Known as hydrophytes, these plants form woodlands, shrublands, and meadows in a curvilinear mosaic affected by the depth to groundwater, soil texture, flood frequency, flood duration, and an ability to survive low oxygen conditions and physical damage from ice flows and beaver cutting. These conditions change with time as the river channel changes, hence the riparian mosaic is constantly shifting.

Riparian trees and shrubs create unusual environments in the landscape. The woodlands cast more shade and have higher humidity and less wind, thereby providing habitat for deer, squirrels, and dozens of species of birds, perches for raptors, and nesting cavities in dead and dying trees for woodpeckers, wrens, racoons, and opossums. Such habitats vary widely in patch size, vegetation structure, and connectivity to other woodland patches. Young willows and scattered trees in narrow, discontinuous strips support few vertebrate nesters, but some are popular with bird watchers looking for yellow warblers, common yellowthroats, and orioles. The cavity nester population increases as trees age and more cavities are available. Bird diversity increases where woodlands cover a large area and include a wide range of contiguous stands of various ages. These conditions are met most often along larger rivers such as the Missouri, Red, Little Missouri, Sheyenne, and White. A total of 105 upland bird species were recorded during the nesting season along the Missouri River between Mobridge and Fort Thompson.[15] Riparian woodlands also are stopover habitat for birds during spring and fall migration. Natural riparian woodlands comprise only 1 percent to 3 percent of the land area in both states, yet their presence increases the total biodiversity by as much as 30 percent to 40 percent.[16]

A characteristic topographic feature of floodplains is the presence of alluvial terraces, also known as benches. They appear as stair steps rising from the river to the adjacent grassland or cropland (fig. 9.9). Dominant tree species change as the elevation of the terraces increases. The frequency of flooding is higher down near the river; soils on the benches above have more fine silt and clay. Thus, along the Red River, for example, cottonwood and willow occupy a narrow strip near the river channel, and American elm (most trees have died from Dutch elm disease), green ash, and boxelder occupy the next bench upslope. Basswood, green ash, hackberry, and bur oak fringe the top. Alexander Henry (the younger) observed the same kinds of trees along the Red River in 1806, writing, "There is an abundance of wood on the banks of the river to answer every purpose . . . for many ages to come."

A similar vegetation mosaic is found along reaches of the Sheyenne River, Big Sioux River, and other smaller rivers in the eastern Dakotas. However, trees are absent in the upper portions of their watersheds, where streamflow is intermittent and less able to create exposed alluvium, where tree seedlings can survive. Historic fires and grazing may be secondary factors limiting the presence of trees in such areas.

The plains cottonwood is the most abundant, widespread, and largest tree in the Dakotas, attaining a diameter of six to eight feet in some places (see fig. 10.14). It is more common in the west, where streamflow is faster and more erosive, and where some eastern riparian tree species cannot grow.[17] The success of cottonwood depends on its abundant light, fluffy seeds that are carried by wind and on the surface of flowing water. The small seeds are deposited on shorelines in a linear pattern as the water level drops, and if the seedlings and young saplings survive subsequent floods, they develop into a narrow band of new trees. As the channel moves, subsequent floods lead to adjacent, curvilinear bands of trees of different ages, which mark where the channel was once located—a conspicuous pattern that clearly illustrates a shifting riparian mosaic (fig. 9.10). The bands disappear as trees

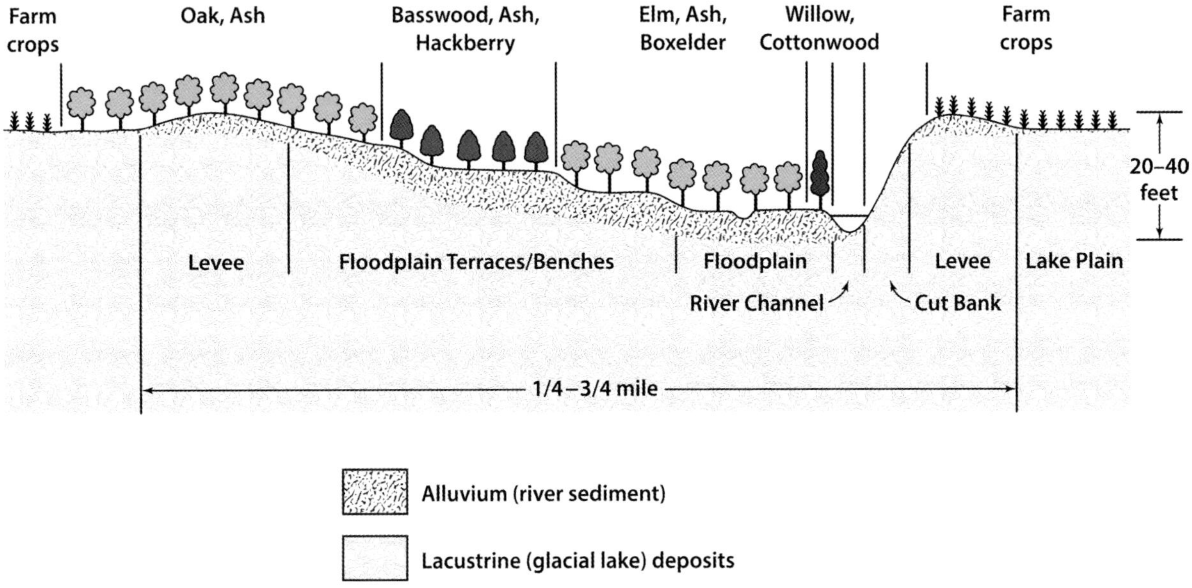

Fig. 9.9. Cross section of the Red River valley showing how predominant trees change from one alluvial terrace (bench) to another. Lower floodplain surfaces are younger, flood more frequently, and have pioneer trees such as cottonwood and peachleaf willow. Higher surfaces are older and have bur oak, green ash, hackberry, and other species that are less well adapted for flooding. Severson and Sieg (2006) review the observations of early explorers in eastern North Dakota woodlands. Adapted from Wanek (1967).

die or are undercut by the river. Peachleaf willow and sandbar willow—a shrub—become established in the same places as cottonwoods. Once riparian woodlands are thick enough and old enough to slow floodwaters, sediment is deposited. Over time, many feet of sediment accumulate, thereby reducing flood frequency and duration on the new bench surface and increasing the opportunity for less flood-tolerant trees and shrubs to colonize the area.

A common observation is of trees toppling into the river as a cut bank is eroded and root systems are exposed (fig. 9.11)—a testament to the power of flowing water and how rivers move across or widen their floodplains. The ecologist Ben Everitt used tree rings to age trees and determine how rapidly the Little Missouri River channel shifted during the past few centuries.[18] Aerial photographs were only available for 75 years or less, not enough time to see much change. Because cottonwood trees establish almost exclusively in even-aged bands on point bars, tree age provides an estimate of the number of years since the channel was located in that place (fig. 9.12). Everitt aged hundreds of cottonwood trees, using small cores from the trunk, to determine the ages of different bands. Some were more than 200 years old. He found that nearly all parts of the floodplain are reworked by the channel in a century or two. One outcome of rapid channel movement is domination of the floodplain by cottonwood, the tree species best adapted to reproduce, grow, and survive in dynamic rivers. Cottonwood is less common where the banks are more stable and the channel changes more slowly, such as along the Red River.

To the surprise of seasoned cottonwood researchers, Jonathan Friedman and his team found slow-growing, living cottonwood trees up to 370 years old on the Little Missouri River floodplain in the North Unit of Theodore Roosevelt National Park.[19] This discovery contradicts the long-held belief that cottonwood trees grow fast but die young. The 370-year-old tree, the oldest known plains cottonwood, had escaped the assault

Fig. 9.10. Banding of cottonwood forests is apparent on some actively meandering rivers, such as the Little Missouri River in Theodore Roosevelt National Park. The right bank has a point bar on the inside of the river bend. Tree age is youngest on the most recently deposited sediment nearest the channel. On the left side of the river, trees topple into the river as the cut bank is eroded. Photo by Jason Lindsey/Alamy Stock Photo.

Fig. 9.11. Erosion of outer cut banks with mature vegetation creates space for river channels to move, such as here where a cottonwood is about to topple into the Missouri River. The eroded sediments are deposited downstream as sandbars, on which young pioneering trees such as cottonwood and willow grow.

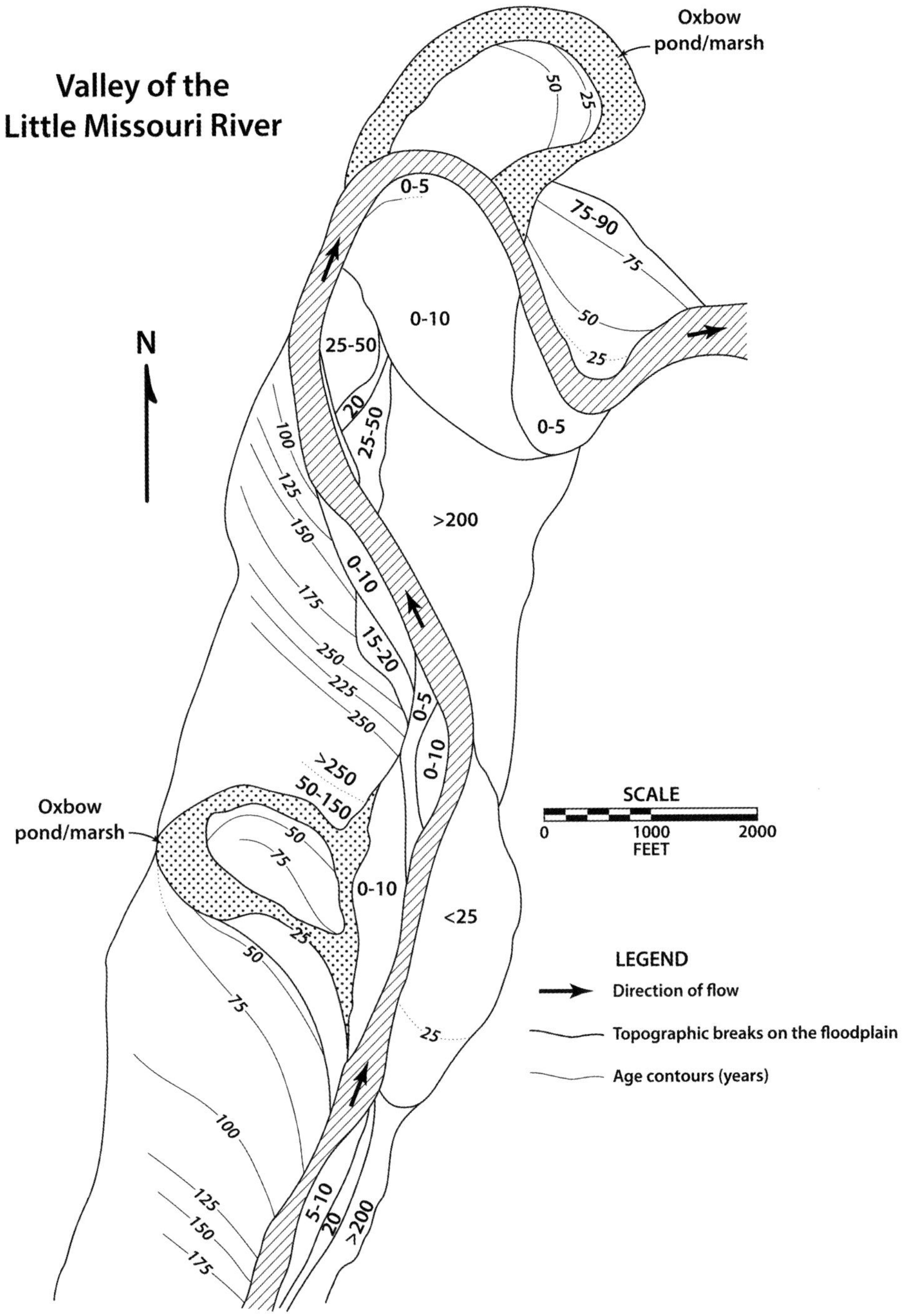

Fig. 9.12. Map of cottonwood ages on a 1.5-mile-long segment of the Little Missouri River floodplain in the North Unit of Theodore Roosevelt National Park. Even-aged bands of cottonwoods establish quickly on sandbars, hence ages obtained by coring trees represent the approximate time in years since the channel was located in that place. The map portrays a rapidly meandering river with young trees located near the channel and older trees away from the channel, nearer the edge of the floodplain. Some parts of the floodplain escaped erosion by the channel for two or more centuries. Adapted from Everitt (1968).

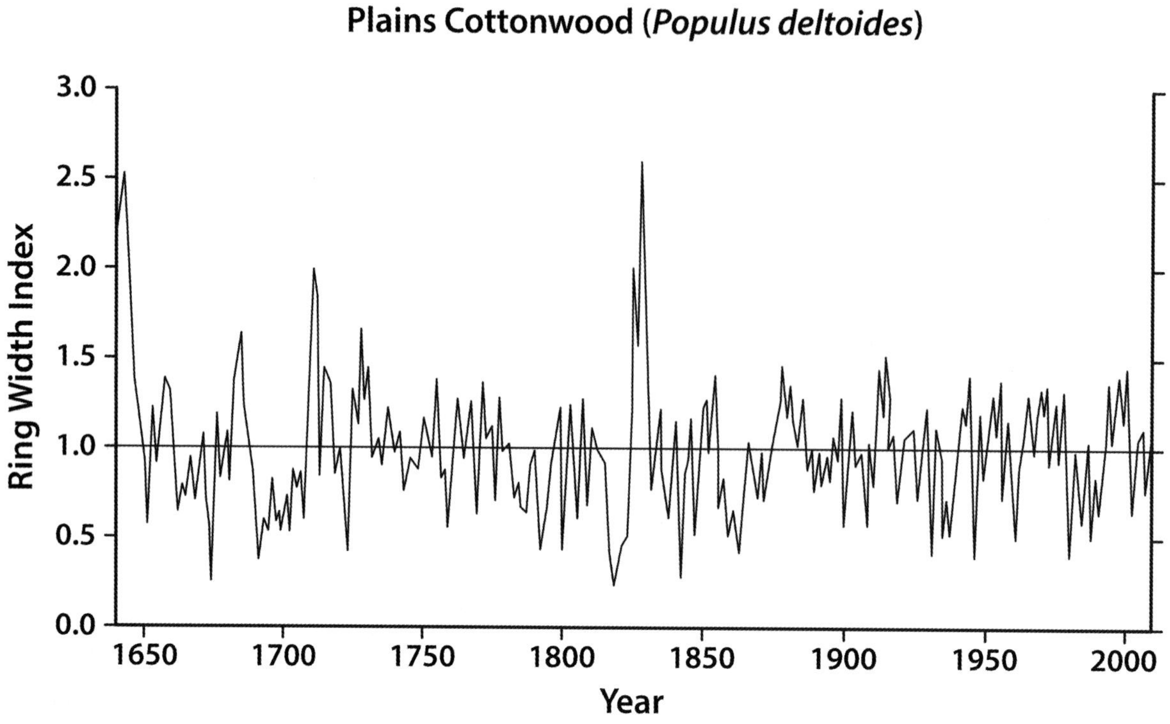

Fig. 9.13. Long-term natural variation of plains cottonwood growth rate (Ring Width Index) in response to high and low streamflow on the Little Missouri River in Theodore Roosevelt National Park from 1643 to 2010. High growth rates indicate flooding, low rates indicate low flows during drought. Five periods of major flooding are apparent. Adapted from Edmondson et al. (2014).

of the meandering river for nearly four centuries. Friedman was able to produce a tree-ring chronology spanning a 367-year-period (1643–2010). Wide tree rings indicated higher precipitation and streamflow, and narrow rings the opposite (fig. 9.13). A prolonged period of slow growth from 1816 to 1823 corresponded to the "sand blowing years" described by the Lakota. This drought period and another in the 1860s were more severe than any during the 1900s, including the Dust Bowl years. In another study, the dendrochronologist Peggy Reily compared tree cores from the Little Missouri and Missouri Rivers and found that the sharply reduced growth of cottonwood trees on the Missouri is correlated with the cessation of flooding and lowering of the floodplain water table by dams.[20] The research of Friedman and Reily, which links climate to tree growth, serves as a baseline from which to detect the onset of climate change.

Fish

The diversity of aquatic life in warm, turbid, slow-moving rivers is no less amazing than in lakes and reservoirs. Fifty or more species of fish are found in Dakota rivers, more than many expect considering the frequency of pollution and intermittent flows.[21] The list includes sport fish such as channel catfish, sauger, and northern pike, along with introduced species such as the common carp. In addition, there are hundreds of species of algae and other aquatic plants, a multitude of invertebrates that include insects and clams, and vertebrates such as amphibians, beavers, muskrats, shorebirds, waterfowl, common snapping turtles, and western painted turtles. About a dozen species of clams are now found in the Big Sioux River, many fewer than decades ago.[22]

Through natural selection, fish have evolved to occupy different ecological niches, with each species using different portions of the available habitat at different times. There are herbivores, omnivores, and carnivores in all size classes; some are bottom feeders, and others feed on plankton, smaller fish, or other swimming organisms. And they have different reproductive strategies. Some spawn only in small tributaries. As in terrestrial ecosystems, introduced fish can reduce the abundance of native species.

Special adaptations are required for fish to survive in turbid rivers. For example, bullhead, catfish, and carpsuckers have ventral mouthparts that enable the suction of silty bottom sediments from which algae, plankton, and other live and dead organic materials are filtered for food. Their bodies tend to be somewhat flattened or humpbacked, and they have stout, sometimes spiny dorsal and pectoral fins that deter predators. Living in murky water and often feeding at night, their eyes are small and not well developed. To compensate, bottom-feeding fish have appendages with taste receptors (barbels) near their mouth (fig. 9.14). In addition, bullheads and other catfish have taste receptors on their skin. Some fish are able to slow their metabolism for extended periods, essentially aestivating when oxygen becomes limiting or during the winter when streamflow is slow and food less available.

Fish habitat is highly variable across a river channel and up and down the river. Shade and concealment cover are provided by plants, boulders, overhanging banks, driftwood, or submerged trees (snags) toppled from eroding cut banks. Some species spend most of their time in certain parts of a river, such as riffles, eddies, or pools, while others depend on a broad range of sites, moving among them as needed to survive and reproduce. Fish move upstream or downstream in free-flowing rivers to find more suitable conditions.[23] Habitat complexity and species diversity often increase at the mouth of a tributary. The location of different kinds of fish habitat can change rapidly after flood events. Floods also connect rivers to lakes and marshes temporarily, thereby providing opportunities for fish migration to or from the stream.

Fig. 9.14. The common channel catfish (above) and the endangered pallid sturgeon (below) both have sleek body shapes and unusual sensory organs—barbels and whiskers—that facilitate survival in turbid rivers. Photos by Sam Stukel, U.S. Fish and Wildlife Service.

From time to time rivers change in ways that lead to fish mortality. Most dramatic is when streamflow ceases and no pools persist or are accessible. High mortality also is caused in winter when oxygen levels in the water drop below critical levels or when some rivers freeze nearly to the bottom. When streamflow resumes, fish typically immigrate from portions of the river or its tributaries that retained sufficient water and oxygen for remnant populations. Generally, larger, deeper, and cleaner rivers have higher fish diversity, more fish biomass, and larger fish than smaller or polluted streams.

Beaver, Bison, and Livestock

Of all the animals found along creeks and rivers, the seldom-seen beaver is best known and has the greatest impact on the landscape. Beaver are regarded as ecological engineers and keystone species because they greatly modify the riparian ecosystem, creating habitat for other species. A beaver can topple a 5-inch diameter tree with its sharp incisors in less than three minutes.[24] Their dams, made of interwoven branches packed with mud, create ponds in which to build and protect their dome-shaped lodges, which benefit fish and other organisms (figs. 9.15 and 9.16). Where damming is not possible, such as on lakes or on wide and fast-flowing rivers, beavers live in dens dug into banks. Without maintenance, beaver dams fail and the riparian landscape experiences another episode of rapid change.

Beaver are found throughout the Dakotas, though in relatively low numbers compared to historical times. A strong European market for beaver pelts to make felt hats for fashionable European gentlemen brought French and English fur traders to North America early in the 1700s.[25] Pelts sold for nearly one hundred dollars in today's

Fig. 9.15. Beaver lodge at the confluence of the White and Missouri Rivers. Lodges protect beavers from predation and are food caches. Photo by Malia Volke.

Fig. 9.16. Beaver dam built on an intermittent stream.
Photo by Carol A. Johnston.

currency. Once beaver numbers were depleted in eastern North America, French-Canadian traders pushed farther west, traversing the prairies of the Northern Great Plains. Alexander Henry, a partner in the North West Company, with a post in Pembina, reported rapidly diminishing beaver numbers in the first decades of the 1800s. By 1820, the North American beaver was in danger of extirpation, but was saved by a fashion shift toward silk hats. For a time, beaver were still common in the

Black Hills, as the Custer expedition in 1874 had trouble finding places to cross beaver-impounded streams.[26] But beaver continued to be trapped, for both their fur and sometimes to eliminate their ponds. They became increasingly rare, and by 1887 trapping was outlawed for a period of five years. Stocking eventually led to their reestablishment throughout the region, but population sizes are much lower than they were at the beginning of the nineteenth century.

History shows that humans have a love-hate relationship with the beaver. On the negative side, beavers sometimes plug road culverts and irrigation ditches. Crossing dammed streams with motorized vehicles is difficult or sometimes impossible. Beaver dams also can impede upstream migration of fish, and the potent greenhouse gas methane is released to the atmosphere from their ponds, as is methylmercury. However, the establishment and abandonment of beaver ponds increases channel complexity that supports a higher diversity of organisms, including muskrats, otter, fish, waterfowl, and amphibians. Their dams and ponds raise water tables, recharge groundwater supplies, filter water, capture sediments, and enable denitrification during periods when bottom sediments are anaerobic. Overall, water quality improves and the riparian zone is subject to less erosion.[27]

Perennial streams are preferred by beaver, but they can survive as well in more abundant intermittent streams.[28] For example, beaver successfully colonized a creek in western South Dakota with daily flows ranging widely from 0.3 to 193 cubic meters per second, converting the creek into a series of stair-stepped wetlands that retained water in summer while the rest of the creek went dry. Beaver ponds in prairie are less spectacular than those in the boreal forest, but ranchers living through major droughts credit beaver ponds for saving their cattle business. Some observers claim that the extirpation of beaver a century or more ago in the rugged topography of the western Dakotas initiated soil erosion in woody draws that continues today.[29] Reintroducing beaver is now viewed as a way of moderating the adverse effects of droughts and climate change.[30]

Large herbivores also affect the riparian zone. Bison moved across the Great Plains in large herds that apparently grazed heavily before moving on. Historical accounts report significant effects of bison on riparian zones. In 1800 on the Red River, Alexander Henry wrote: "The willows are entirely trampled and torn to atoms, even the bark of the smaller trees are [*sic*] in many places totally rub'd off by the Buffalo. . . . The numerous paths . . . and the vast quantity of dung . . . gives this place the appearance of . . . where Cattle have been kept for many years."[31] About the Park River floodplain, Henry wrote, "The few spots of wood which is a long [*sic*] this river appear to have been ravaged by the Buffaloe, none but the large trees are standing. . . . The ground is bare and more trampled . . . than the gate of a farmyard." Explorers frequently mentioned buffalo paths through riparian woodlands.[32] Excessive numbers of domestic livestock confined to floodplains can cause similar effects:

- Streambanks that have been broken down by trampling, thereby accelerating erosion and widening the streambed, reducing water depth, and destroying overhanging banks important to fish
- Dense, pasturelike riparian meadows with relatively few trees and shrubs, due to trampling and browsing
- Trails leading to water, and abundant animal waste
- A simplified ecosystem that sustains less plant and animal diversity
- Abundant Eurasian weeds that displace native riparian species

Two candidates for the least tamed, most wild rivers in the region are the White River in South Dakota and the Little Missouri River in North Dakota (box 9.1). Both rivers flow into Missouri River reservoirs but neither has a mainstem dam. Also, both are located in sparsely populated areas

Box 9.1. Two Wild Rivers

The White River is one of the longest undammed rivers in the United States, flowing about 580 miles from its source in Nebraska to Lake Francis Case, a reservoir on the Missouri River behind Fort Randall Dam. Summer flow rates are affected only by modest diversions for agriculture. The downward slope of the entire river channel averages about 5.8 feet per mile (three feet per mile over the last 70 miles). The White and its tributaries created the rugged topography of Badlands National Park. Cottonwood and peachleaf willow of various ages are abundant on its floodplain, an indication that channel shifting is sufficient to maintain these riparian species. Forty-one native fish species and eight introduced species have been found in the river, along with two native species of turtles and five native amphibians. Large mammals include white-tailed deer, mule deer, raccoon, porcupine, red fox, and coyote. Big Horn sheep are found in the badlands. Sixteen miles of the White River were lost when Lake Francis Case filled, but its heavy sediment load created a delta that now has cottonwood, willow, and other riparian species (fig. 9.17). Sturgeon, sauger, tern, and plover benefit from the shallower, warmer water of this emerging delta, assisting in the recovery of riverine species jeopardized by damming the Missouri.

The Little Missouri River flows about 560 miles northward from northeast Wyoming (fig. 9.18). It has a gentler average slope than the White River, but it too has carved badlands that are now protected in a national park, this one named for Theodore Roosevelt. Wildlife in the park include bison, bighorn sheep, elk, white-tailed deer, mule deer, pronghorn, and feral horses. Twenty-two species of native fish and four species of introduced fish have been found in the river, and roughly half of North Dakota's 1,300 plant species can be found in the area, the driest part of the state. Historians attribute President Roosevelt's conservation initiatives to his experiences in the badlands. Garrison Dam on the Missouri created a reservoir named Lake Sakakawea, which shortened the Little Missouri by approximately 40 miles. A delta has formed there as well. Like the White River, the Little Missouri is sparsely populated and has a shifting floodplain with riparian woodlands of plains cottonwood and willow. The river has relatively natural flows upstream of the national park, with one small diversion near Medora. Temporary surface-water permits for uses other than recreation and agriculture are prohibited in this part of the watershed, which covers the majority of the river's length. However, hundreds of temporary water permits have been issued downstream from the park's North Unit for the Bakken oil fields.[a]

[a]Volke et al. 2015; Shuh et al. 2017.

and flow through large areas of public and tribal lands protected to varying degrees from development. Because the White and Little Missouri Rivers have retained many of their natural characteristics, they are highly valued by conservation biologists. Also, except for the absence of large bison herds, the viewshed from many high points along both rivers has changed little from the descriptions recorded by early Euro-American explorers.

Restoring Riparian Landscapes

Some heavily damaged riparian habitats can be rehabilitated. However, because creeks and rivers are interconnected and penetrate all parts of their watershed, the challenges are much greater than restoring a tract of prairie. How can a river segment be restored when there are so many upstream factors that cannot be controlled? How many miles must be restored to achieve objectives that are considered desirable? And who decides and who pays for the work?

Such questions are beyond the scope of this book, but consider the following:

- A community-wide effort is required throughout the watershed. Towns and industrial complexes can reduce their effluents; farmers can minimize

Fig. 9.17. White River delta at flood stage. Water flows from right to left through the delta into Lake Francis Case on the Missouri River. The linear cottonwood woodlands on the natural levee, on the far side of the channel, are similar to those destroyed when the reservoir was filled. The silver-colored trees are Russian olive, an invasive tree. Photo by Malia Volke.

Fig. 9.18. The Little Missouri River meanders through Theodore Roosevelt National Park and associated river bluffs and badlands. The Little Missouri is the most natural river in North Dakota. Note the meanders, point bars, and cottonwood forest on the broad floodplain. Rocky Mountain juniper is the most common tree on the uplands.

the loss of soil and nutrients through adopting conservation tillage practices (see chapter 5), and wetlands can be restored to capture runoff and filter water. Increased groundwater quality and quantity can result (see chapter 11).

- Where possible, control livestock grazing, foot and vehicle traffic, and the spread of invasive plants and animals; avoid trampling soils and vulnerable plants.[33]
- Enable floods and meandering channels on reaches where property damage can be avoided; where feasible, remove buildings from the floodplain.
- Where benefits are possible, remove wetland drainage tiles and reintroduce beavers.
- Avoid introducing new diseases and invasive organisms (box 9.2).

The farm-to-city connection is illustrated by a lawsuit filed by a city in Iowa against drainage districts to compensate the costs of removing toxic nitrates in runoff from farm fields that entered the Des Moines River, a source of the city's drinking water.[34] A slim majority of the Iowa Supreme Court did not hold the districts liable. However, more legal actions involving urban and rural environmental issues are likely in the future unless conservation measures are implemented on farmland. The state of Minnesota's solution was to require producers to plant grass strips along the Minnesota River to absorb nitrates and other pollutants, as a means of lowering their concentrations in river water used downstream by towns and cities.[35] State and federal programs in the Dakotas offer lower tax rates or payments if perennial grasses and forbs are planted in buffer strips along streams, lakes, ponds, and wetlands.

There is proof that riparian restoration works, although it may take time. Pristine conditions are not attainable, but water quality and wildlife habitat can be improved. For example, the modern Mortenson Ranch in western South Dakota was pieced together after the Dust Bowl from about 40 abandoned homesteads.[36] Every acre had been overgrazed or farmed with traditional methods. Restoration focused on Foster Creek, a once tree-lined tributary to the Cheyenne River. By the time homesteaders left, the creek was a deep gully

Box 9.2. Aquatic Invasive Species and Hybrid Vigor

Invasive organisms threaten water quality, fish populations, and the biological diversity of creeks, rivers, wetlands, and lakes. They include quagga mussels, zebra mussels, and New Zealand mud snails, all of which displace native clams and snails and alter water chemistry. Other invasive species include various fish, including carp and snakeheads, that make life difficult for game fish. Plants like purple loosestrife, Eurasian watermilfoil, and common reed change the character of aquatic ecosystems dramatically. All are difficult to control, and the costly problems they create are sufficiently severe that state legislatures have passed laws intended to curb their spread. It is now routine for law enforcement officers to inspect boats before they are launched, thereby reducing the chances that such plants and animals will be inadvertently introduced. Also, shoreline anglers are asked to inspect their waders, live bait often is prohibited, and emptying minnow buckets into the lake after fishing is discouraged. A day of fishing—for many a time of freedom and relaxation—is now a bit less so because of organisms introduced from other continents. Notably, some of the problematic plants are hybrids between a native species and an introduced European species. Well-known examples are common reed and reed canarygrass. Hybrid cattail is similar, except it is the progeny of two native cattail species, the broadleaf cattail (*Typha latifolia*) found historically in the Dakotas and the narrowleaf cattail (*T. angustifolia*) found most commonly in eastern North America, though now spreading westward. The hybrid tolerates deeper water, which leads to the loss of open water. State game and fish departments now maintain websites that describe how to recognize and control invasive species.

without a tree in sight. A landscape that once functioned like a sponge was eroding rapidly.

A two-pronged approach was adopted to reclaim the land: first, runoff was reduced in uplands by reconfiguring pastures and adopting a rest-rotation grazing system, and second, floodplains and shallow aquifers were rebuilt by constructing temporary dams to slow water, capture sediment, and enable water seepage to the riparian zone. The dense grass and forb cover of the riparian zone was used as temporary calving pasture in early spring, but was rested during the remainder of the growing season. Most dams, built during the 1980s to the 1990s, did their work in a few years by capturing sediments and creating mudflats on which thousands of cottonwood and willow began to grow naturally. The deep gullies no longer drained groundwater from the fringes of the floodplain, and the subirrigated floodplain filtered the water and promoted the growth of native range grasses. Beaver probably had done this kind of work previously, providing an ecosystem service that is now costly to reproduce with machinery.[37]

About 30 years have passed since the Mortenson's grazing system was adjusted and dams constructed. Some reaches are now lined with young cottonwood and willow; others are dominated by prairie cordgrass, a long-lived perennial native grass that forms dense patches (figs. 9.19 and 9.20). In winter and early spring, the mulch-covered sod insulates resting cattle from the cold ground, thereby improving calving success. Yellow warblers and common yellowthroats nest in thickets of sandbar willow. Complete restoration can be a

Fig. 9.19. Restoring grassland can facilitate recovery of riparian woodlands, as shown here on the Mortenson Ranch. Prairie sandreed grows in the foreground, green ash lines the intermittent stream and upland draws. Oahe Reservoir on the Missouri River is visible in the distance.

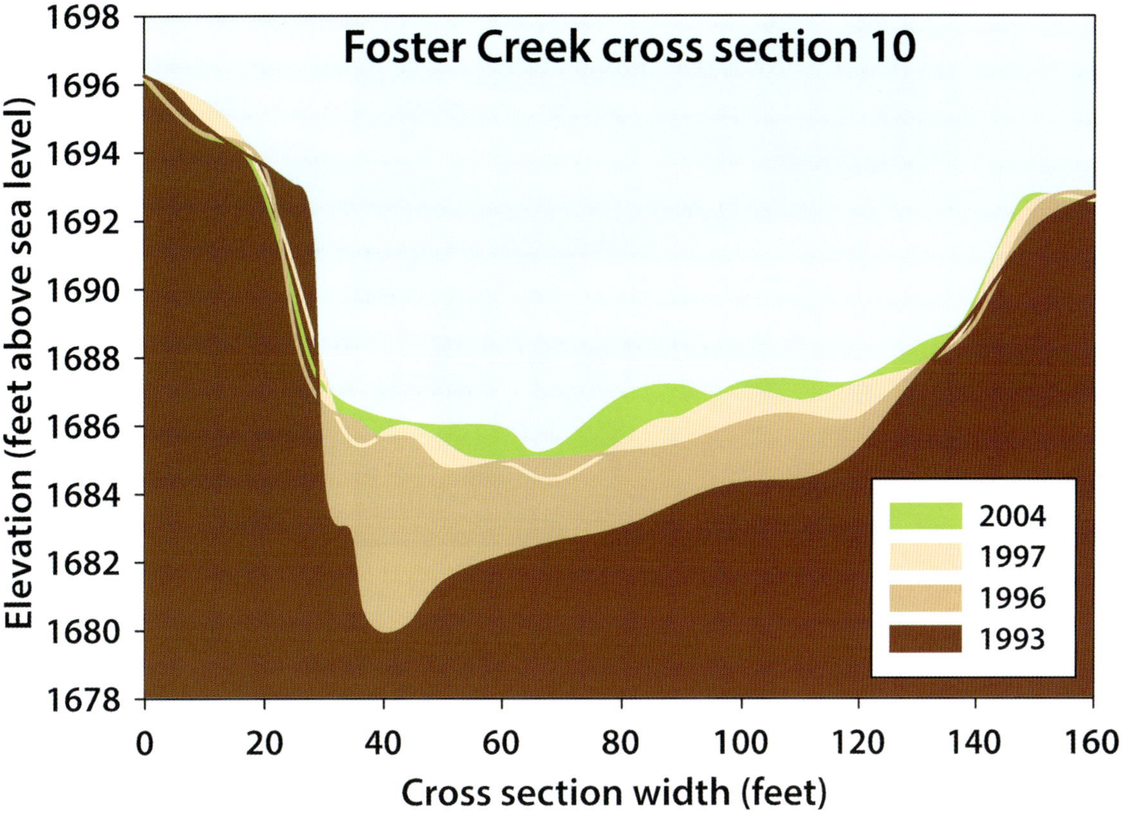

Fig. 9.20. Repeated measurements of sediment accumulation behind three shallow dams on Foster Creek, Mortenson Ranch, 1993–2004. Sediment deposition was greatest early in the record, shortly after earthen dams were built. Seven feet of sediment accumulated over the old, degraded stream channel. The purpose of the dams was to raise the floodplain by sedimentation, not to maintain a pond. The stream now flows in a shallow channel across the surface of the accumulated sediment. The entire surface of the sediment is inundated during floods.

long-term process, but rewards are achievable after only a few years.

The Dam Conundrum

Mention dams in the Dakotas and the Missouri River comes to mind, but many smaller rivers and streams also have been impounded. Nothing has a more adverse effect on riparian ecosystems than flooding them permanently with water or, below the dam, depriving floodplains of periodic floods and sediment. Many reservoirs are beginning to show their age by filling with sediments. Already, some no longer meet their intended purpose, whether water-based recreation, stock watering, or flood retention. Failing dams are sometimes removed because they are too expensive to repair or the accumulated sediment too difficult to remove. Notably, about a thousand dams were removed in the United States between 1990 and 2015—two in North Dakota and five in South Dakota.[38]

There are three major kinds of dams: storage, run-of-the-river, and low-head (not including sediment capture types described above). Unless modified with fish ladders, storage and run-of-the-river dams block upstream passage of fish and other aquatic organisms. Run-of-the-river dams

Fig. 9.21. Low-head dams on the Big Sioux River near Sioux Falls (top) and the Sheyenne River south of Devils Lake (bottom). These dams block fish passage during low to average flows but not at times of high water. Sioux River photo by Greg Latza, People-Scapes Photography.

have small reservoirs and only moderate effects on streamflow. Storage dams have the largest effect because they capture large volumes of water that can be released as desired, such as for flood control, irrigation, and hydroelectric power. Examples of storage dams in South Dakota are Shadehill Reservoir on the Grand River, Pactola Reservoir on Rapid Creek, and Angostura Reservoir on the Cheyenne River. Those in North Dakota include Lake Darling on the Souris River, Arrowwood Lake on the James River, and Lake Tschida on the Heart River.

Low-head dams are very common on Dakota rivers and allow water to flow over the dam when streamflow is high (fig. 9.21), permitting some fish to migrate upstream. Ponding behind low-head dams during droughts keeps the channel from completely drying out. Thus, aquatic organisms survive and recolonize the channel when normal flows return. Their original purposes, however, were largely commercial—for lumber and grist mills, small hydropower plants, ice harvesting, and livestock watering. About ten low-head dams less than 15 feet high were built along the Big Sioux River and have become popular fishing spots. The dam at Klondike was famous as a place to catch large catfish until it was removed in 2013 to facilitate safer canoe and kayak recreation. It was replaced by a shallow rock ramp for fish passage and to recharge local groundwater supplies.

The ecological effects of low-head dams on the Yellowstone River are more consequential. The Yellowstone is 692 miles long, the longest river with no storage dams in the United States, but six low-head dams were built on the main channel to facilitate water diversions for irrigation in eastern Montana. These dams have become controversial because of the effect they have on upstream migration and spawning of the endangered pallid sturgeon. The turbid, warm-water, lower Yellowstone is a refugium for many of the Missouri River's native fish that cannot consistently tolerate the clear, cold water of Lake Sakakawea located just downstream of the Yellowstone's confluence with the Missouri (see fig. 1.9). The sturgeon and other fish swim upstream in the Yellowstone but often are entrained in irrigation canals and killed. Experts still regard the lower Yellowstone as the most favorable place on the upper Missouri River for the sturgeon to successfully reproduce, but low-head diversion dams block most adults from progressing upstream. To solve this problem, the Corps of Engineers is considering a diversion canal that may allow fish to circumvent the first dam.[39] Rather than canals, environmental groups favor pumping water for irrigation from the channel so that the diversion dam can be removed, thereby providing unobstructed upstream migration of adults and downstream drifting of pallid larvae. A decision on which plan to adopt has not yet been made.

The construction of small and large dams of all types has created a heavily engineered landscape in places, one that has improved the lives of many people. Notably, some municipalities, agencies, and landowners have already collaborated in taking steps to reclaim some of the historical benefits provided by rivers and their riparian ecosystems, but even with improved management many dams will outgrow their usefulness. A major challenge will be to find the funds and minimize the short-term damage to the river when removing or disabling dams. This process has started, but most studies have been short (less than 5 years) and do not adequately represent the diversity of types of dams, watershed conditions, and dam removal methods.[40]

Rivers in a Changing Climate

As described in chapter 3, the watersheds of Dakota rivers are likely to experience more climate warming during this century. Among the effects will be earlier melting of snow and reduced snowpack, longer summers, and greater evapotranspiration (ET). Mean annual precipitation is projected to increase more in the east, partially compensating for increased ET. More intense precipitation in the cooler months may produce more severe floods, while the higher ET in warmer months may lead to more severe droughts and lower base flows of rivers.[41] First- and second-order streams are likely to have a more variable flow regime with more floods and more droughts. Large rivers fed by mountain snowpack, such as the Missouri and Yellowstone, are likely to be most affected because of greatly reduced snowpack and earlier melting expected in the future. Hydrographs in the Northern Rockies show that April snowpacks have declined from 20 percent to 80 percent over the period 1955–2016.[42] Earlier melting of a thinner snowpack contributes less water to stream flow, and it runs off earlier in the spring, potentially reducing water for municipalities, industry, agriculture, hydropower, and recreation far downstream. The western Dakotas are projected to experience more severe effects of climate change than areas in the east.

Native plants and animals are adapted to such fluctuations as long as they are within the range of variability experienced during their evolutionary history. If not, they are challenged with migrating northward. Fortunately, riparian ecosystems provide a corridor through which wildlife can move more easily than through heavily farmed uplands. Rivers such as the James, Big Sioux, Red, and parts of the Missouri are oriented north and south, providing organisms a direct northward route to elude new stresses from warming. Maintaining free passage of organisms northward, along with river breadth and intact riparia, should be an essential component of any comprehensive plan to mitigate the adverse effects of climate change.

Aquatic organisms are particularly vulnerable to extirpation in the future climate. Those that occur in streams or rivers with dams or in isolated watersheds may be unable to migrate to escape warmer water. Trout in the generally shallow Black Hills streams would be especially vulnerable to warmer stream water. Warmer water not only holds less oxygen, but the oxygen demand for fish and many other species increases as their body temperature increases. In this way, climate change puts both cold- and warm-water fish in a "thermal squeeze." Trout are not native to the Black Hills but have done well in the past through stocking and habitat creation in streams. Trout fishing there is an important source of tourist income. In the past few years, some trout streams in nearby Montana have been closed to fishing due to increasing water temperatures that cause fish to die after experiencing the stress from being caught, even though they are quickly released back into the stream.

The British limnologist Hugh Hynes wrote, "You can't divorce a stream from its valley." A stream is the product of many factors including climate, soils, topography, upland and riparian vegetation, land management, natural disturbances, and water usage, among others. The challenge of identifying the singular effects of climate change on streams is to be able to disentangle one variable from all of the others. Especially challenging are detecting the effects of climate over a large watershed. The stream hydrograph—seasonal streamflow patterns—for the lower reaches of the Missouri River, for example, is affected by climate change in the northern Rocky Mountains as well as throughout much of the Northern Great Plains. Hundreds of stream gages exist in its valley, some with nearly a century of data. Changes in hydrographs outside of historic patterns can potentially indicate the effects of climate change. Continued long-term monitoring and analysis of numerous variables is essential.

Chapter 10 The Missouri River

The Missouri is the longest river in North America, starting in the mountains of Montana and ending where it joins the Mississippi River just north of St. Louis—a distance of 2,546 river miles. In the Dakotas, it forms the border between young, glaciated landscapes to the north and east and ancient, unglaciated landscapes to the south and west. With mean daily flows of about 29,600 cubic feet per second, the Missouri transports six times more water than the next largest river in the two-state area, the Red River of the North (see table 9.1). Its broad, deep valley has provided shelter and sustenance for people who first arrived over ten thousand years ago.[1] Much later it was the "waterway to the Pacific" used by Captain Meriwether Lewis and Second Lieutenant William Clark. Today the upper Missouri has been largely transformed into a series of large reservoirs. This chapter focuses on how this river was first used by indigenous people, the observations of explorers in the 1800s, and the more recent, dramatic developments that reduce floods, generate hydroelectric power, and provide water for municipalities, industry, agriculture, and recreation.

Early Human History

The Missouri River valley provided a diversity of resources to the first occupants of the region (fig. 10.1). Water, firewood, and shelter were easy to find, and there were numerous edible plants. Fish were abundant, and bison, elk, pronghorn, and deer were found throughout the year. Plains cottonwood, willow, American elm, green ash, boxelder, and bur oak provided wood for fuel and the construction of shelters, farming tools, bull boats, and canoes. As described in chapter 5, fertile areas of the floodplain were farmed, primarily for corn, beans, and squash.

Seven decades before Lewis and Clark started their trip, a French fur trader, Pierre La Vérendrye, learned from an Assiniboine tribe near Lake Winnipeg of people who lived in large villages along a great river and with whom they traded for corn.[2] In 1738, Vérendrye visited Mandan and Hidatsa villages along the Missouri near the Knife and Heart Rivers tributaries. Farther south they visited Arikara villages near the Bad and Cheyenne Rivers (fig. 10.2). Each village was a center of Indian trade at the time. These early farmers cultivated the floodplain of the Missouri and exchanged their crops for products of the chase brought to them by nomadic bison hunters. Vérendrye's visit to the Mandans is the earliest documented European contact with the Northern Plains villagers.[3] Taking into account the large number of villages and the strong market for corn, he must have seen

Fig. 10.1. The Missouri River downstream of Fort Randall Dam. Note the islands, sandbars, riparian meadows, cottonwood groves, and, in the distance, woody draws. Photo by Greg Latza, PeopleScapes Photography.

Fig. 10.2. Mandan village overlooking the Missouri River, with bull boats made from willow branches and bison hides. After Karl Bodmer (Swiss, 1809–1893), Friedrich Salathé, engraver, *Mih-Tutta-Hang-Kusch, a Mandan village,* n.d., hand-colored aquatint and engraving on paper, Joslyn Art Museum, Omaha, Nebraska, Gift of the Enron Art Foundation, 1986.49.517.16.

thousands of acres under cultivation. Agriculture broadened the subsistence base of the villagers and their trading partners to a tripartite system of bison hunting, horticultural crop production, and hunting and gathering. For hundreds of years they held the keys to the granary of the Northern Great Plains. Horses augmented their lifestyle in the mid-1700s.

Plains Indian villages—mostly Mandan, Hidatsa, and Arikara—were concentrated along the Missouri between present-day Chamberlain, South Dakota and the mouth of the Knife River north of Mandan, North Dakota. Their early earthen lodges were rectangular in shape and typically about 25 by 35 feet in size.[4] Entrances were located on the sheltered, south end of the lodge and consisted of a narrow, roofed, antechamber eight feet long. Mats of willow and red-osier dogwood were laid across the frame of the lodges and covered with straw and mud. Communal or ceremonial lodges were often octagonal and about 50 to 60 feet in diameter. Archaeological excavations have revealed that the average diameter of posts at floor level was about seven inches,[5] a size achieved in 20–25 years and which was readily available, easily transported, and could be worked with stone tools. Young cottonwood trees of this size grew naturally on sandbars repeatedly formed by floods. They also used driftwood, originating from trees that fell into the river as the banks eroded.

The largest villages had 200 or more single-family dwellings. Over time, circular structures became more common, perhaps due to greater heating efficiency during the winter (figs. 10.3 and 10.4). Central fire pits provided more uniform heating. Trash

Fig. 10.3. Interior of the earthen lodge of a Mandan chief. After Karl Bodmer (Swiss, 1809–1893), Narciso Edmundo Jose Desmadryl, engraver, *The Interior of the Hut of a Mandan Chief*, n.d., hand-colored aquatint and engraving on paper, Joslyn Art Museum, Omaha, Nebraska, Gift of the Enron Art Foundation, 1986.49.517.19.

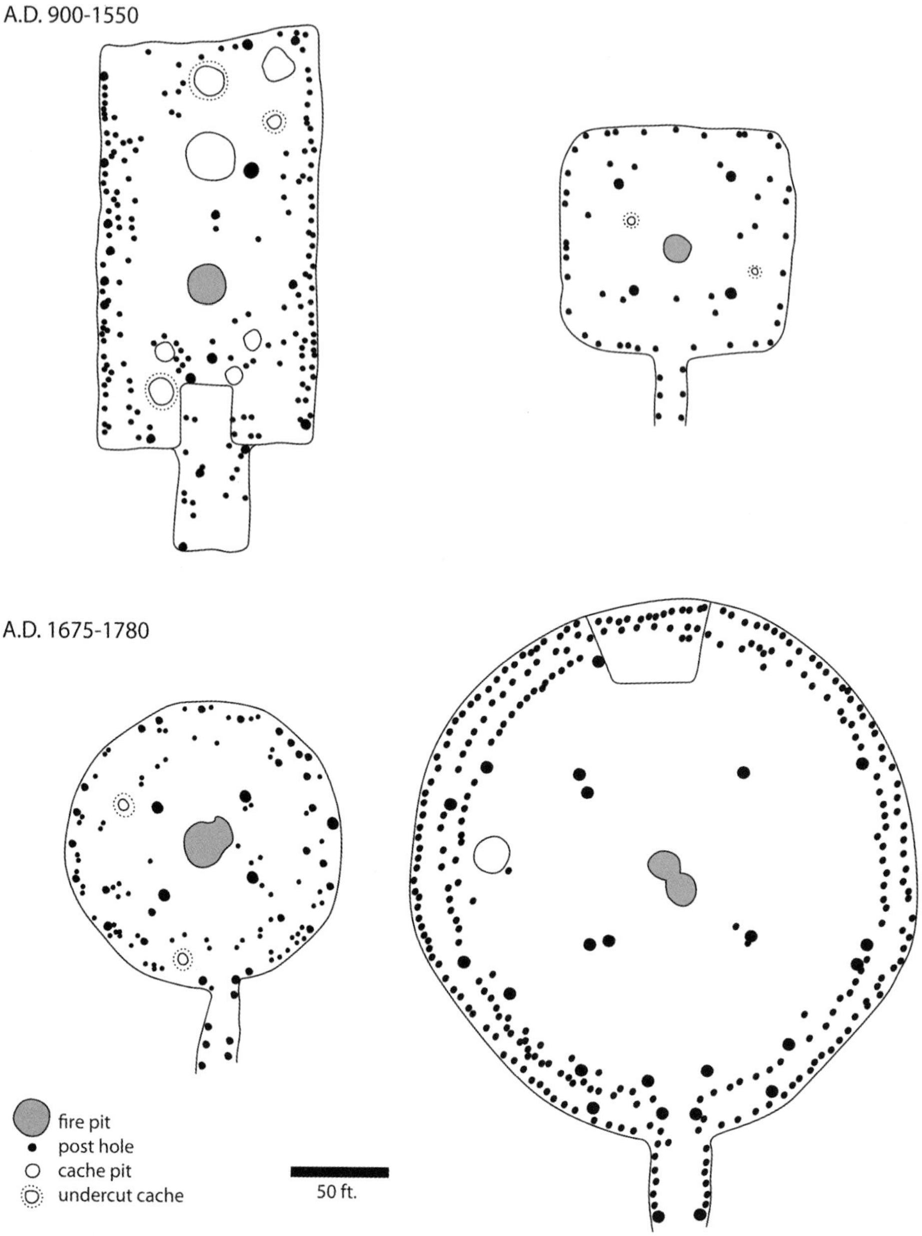

Fig. 10.4. Evolution of lodge design in middle Missouri Villages. The floor plans show location and number of wooden poles and cache pits. Rectangular shapes were typical of early designs in eastern forests and the Plains Woodland period. The square shape was adopted in early Great Plains locations. Round designs were common in recent villages on the Northern Great Plains. The lower right drawing is a ceremonial lodge. Food was stored or cached in underground pits within the lodges (see chapter 5). Adapted from Lehmer (1971).

mounds have provided archaeologists with a treasure trove of cultural information, especially about diets.[6] The enormous quantity of bones, 90 percent from bison, indicates they were meat eaters. Also common were bones from pronghorn, deer, dog, and elk, in addition to badger, beaver, bobcat, coyote, fox, ground squirrel, porcupine, prairie dog, jackrabbit, cottontail, squirrel, skunk, vole, and wolf. The remains of birds and other animals were found, including those of the bittern, crow, eagle, goose, duck, grouse, grackle, hawk, heron, magpie, owl, passenger pigeon, prairie chicken, raven, whistling swan, whooping crane, catfish, gar, frogs, turtles, and mussels. Plant remains included sunflower, chokecherry, blackberry, cherry, grape, plum, goosefoot, marsh elder, rose, dogwood, buffaloberry, and dock. Clearly, people were living off the land.

Archaeologists have suggested that local exploitation of timber resources was a factor in decisions to move whole villages closer to new supplies of wood.[7] Likely causes of shortages included overexploitation, channel realignment during floods that eroded woodlands, growth of local trees beyond the sizes useable or desired, and shortage of driftwood during periods of low streamflow. Other motivations for relocation were warfare, disease, tribal reorganization, and climate change. Timber supplies along the Missouri River were depleted even more rapidly in the 1800s by immigrants of European descent, who constructed large forts and burned wood in their steamboats.[8]

Village prosperity was dealt a tragic blow by diseases from Europe.[9] Collectively called smallpox but including measles, chickenpox, and cholera, high rates of mortality left survivors with a disorganized remnant of their former culture. Four epidemics have been identified (1780–1781, 1801–1802, 1837–1838, 1856), but earlier ones are suspected. Mortality rates in 1837 alone were an estimated 33 percent of the Hidatsa, 50 percent of the Arikara, and a staggering 98 percent for the Mandan. The cumulative effects of epidemic diseases, coupled with the rising power of the horse tribes who suffered lower losses from disease, particularly the Dakota, and the economic competition from trading posts stimulated by the Lewis and Clark expedition, brought an end to the villagers' widespread culture. In 1804, Lewis and Clark recorded many abandoned villages along the river.[10] The last village was abandoned in 1886, ending a 900-year legacy of lodge-dwelling village life in the Northern Plains.

Various lines of evidence led to the conclusion that the sedentary tribes and their trading partners had considerable impact on their natural resources. Bison numbers probably declined during the first two-thirds of the eighteenth century when the villages and nomadic tribes rode a wave of good health and economic prosperity.[11] The arrival of steamships brought disease, but they also facilitated the export of pelts and expanded the need for large quantities of fuelwood harvested from the floodplain forests. Still, despite growing resource exploitation, the river retained the capacity to meander and to produce new crops of trees and associated wildlife. Undiminished grasslands continued to support large herds of bison.

Observations by the Corps of Discovery

Little of the upper Missouri River region had been explored by European Americans prior to 1804, the year Meriwether Lewis and William Clark arrived with their crew of 45 men and one dog. They rowed, sailed, pushed, pulled, and paddled their heavy, metal-framed keelboat and several swifter pirogues and canoes up the river.[12] As described in chapter 2, the primary objective was to determine if a water route could be found to the Pacific Ocean, but they also were charged with learning as much as possible about the natural resources of land acquired with the Louisiana Purchase. Lewis was the first person with formal education in science to report on the plants and animals.[13] However, the captains received vital information from the people they encountered along the way, people who knew much about the ecology of the land where their ancestors had lived for so long—the location of salt deposits, plants to use for medicine, and where, when, and how food and construction materials could be obtained. Lewis and Clark and other explorers recorded elements of this traditional knowledge in their diaries, journals, drawings, and paintings; and later, scientists benefited from interviews with elders who had learned much about plants and animals from their predecessors.

With increasing frequency as the Corps moved

upstream, the captains wrote about animals that were new to them. Near the mouth of the Platte River, least terns and white pelicans were collected and described. A coyote, which they called a "prairie wolf," was pursued but, not surprisingly, escaped collection. Prairie dogs and pronghorn were described soon after entering present-day South Dakota, and bison, deer, and elk became abundant upriver. At Spirit Mound, near where the town of Vermillion would be founded, and one of the few actual spots where the expedition was known to have walked, Lewis wrote "we beheld a most butifull landscape; numerous herds of buffalo were seen feeding in various directions." Lewis later wrote, "I do not think I exaggerate when I estimate the number of Buffaloe . . . at one view to amount to 3000." (September 17, 1804, near the mouth of the White River).[14]

Farther upstream, the Corps encountered Arikara villages and fields that had been largely abandoned, possibly due to the smallpox epidemic of 1781. Survivors sold them corn, beans, pumpkins, watermelons, squashes, and tobacco, plus hog peanut or ground bean, a native legume that grew in riparian woodlands and produces large nutritious beans, some above- and some belowground (fig. 10.5). The Arikara described how they raided small mammalian caches of these beans but left a few or added corn to the cache, not wanting to take all of the food stored by the animals.[15] The travelers were intrigued by riverworthy bull boats (see fig. 10.2 and box 10.1). After reaching present-day North Dakota, Clark wrote in his journal about "emence herds" of pronghorn crossing the "wide" river in October, apparently migrating to the vicinity of the Little Missouri badlands and Killdeer Mountains for the winter.[16] On October 20 he wrote, "the Countrey thro which I passed this day is Delightful, Saw great numbers of Buffalow Elk Goats [pronghorn] & Deer." The following spring Lewis wrote about seeing "timber" along the river with cottonwood, elm, ash, boxelder, willow, chokecherry, redwood [probably

Fig. 10.5. The hog peanut (*Amphicarpaea bracteata*) and its two types of seed pods, one that developed above ground and the other in the soil, like a peanut. The Mandan called the plant "mouse-bean" because mice and other rodents cached the seeds. Photos by Jerry D. Davis.

Box 10.1. Bull Boats

Bull boats were an important form of shallow water transportation for some upper Missouri River tribes. The circular frame was fashioned from willow saplings and covered with a single hide from a bull bison, hence the boat's name. Stability and abrasion resistance were improved by having the fur on the outside. A single paddle was used to pull the boat forward, or a boat with passengers or cargo would be pulled by a swimmer. Explorers witnessed fleets of 80 or more bull boats moving provisions to villages. Fur traders expanded the concept by building larger bull boats, some 30 feet by 12 feet, to move pelts down shallow rivers. These may have been the lightest and shallowest-draft vessels ever constructed for their size and use.

red-osier dogwood], and redberry [probably silver buffaloberry].

Overwintering with the Mandans

For a distance of 20 miles upstream from present-day Bismarck, the Corps of Discovery noted numerous abandoned villages before finding two that were occupied by members of the Mandan tribe and three by the Hidatsa. The five villages had a combined population of about 4,000 people. The Corps built their own winter quarters, Fort Mandan, on the left bank of the Missouri (about 14 miles west of the present-day city of Washburn), directly across the river from the Mandan village. The historian Stephen Ambrose summarized their activities that winter, writing, "The winter at Fort Mandan involved hunting, trading, keeping fit, dealing with the cold, doing extensive repairs to old equipment and building new canoes, visiting the Indians, and much more. But for Meriwether Lewis it was primarily a winter of scholarship, of research and writing."[17]

Lewis and Clark spent a total of about nine consecutive months in the Dakotas on their trip westward. Large amounts of corn were received from the Mandans in exchange for battle axes forged by the expedition's blacksmith. Historians have speculated that the Corps would not have survived their first winter if it were not for Mandan corn and other vegetables. The captains wrote their first major report to President Jefferson during that winter. It included Clark's map (fig. 10.6), which, in the spring, was shipped downstream to St. Louis on the cumbersome keelboat along with 108 mounted plant specimens that included the roots of purple coneflower, the seeds of tobacco, and several varieties of Indian corn. Also on the boat were rock specimens, skeletons, horns, the skins of mammals, and a live prairie dog, magpie, and sharp-tailed grouse. The prairie dog and magpie were still alive when they arrived in Monticello.

Leaving Fort Mandan, laboring against the swift current required a large amount of food—an estimated nine pounds of meat per day per person.[18] The number and variety of animals consumed during their expedition is testimony to the wealth of game they encountered (table 10.1). Near present-day Montana, Lewis wrote on April 17 [verbatim], "we saw immence quantities of game in every direction, around us as we passed up the river; consisting of herds of Buffaloe, Elk, and Antelopes with some deer and woolves." Five days later, he added, "I ascended to the top of the cutt bluff this morning, from whence I had a most delightful view of the country, the whole of which except the vally formed by the Missouri is void of timber or underbrush, exposing to the first glance of the spectator immence herds of Buffaloe, Elk, deer, & Antelopes feeding in one common and boundless pasture."

Grizzly bears were seen also, and Lewis described the snow geese they saw. Geese were observed nesting in the top of a large cottonwood. After climbing to a hilltop near the mouth of the Yellowstone River, he wrote, "from whence I had a most pleasing view of . . . the wide and

Fig. 10.6. A section of William Clark's map that illustrates evacuated Indian villages along with channels, islands, bars, and other riparian features. Source: Prince Maximilian (German, 1782–1867), Lewis and Clark Expedition Map, Clark-Maximilian Sheet 18, Route about October 23–November 1, 1804, Joslyn Art Museum, Omaha, Nebraska, Gift of the Enron Art Foundation, Photograph © Bruce M. White, 2019.

fertile vallies formed by the missouri and the yellowstone rivers, which occasionally unmasked by the wood on their borders disclose their meanderings for many miles in their passage through these delightfull tracts of country. . . . the whol face of the country was covered with herds of Buffaloe, Elk & Antelopes, deer are also abundant, but keep themselves more concealed in the woodland. the buffaloe Elk and Antelope are so gentle that we pass near them while feeding,

Table 10.1. Number of mammals and birds harvested by the Lewis and Clark expedition, primarily for food, from the time of leaving St. Louis on May 14, 1804, to the day it ended on September 23, 1806[a]

Deer	1,001
Elk	375
Bison	227
Pronghorn	62
Bighorn sheep	35
Grizzly bear	43
Black bear	23
Gray wolf	18
Beaver	113
River otter	16
Waterfowl	149
Grouse	46
Turkey	9
Plover	48
Indian dog	190
Horse	12

[a]Source: Burroughs 1961.

without apearing to excite any alarm among them" (April 25).

Lewis measured the widths of the Missouri and Yellowstone Rivers at their confluence and found them to be nearly identical. Some later explorers were not sure which channel was that of the Missouri. On or about May third, the Corps of Discovery entered what is now Montana. Two days later Lewis wrote about wolves: "we scarcely see a gang of buffaloe without observing a parsel of those faithful shepherds on their skirts in readiness to take care of the mamed & wounded."

The Missouri as Riverine Highway

The Missouri provided a new gateway for accessing wildlife resources from the south and east, rather than from fur-trading companies and the Assiniboine to the north in Canada. At the start, keelboats, mackinaws, and pirogues loaded with trade goods were laboriously sailed and paddled upstream against the current, and then, some months later, drifted downstream laden with furs, hides, and other products. Moving people and goods in relatively small boats against the current was slow and arduous, but that changed in 1819 with the first Missouri River steamboat—thirteen years after Lewis and Clark completed their expedition.[19] Hiram Chittenden, biographer of steamboat captain Joseph La Barge, wrote: "The river is like a great spiral stairway leading from the ocean to the mountains. A steamboat at Fort Benton [Montana] is 2,565 feet . . . above the level of the sea; yet so gentle is the slope nearly all the way that, in placid weather, the water surface resembles that of a lake. This wonderful evening-up of the slope of the river by the extreme sinuosity of its course is a fact not only interesting as a natural phenomenon, but of the utmost importance in the behavior and use of the stream."[20]

The fate of bison, the rise of steamboats, and fuelwood harvesting were inextricably linked. Millions of hides acquired at remote forts and trading posts could now be shipped to St. Louis and beyond, but massive amounts of fuelwood had to be cut or collected every day or two.[21] One of the boats, *The Yellow Stone*, reportedly used 10 or more cords of wood daily, an amount that would fill a railroad flatcar.[22] Able passengers were sometimes obliged to assist the crew in wood collecting. Over time businesses developed whereby "wood hawks" would cut, stack, and dry wood on the riverbank, to be sold to passing steamboats. The cutting of so many trees led to localized depletion of the riparian forest, especially near navigable channels.

One of the first steamboat passengers to travel up the Missouri River was the artist George Catlin, who in 1832 traveled on *The Yellow Stone*'s maiden voyage to Fort Union, 2,000 miles from St. Louis. Their return cargo of buffalo robes and furs started the export of massive volumes of wildlife products from the Northern Great Plains. Catlin noted

Fig. 10.7. A consequence of rapid channel meandering during floods is the transport of whole trees and woody debris from the floodplain into the river channel, making navigation treacherous. After Karl Bodmer (Swiss, 1809–1893), Lucas Weber and Johann Hürlimann, engravers, *Snags (Sunken Trees) on the Missouri*, hand-colored aquatint and engraving on paper, Joslyn Art Museum, Omaha, Nebraska, Gift of the Enron Foundation, 1986.49.517.6.

that the voyage had "almost insurmountable difficulties which continually oppose the voyageur on this turbid stream." He wrote at length about the amount of dead wood present in the channel (fig. 10.7), dreaded by the captains of steamers but now regarded by biologists as important habitat for fish. About 400 steamboats were sunk or disabled by snags on the Missouri River. The average lifespan of a steamboat was five to seven years.[23]

Hiram Chittendon wrote that "the shores of this river (and, in many places, the whole bed of the stream) are filled with snags and raft, formed of trees of the largest size, which have been undermined by the falling banks and cast into the stream; their roots becoming fastened in the bottom of the river, with their tops floating on the surface of the water, and pointing down the stream, forming the most frightful and discouraging prospect for the adventurous voyager. . . . Almost every island and sand-bar is covered with huge piles of these floating trees, and when the river is flooded, its surface is almost literally covered with floating raft and drift wood which bid positive defiance to keel-boats and steamers, on their way up the river." According to Stephen Ambrose, "The current ran at five miles per hour usually, but it sped up when it encountered encroaching bluffs, islands, sandbars, and narrow channels. Incredible to behold were the obstacles—whole trees, huge trees . . . that had been uprooted when a bank caved in; hundreds of large and thousands of smaller branches; sawyers, trees whose roots were stuck on the bottom and whose limbs sawed back and forth in the current, often out of sight; great piles of driftwood clumped together, racing downriver, threatening to tear holes in the sides of the boat; innumerable sandbars, always shifting, swirls and whirlpools beyond counting."[24]

Catlin spent a month at Fort Union and three weeks at the Mandan villages. His paintings of people (fig. 10.8), riverine landscapes, and villages number in the hundreds. He was aware that the villagers of the upper Missouri were in decline because of diseases and increasing raids by nomadic tribes with horses, which had been pushed westward by Europeans. Catlin predicted that all the bison would be killed, devastating the tribes.[25] His paintings provide glimpses of the land and its people, and he made a rare conservation appeal for that time period, urging Americans to establish the Great Plains as a park in which nature and the Indian way of life would be saved. That did not happen, and during the following 40 years the bison were driven precariously close to extinction (see chapter 13).

A year after Catlin's trip, in 1833, Swiss artist Karl Bodmer accompanied the German scientist Prince Maximilian of Wied-Neuwied up the Missouri, traveling on the steamboat *Assiniboine* to Fort Union, and from there by keelboat to Fort McKenzie (near the Marias River, in present-day Montana). Unfortunately, the *Assiniboine* burned on the return trip. Its entire cargo was lost, including furs, hides, and several cases of Maximilian's bird and animal specimens. Bodmer's paintings are renowned for their accuracy and sensitivity to the nuances of the people and places of frontier America.[26] They depict not just the natural character of the Missouri and its valley but also the impact of people on the landscape. One painting portrayed the bleakness of the Missouri in the dead of winter, bereft of wood or other vegetation (fig. 10.9). In general, his art supports the conclusion by contemporary archaeologists that woodland resources could have become exhausted in the vicinity of villages, sometimes leading to relocation. Notably, some Indians moved seasonally to forests lower in the topography during winter to avoid the bitterly cold winds in higher, more exposed locations (fig. 10.10).

The famous ornithologist and painter John James Audubon began his steamboat trip up the Missouri on April 25, 1843. The 58-year-old naturalist kept a voluminous journal that included graphic drawings of people, military posts, pioneer settlements, bird life, and big game.[27] He first saw bison near the mouth of the James River in present-day South Dakota (May 20), writing that the ground was literally covered with their tracks and the bushes thick with their hair. At Fort Union he wrote that bison were "all around" and that wolves could be seen from the fort. Audubon described the sounds: "Wolves howling, and bulls roaring, just like the long-continued roll of a hundred drums. Buffaloes all over the bars and prairies, and many swimming; the roaring can be heard for miles."

Traveling along the Yellowstone River valley, Audubon lamented the slaughter of bison:

> What a terrible destruction of life, as it were for nothing, or next to it, as the tongues only were brought in, and the flesh of these fine animals was left to beasts and birds of prey, or to rot on the spots where they fell. The prairies were literally covered with the skulls of the victims, and the roads the Buffalo make in crossing the prairies have all the appearances of heavy wagon tracks. . . . One can hardly conceive how it happens, notwithstanding these many deaths and the immense numbers that are murdered almost daily on these boundless wastes called prairie, besides the hosts that are drowned in the freshets, and the hundreds of young calves who die in early spring, so many are yet to be found. [A late heavy snow was considered the cause of a massive die-off of buffalo calves that year.] Daily we see so many

Fig. 10.8. Four Bears, a Mandan chief who interacted with the German naturalist Prince Maximilian of Wied-Neuwied and artist Karl Bodmer. Painting by George Catlin (American, 1796–1872), William Day and Louis Haghe, lithographers, *Mah-To-Toh-Pah*, hand-colored lithography on heavy wove paper, Joslyn Art Museum, Omaha, Nebraska, Gift of the Walter Scott Jr., Foundation, 1995.12.30.

Fig. 10.9. Winter on the Missouri River with Fort Clark in the distance. Note the scarcity of wood. After Karl Bodmer (Swiss, 1809–1893), Friedrich Salathé, engraver, *Mih-Tutta-Hang-Kusch, a Mandan village*, n.d., hand-colored aquatint and engraving on paper, Joslyn Art Museum, Omaha, Nebraska, Gift of the Enron Art Foundation, 1986.49.517.16.

Fig. 10.10. Winter village in floodplain forest along the Missouri River. After Karl Bodmer (Swiss, 1809–1893), Narciso Edmundo Jose Desmadryl, engraver, *Winter Village of the Minatarres*, 1841, hand-colored aquatint and engraving on paper, Joslyn Art Museum, Omaha, Nebraska, Gift of the Enron Art Foundation, 1986.49.542.26.

that we hardly notice them more than the cattle in our pastures about our homes. But this cannot last; even now there is a perceptible difference in the size of the herds, and before many years the Buffalo, like the Great Auk, will have disappeared.

The importance of the work of Catlin, Bodmer, and Audubon cannot be overstated in documenting the natural history of the upper Missouri River and the culture of the people living there. The work of Catlin and Bodmer was done three years before the smallpox epidemic of 1837. Audubon arrived after the epidemic but in time to see and record the wildlife before much of it disappeared. Collectively, their wildlife and landscape paintings provide a benchmark against which to measure even more dramatic transformations of the river yet to come. Among the many birds that Audubon described was the western meadowlark, nearly identical to the eastern species except for its song (fig. 10.11).

Steamboat traffic on the Missouri River boomed and diversified during the following few decades because of new boats with shallower drafts and more powerful engines. This modern technology stimulated a brisk fur and hide business, transported army troops, and supplied Indians with goods promised as part of treaties (box 10.2). The export of wildlife was massive. In 1848 Joseph La Barge made the round trip from St. Louis to Fort Union in only 65 days. His cargo on the return trip included 1,700 bales of buffalo robes, 260 packs of furs, quantities of salted buffalo tongue, and a menagerie of live native animals, including bison, bear, antelope, elk, and beaver tethered or penned on the main deck.[28]

Dakota Territory was formed in 1861, with Yankton as its first capital. Fur merchants in Sioux City reported that hunting and trapping by Euro-Americans were replacing the traditional Indian food-gathering system. Exports became a mix of wild and pastoral products, signifying a shift in local economies. In 1882 one shipper loaded wool, sheep pelts, cattle hides, wolf pelts, and the hides of bison, pronghorn, deer, elk, beaver, and mountain sheep.

Box 10.2. Unintended Consequences

People were not the only passengers on boats traveling upstream from St. Louis to the Mandan villages and beyond. Norway rats introduced to America from Europe disembarked an army keelboat at the Knife River villages in 1825, and more rats arrived on boats that came later. Initially these newcomers were welcomed by the Mandan because they killed the native deer mice that plagued their lodges. However, the rats could dig deep tunnels that enabled them to raid buried food caches. Soon there were too many rats. A fur trader living at nearby Fort Clark killed 1,686 rats in one year, from June 1836 to May 1837. Swapping a native mouse for an introduced rat was a bad deal, as rats were eating 250 pounds of corn per day. By 1837 the Mandan faced three life-threatening problems: a scarcity of bison, disease, and lean corn reserves.[a]

[a]Wood et al. 2011; Fenn 2014.

Box 10.3. Embalming with Cottonwood

The most successful voyage led by the indomitable steamboat captain Joseph La Barge was marred by the tragic death of passenger Captain W. D. Speer, a British Army officer on furlough from India. The officer was mistakenly shot dead in the dark by a sentinel as he ascended the steps to the hurricane deck of the steamship *Octavia*. La Barge was worried that the body would decompose in the warm summer weather before it could be returned to St. Louis. In La Barge's words: "I put off the remains of Captain Speer at Fort Buford [near present-day Williston] to await my return. I asked the commanding officer if he could suggest any way of embalming the body. He advised the construction of a large box and the filling of it with green cottonwood sawdust. The experiment seemed to work well, although I had never heard of such a thing before." The sawdust apparently slowed the decomposition of the corpse sufficiently for La Barge to cruise onward upstream to Fort Benton, return to pick up the coffin, and deliver the Captain's remains weeks later to authorities in St. Louis.[a]

[a]Chittenden 1962.

Fig. 10.11. The western meadowlark was the first new bird species discovered by Audubon in the Dakotas. The meadowlark is the state bird of North Dakota. Shown is Audubon's painting of the eastern meadowlark, virtually indistinguishable from the western species except for its song. *Meadow Lark from Birds of America* (1827) by John James Audubon (1785–1851), etched by Robert Havell (1793–1878). Digitally enhanced by Rawpixel, September 21, 2017. Licensed by Creative Commons Attribution-Share Alike 4.0 International license.

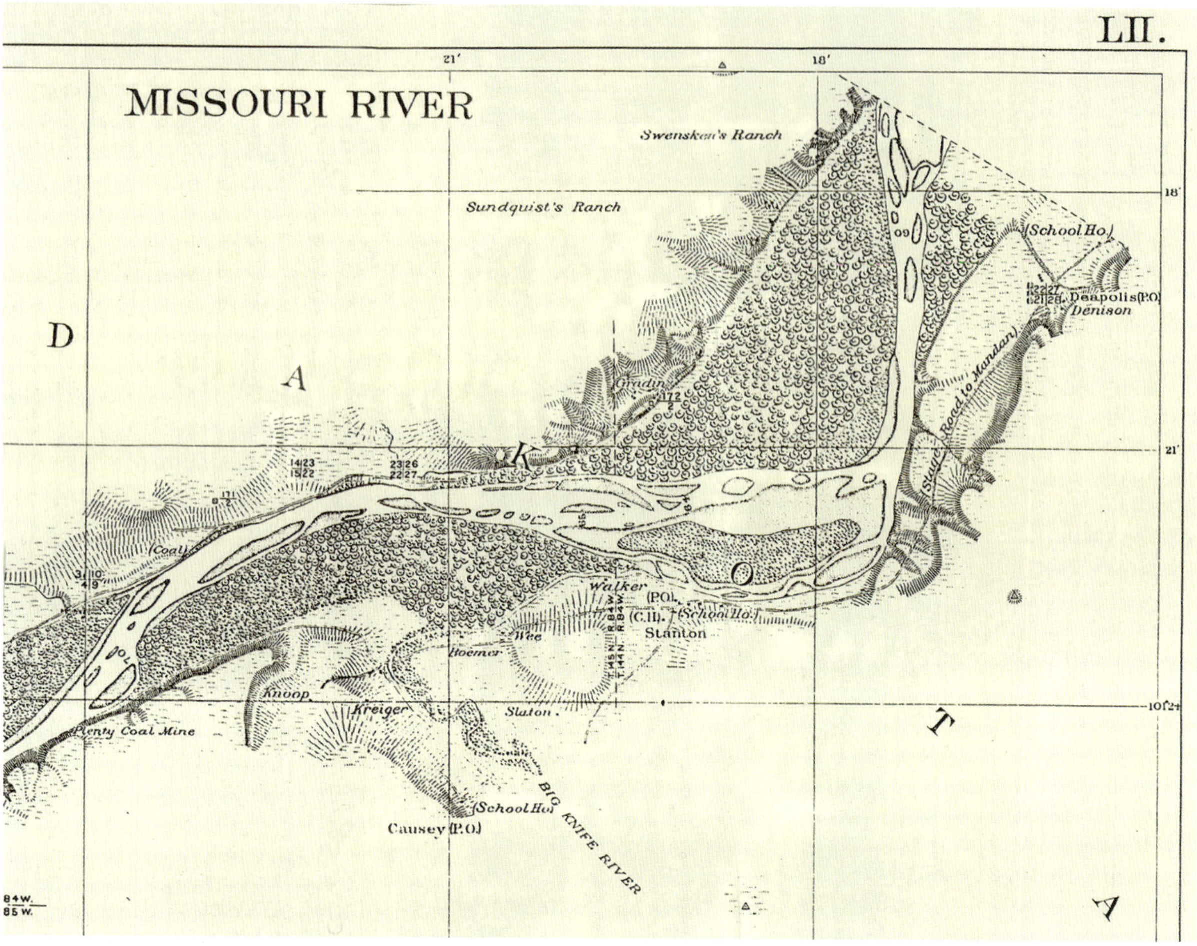

Fig. 10.12. Section of the Army Corps of Engineers 1889 map of the Missouri River valley, including the historically significant Big Knife River tributary. The age of the extensive floodplain forest is shown by different symbols: fine stippling—cottonwood and willow seedlings on new islands and sandbars; small to large circles—young to old forests dominated by cottonwood and peachleaf willow. Note the coal mine on the left. Source: Missouri River Commission 1894.

Rising land values, the completion of railroads, and the advance of Euro-Americans accelerated the killing of wild animals. Contributing to the near extinction of large mammals were drought, severe winters, cattle-borne diseases, and the construction of fences that interfered with movements of nomadic species, especially bison.

Responding to the hazards of river travel, Congress passed the Rivers and Harbors Act in 1882, which led to separate management plans for the Missouri above and below Sioux City. The Missouri River Commission was established in 1894 and published the first professional maps of the river from St. Louis to Three Forks, Montana (fig. 10.12). In 1912, work began to stabilize riverbanks, remove snags, and dredge and maintain a six-foot-deep navigable channel to improve commercial shipping between Sioux City and St. Louis. The Missouri in the Dakotas was not channelized, but the era of steamboats and wildlife export ended when more efficient railroads and gas-powered boats arrived.

Trees and Floods

Lewis and Clark's journals confirm that the Missouri River floodplain was generally well timbered throughout the Dakotas. Exceptions were near Indian villages, where forests had been cleared for cropland or harvested for fuel and construction material. Again, quoting from Meriwether Lewis's journal: "The country [northern South Dakota] generally consists of low, rich, timbered ground." A month later, in central North Dakota, he wrote that timber was scarce near the villages and "was supplied from the opposite side of the river, where it was and still is abundant" (October 1804, present-day central North Dakota).

Of the tree species recorded in the explorers' journals, the cottonwood was mentioned by far the most often, revealing its dominance. This tree was critical to the success of the expedition because it was the only one large enough and with light enough wood to make canoes to navigate the shallow, upstream sections of the river. It was also used to build and heat Fort Mandan in the bitter winter of 1804–1805. Meriwether Lewis wrote, "The cottonwood which is so abundant in this country . . . arrives at a great size, grows extremely quick the wood is of a white colour, soft spngey [*sic*] and light preogues [pirouges] are mostly made of these trees."

The cottonwood also was heavily utilized by American Indians for lodge poles and as winter food for their horses, which would eat the soft, corky branches. To the surprise of the captains, horses fed cottonwood branches could endure long periods of bison hunting without consequences. The ethnobotanist Melvin Gilmore wrote in 1914 about mystic properties attributed to the cottonwood, a tree that appeared "always so self-reliant, showing such prodigious fecundity, its lustrous young leaves in springtime by their sheen and by their restlessness reflecting the splendor of the sun . . . it is said the air is never so still that there is not motion of cottonwood leaves. Even in still summer afternoons, and at night when all else was still, they could ever hear the rustling of cottonwood leaves by the passage of little vagrant currents of air. And the winds themselves were the paths of the Higher Powers" (fig. 10.13).[29]

The Missouri River in its natural condition flooded frequently and meandered across its floodplain on a massive scale (fig. 10.14). Floods created considerable heterogeneity on the floodplain, including point bars, meander scars, oxbow lakes, and cut banks. Typically, two surges in natural streamflow occurred each year, the first in April during local snowmelt and rainfall on the plains and the second in June, fed largely by melting snow in the Rocky Mountains. Lewis and Clark were keen observers of nature and often noted how channel shifting and sandbar formation favored the reproduction and dominance of cottonwood and willow on the floodplain. On the return trip two years later, Clark wrote: "I observe a great alteration in the Corrent course . . . where there was Sand bars in the fall 1804 at this time the main Current passes, and where the current then passed is now a Sand bar—Sand bars which were then naked are now covered with willow Several feet high. the enteranc of some of the Rivers & Creeks Changes owing to the mud thrown into them, and a layor of mud over Some of the bottoms of 8 inches thick" (August 20, 1806).

The first complete inventory of floodplain vegetation along the Missouri in the Dakotas was that of the General Land Office Survey (1872–1881). Thousands of "witness" trees along survey lines and at section corners were identified by species, their diameters measured, and locations specified to facilitate relocating survey lines if the surveyors' markers were later removed or destroyed. Earlier inventories of specific sections of the river also are available.[30]

These survey notes and plat maps confirmed what Lewis and Clark saw, namely that most of the bottomland was densely wooded. Several types of forests were identified. About 80 percent

Fig. 10.13. An unusually large plains cottonwood (20 feet in circumference) along the Missouri River near the Cross Ranch in North Dakota, and a close-up photo of its leaves.

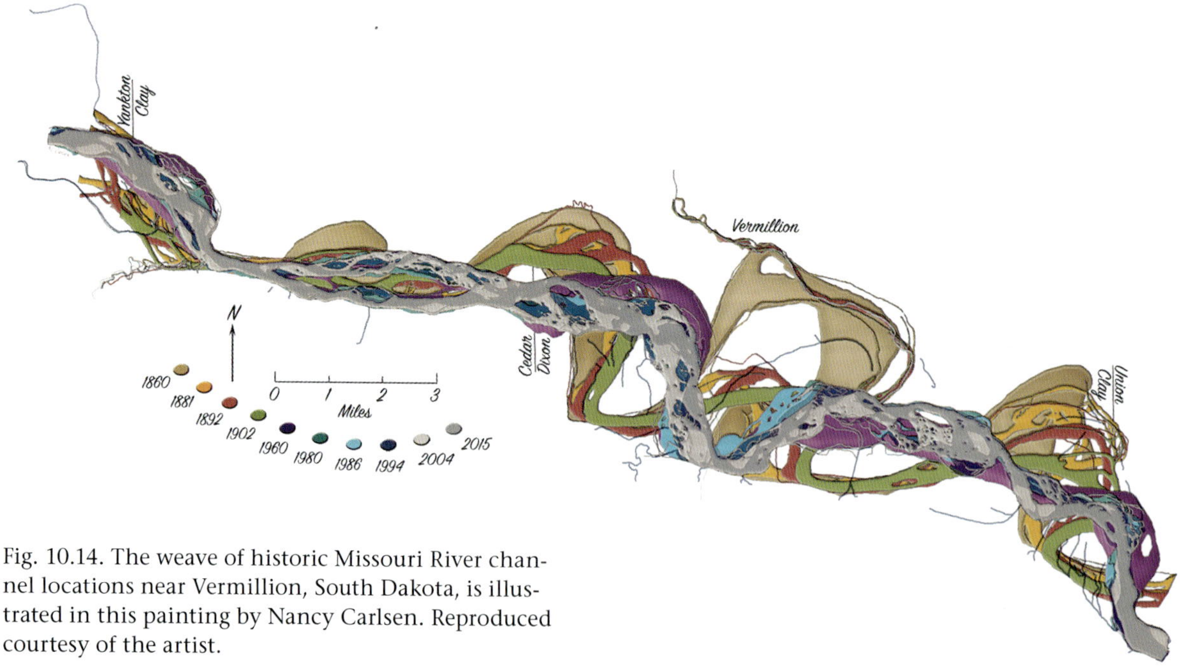

Fig. 10.14. The weave of historic Missouri River channel locations near Vermillion, South Dakota, is illustrated in this painting by Nancy Carlsen. Reproduced courtesy of the artist.

of the floodplain was dominated by plains cottonwood, peachleaf willow, and various willow shrubs, ranging from young, dense stands on low benches nearest to the river channel to old stands of giant cottonwoods four to six feet in diameter on higher benches. Other trees included American elm, green ash, and boxelder. Bur oak occurred in upland woody draws but rarely on the active floodplain. The elm, ash, and boxelder forest type occurred on high benches that had escaped erosion by the river for a century or more.

Research has determined that the forests along the wild Missouri can change in a successional sequence, though one that could be interrupted. Pioneering stands of young cottonwood and willow on shorelines sometimes persisted long enough to be replaced by American elm, green ash, and boxelder. Cottonwood gradually became less common because its seedlings do not normally survive in the understory environment, unlike the seedlings of the other species. However, the channel is constantly shifting, often toppling trees into the river as cut banks erode. If this happens, the cottonwood becomes established again, or meadows or shrublands develop. Thus, the movement of the river across its floodplain creates a continuously shifting mosaic of woodlands, shrublands, and meadows of different ages.[31]

Aquatic Life

The wild river's reworking of the floodplain brought leaves, branches, and even whole trees into the channel, which provided food and habitat for invertebrates, fish, and other aquatic life.[32] Toppled trees that lodged downstream created riffles, quiet water, and shelter beneficial to different species. As the floodplain mosaic shifts, so does the mosaic of the river itself—measured in terms of channel depth, streamflow velocity, the texture of bottom sediments, and overhanging banks, logs, and snags. Habitats range from a deep, fast-flowing, changeable, and turbid main channel to shallower off-channel oxbow lakes, sloughs, and back-water marshes. Occupying these habitats are about 50 different kinds of fish.[33] Top predators are the channel catfish and pallid sturgeon (fig. 10.15), both growing to over 100 pounds and adapted for catching prey in murky water. The shovelnose sturgeon, sauger, and paddlefish feed on bottom-dwelling insects, minnows, and plankton. Some species feed primarily on plants, such as the river carpsucker, bigmouth and smallmouth buffalo, freshwater drum, white sucker, and redhorse sucker. The quillback sucker, goldeye, and longnose and shortnose gar are found in relatively clear water. Dozens of small fish, primarily in the minnow family, occupy shallow, off-channel habitats. The flathead chub is the most common minnow, but other species include the plains minnow, brassy minnow, silvery minnow, sicklefin, and sturgeon chub.

Many species of fish have morphological adaptations for life in rivers. Bottom-feeding fish adapted for fast, turbid waters have flattened heads with mouths pointing downward, a humpbacked body shape, enlarged pectoral fins, reduced eyes, and well-developed chemosensory, whiskerlike organs known as barbels (sturgeon, chubs, buffaloes, carpsuckers, blue suckers, catfishes, burbot, and freshwater drum). Clear water in the present-day river, the result of sedimentation in the calm water of reservoirs, has negatively affected many native riverine species, such as paddlefish, burbot, silver chub, plains minnow, western silvery minnow, blue catfish, and sauger. Other fish have become more common, especially sight-feeding species, such as skipjack herring, gizzard shad, white bass, bluegill, walleye, white crappie, emerald shiner, and red shiner. Of particular concern are new invaders from Asia, for example, bighead and silver carp, which are rapidly appearing in the lower Missouri River, including southeastern

Fig. 10.15. The pallid sturgeon is a rare and endangered species in the Missouri River. Photo by Sam Stukel, U.S. Fish and Wildlife Service.

South Dakota. These invaders are likely to have impacts on the fish and invertebrate community, as they have on other rivers.

Overall, the wild and dynamic Missouri River provided "something for everyone," such as sandbars required for nesting by the least tern, piping plover, and killdeer; young and middle-aged cottonwoods and willows used by people as well as the eastern and spotted towhee, Baltimore oriole, and yellow warbler; and forests for the scarlet tanager, great crested flycatcher, and blue jay.[34] About 35 species of mammals are frequently found in the valley, with the meadow vole, least shrew, and meadow jumping mouse found in young woodlands. Older trees attract the deer mouse, white-footed mouse, short-tailed shrew, fox squirrel, white-tailed deer, raccoon, and bobcat. The beaver, long-tailed weasel, and mink are associated with woodlands near water. Bison, pronghorn, and elk would spend most of their time in upland prairie, but like many species they depended on the river as well. People also found food and shelter on or near floodplains. Of the floodplain forest that remained after prehistoric farming and the cutting of wood for steamboats, 38 percent was cleared and farmed between 1881 and 1938.[35]

About 70 percent of the remnant floodplain has been cultivated.

Taming the Missouri River

Economic and ecological catastrophes early in the twentieth century, including the Dust Bowl, led to massive federal engineering projects that would transform the Missouri River valley and the lives of its residents. Six earthen dams and their reservoirs (fig. 10.16 and 10.17, table 10.2) were constructed by the Army Corps of Engineers to achieve the following objectives: reduce flood damage to farmland, towns, and cities; store water to irrigate crops; store water upstream that could be released to improve navigation downstream; generate hydropower to stimulate the region's economy; provide construction jobs for returning veterans from World War II; and improve the river for what, at the time, seemed to be more desirable opportunities for outdoor recreation.[36] The six reservoirs have been compared to a string of pearls: deep, cold-water artificial lakes connected by strings of original floodplain now mostly protected from floods. About 75 percent of the combined river length in the Dakotas is now reservoir, 630 miles of the 853 total miles of river. Towns and cultural artifacts were destroyed; thousands of people were displaced. Garrison Dam submerged 156,000 acres of the Fort Berthold Reservation; Oahe Dam submerged 160,000 acres of the Standing Rock and Cheyenne River Reservations.[37]

The reservoirs are novel ecosystems. Most fish species either were introduced for sport fishing, such as the Chinook salmon, or were formerly uncommon native species that thrived with the

Fig. 10.16. Fort Randall Dam and its reservoir (Lake Francis Case), one of six impoundments on the upper Missouri River built by the U.S. Army Corps of Engineers. Photo by Darrell Napton.

Fig. 10.17. The famous big bend of the Missouri River in central South Dakota, now part of Lake Sharpe behind Big Bend Dam. In the 1800s, steamboat passengers would disembark to walk across the narrow neck of land, thereby lightening the boat so it could move upstream more rapidly along this part of the river. Photo by Greg Latza, PeopleScapes Photography.

Table 10.2. Dams and reservoirs on the Missouri River[a]

Dam	Miles upstream	Reservoir name	Began	Closed	Filled	Storage (acre-feet)
Canyon Ferry	2250	Canyon Ferry Lake	1949	1953	1955	1,997,900
Fort Peck	1767	Fort Peck Lake	1933	1937	1940	18,700,000
Garrison	1387	Lake Sakakawea	1946	1953	1954	23,000,000
Oahe	1071	Lake Oahe	1948	1958	1962	23,500,000
Big Bend	987	Lake Sharpe	1959	1963	1966	1,900,000
Fort Randall	878	Lake Francis Case	1946	1952	1953	5,500,000
Gavins Point	811	Lewis & Clark Lake	1952	1955	1957	492,000

[a] Source: National Research Council 2011 (except for the Canyon Ferry Dam).

aid of stocking, such as the walleye. The spottail shiner and rainbow smelt are exotic species introduced to be forage fish for salmon and walleye. Native warm-water species such as the shovelnose sturgeon, pallid sturgeon, and sauger are largely absent from reservoirs except in the warmer, shallow water found near deltas, which form in the upstream portions of the reservoirs and at the mouths of tributaries. The sharp decline in the abundance of the pallid sturgeon following dam construction caused this fish to be listed federally as a rare and endangered species. It is not known

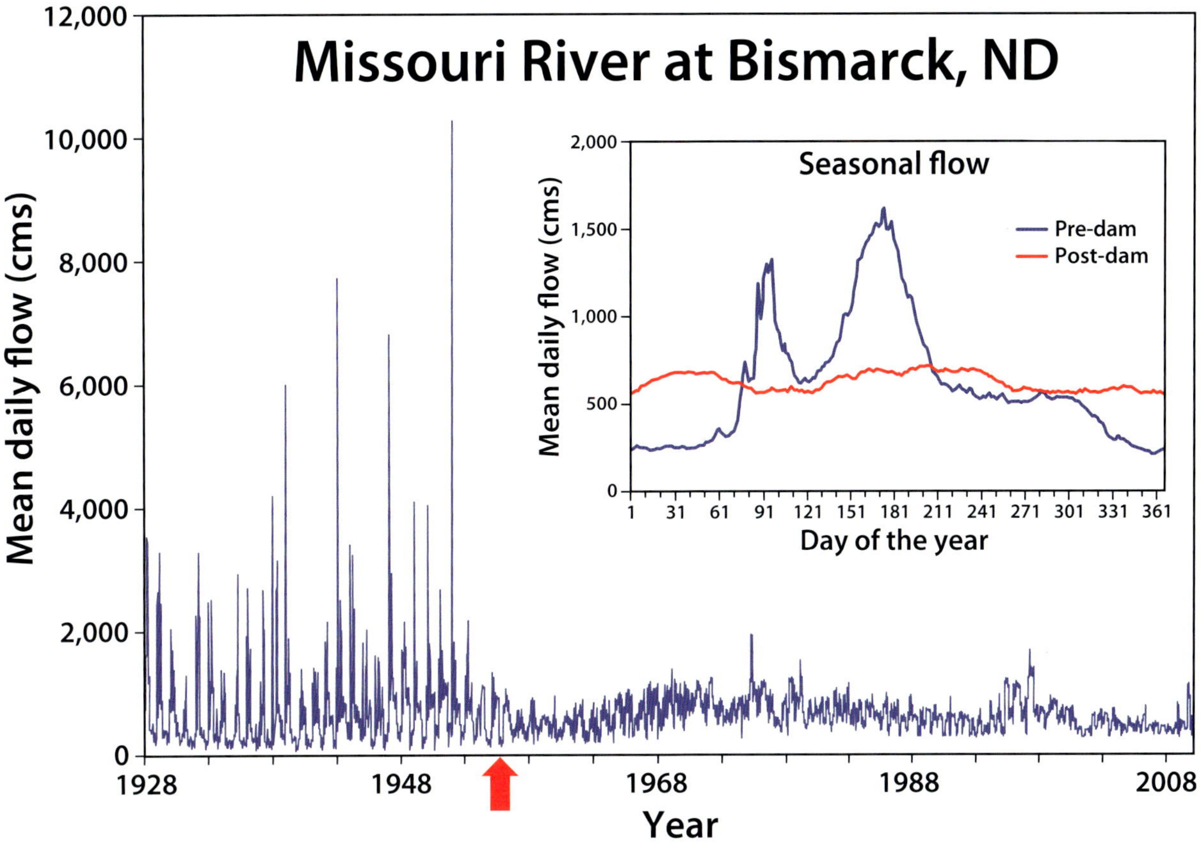

Fig. 10.18. Mean daily streamflow (cubic meters per second) per year for the Missouri River at Bismarck, North Dakota, before and after construction of the Garrison Dam. The arrow marks the time of dam closure. The inset shows mean daily flow each day of the year before and after dam closure. Adapted from Johnson et al. (2014).

if any fish species were extirpated, as no surveys were done prior to dam construction.[38]

The outward appearance of the remnant reaches, with riparian forests and shrublands, leaves the impression that what is left of the river is healthy and self-perpetuating. Nothing could be further from the truth.[39] The lifeblood of a river is the natural flow, floods, and sediment regime. All have been drastically altered by the presence and operation of the dams (fig. 10.18). Just before the large dams were built, about 125 million metric tons of sediment per year were transported by the river past Yankton. The sediment load of the Missouri since the dam has been built is only 1 percent of that in the wild river, which causes channel degradation and incision in the remnant reaches.[40] An incised channel moves less, even with high flows, and more flow is required to overtop its banks to irrigate the floodplain. Channel movement is further limited by bank stabilization using riprap and revetment. The effects of flood suppression and sedimentation in reservoirs has far-reaching downstream effects, as sediment is vital to the land-building process in the Mississippi River delta.

The environmental changes caused by the dams are consequential to all native riverine species. The clear water that is released below dams is too cold for the native warm-water fish, especially those species that are adapted to forage in turbid water. With flood suppression, there is a loss of the sandbars required for successful cottonwood establishment and the nesting of birds such as the threatened piping plover and endangered interior least tern. Other significant impacts are caused by interrupting the exchanges of water, nutrients, and woody debris between river channel and floodplain that historically maintained the high diversity and productivity across a wide range of habitats. The now-dominant cottonwood community, a legacy of floods from before the dam, will be greatly diminished in less than a century as a result of the near absence of sandbar habitat for its reproduction. In its place could grow the invasive eastern redcedar, Russian olive, and saltcedar.[41]

The first comprehensive study of a riparian forest remnant on the Missouri was carried out in 1969–1971 in central North Dakota along a 100-mile long reach between Lake Sakakawea and Lake Oahe.[42] After examining 34 study areas, two predictions were made about the long-term effects of dams: first, cessation of flooding would slow or stop channel meandering and the creation of sandbars required by cottonwood and willow for seedling establishment; and second, green ash would become dominant because it was reproducing more readily than cottonwood, willow, American elm, and boxelder, leading to a future forest of lower diversity.

These predictions were evaluated some 40 years later by comparing aerial photographs and remeasuring the vegetation of the 30 remaining plots.[43] The data supported the first prediction: aerial photographs showed clearly that the river channel had moved very little in the 65 years since the closure of Garrison Dam, causing a dramatic reduction in sandbar formation and cottonwood reproduction. Aging cottonwood trees were being replaced by other species. The second prediction also was supported, but two introduced beetles have cast doubt about the future of green ash. The emerald ash borer kills ash trees and was introduced to Michigan from China in 2002 (see chapter 7). At this writing, the borer is expanding westward, having entered southeastern South Dakota in 2017. The second problem, Dutch elm disease, spread by another beetle, was not present in 1969 but by now has killed most older American elm trees in the study area. American elm was once second to cottonwood in abundance and size along the Missouri. Boxelder seedling survival has declined also, probably because of the absence of moisture from spring floods. All dominant tree species are or could be affected adversely in the future. Lewis and Clark would scarcely recognize the river they wrote about 220 years ago.

While the remaining sections of the historical Missouri River ecosystem are deteriorating, the reservoirs are heavily visited by anglers and others seeking water-based recreation. The walleye is now the most sought-after fish. Thousands of anglers and tourists visit the reservoirs each year, adding millions of dollars to local economies in the form of lodging, guiding services, fuel, bait, licenses, and food. However, expensive stocking is required to maintain native species as well as those that have been introduced (chinook salmon, kokanee salmon, brown trout, rainbow trout, lake trout, and steelhead trout).[44] Stocking is especially important for the recovery of the endangered pallid sturgeon.

Reclaiming Elements of the Wild Missouri

The dams on the Missouri River were completed before passage of the National Environmental Policy Act (NEPA) in 1969 and the Endangered Species Act in 1972. Consequently, the potential adverse effects were not considered as thoroughly as they would have been today. Engineers knew, of course, that floods would be reduced greatly in frequency and magnitude. That was a primary objective. They also probably knew that the movement of fish and other organisms would be restricted, causing declines in some species. That was a trade-off. Perhaps they did not consider the implications of converting the river from a relatively warm, turbid-water fishery to a cold, clear-water fishery—the result of sedimentation in the reservoirs and cold-water releases from the dams.[45] Also, strong disagreements between upstream and downstream states may not have been expected. Water management priorities in upstream states center on spring flood suppression and maintenance of high reservoir levels to support fish and wildlife. Priorities for the downstream half of the river are to improve shipping during low-water periods in the fall.

Short of removing the dams, it will be impossible to truly restore the Missouri River, but steps have been taken to maintain or reclaim elements of the ecosystem that are now highly valued. A relatively free-flowing reach of the river upstream from Fort Peck Dam in Montana is now protected by the National Wild and Scenic Rivers Act, and two reaches in South Dakota and Nebraska have limited protection: a 39-mile segment from Fort Randall Dam to Lewis & Clark Lake and a 59-mile segment downstream from Gavins Point Dam.[46] More specifically, the Army Corps of Engineers has made attempts to improve the habitat of three species now protected by the Endangered Species Act: the pallid sturgeon, least tern, and piping plover. Also, some riparian woodlands with eagles nesting in large cottonwoods are now protected.[47] On the lower Missouri River, the Corps has constructed warm-water side channels designed to favor the pallid sturgeon and their numbers are augmented with hatchery-raised fish. Thus far, no successful sturgeon reproduction has been documented.[48]

Small-scale restoration projects also have been initiated, such as the building of sand islands for tern and plover habitat, but they have been criticized as inadequate to address the scale of the problem. Other more systemic projects have been proposed, such as prescribing a more natural flow regime, including a higher but still modest spring flood peak, releasing warmer surface water from the reservoirs rather than deep cold water, conducting prescribed flood experiments, passing sediment around the dam, and flushing reservoirs of sediments to supply current shortages downstream. Also, on tributaries, if not the Missouri itself, some bank stabilization structures or dams could be removed to allow the channel to move as it did historically. Unless some of these remedies are implemented, a report published by the National Research Council predicts continued deterioration.[49]

Forty years after dam completion, a surprise development has been the formation of nine sizable deltas where the Missouri and its larger tributaries drop sediment into the mainstem reservoirs (fig. 10.19 and 10.20). New patches of cottonwoods and willows have become established on some of these deltas, mitigating to some degree the massive losses of riparian woodlands when the reservoirs were filled.[50] A downside of mainstream deltas has been backup flooding and property damage in Bismarck and Pierre.

Another unexpected development was a major flood that occurred in 2011, caused by record snowmelt in the northern Rocky Mountains and torrential rains in eastern Montana.[51] Water was released from the reservoirs at an unprecedented rate to protect dam and reservoir shoreline infrastructure, but flows exceeded flood stage on many segments of the river from June to September. Even this flood was not enough to initiate significant channel movement in the remnant reaches, as floods of this magnitude did in the past. Without channel movement the prospects for successful cottonwood reproduction are discouraging. Scientists concluded that the degrading channels below dams need to be elevated to floodplain level and that bank stabilization structures need to be removed before floods can be effective in moving the channel and creating new sandbars. The Army Corps of Engineers could take such steps with earth-moving machinery, but debates

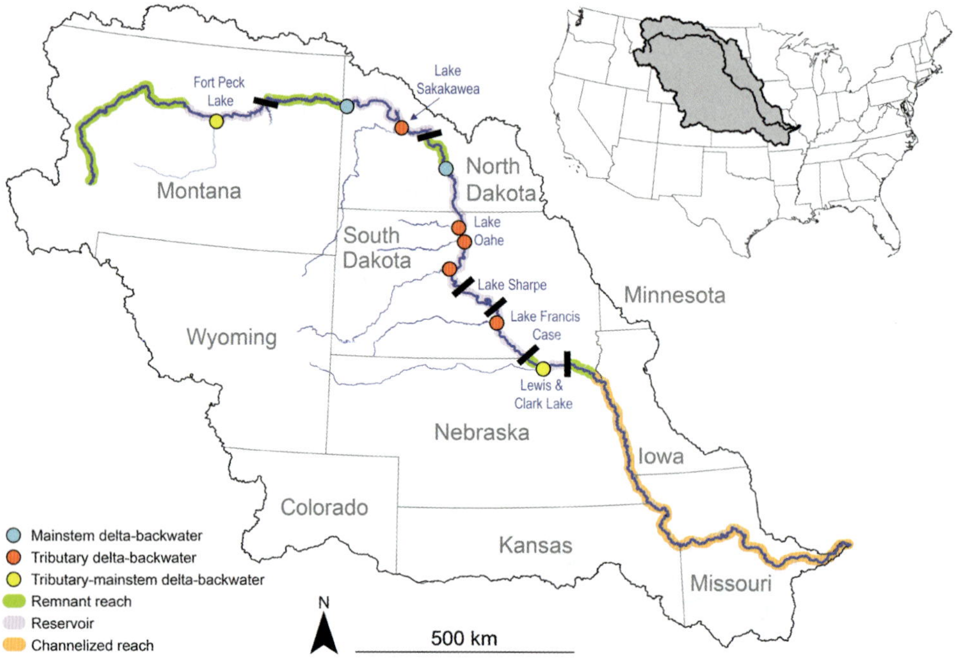

Fig. 10.19. The locations of nine major deltas in the upper Missouri River, which form where rivers flow into reservoirs. Adapted from Volke et al. (2015).

Fig. 10.20. The White River delta with a wooded lobe extending into Lake Francis Case on the Missouri River. The delta extends upstream about 20 miles to where sediment deposition ceases. Dominant trees on the delta are plains cottonwood and peachleaf willow; coyote willow is a common shrub. Photo by Gray Tappan.

among numerous stakeholders will make agreements difficult to achieve.

Climate Change

Rapid climate change now complicates everything. With less snowmelt in the Rocky Mountains and lower inflows to the reservoirs, a higher proportion of the water will be allocated for legislatively mandated uses, namely, hydropower, municipal needs, power-plant cooling, and downstream navigation. Less water will be available to improve habitat for species of concern and outdoor recreation. Slower recovery of reservoir levels following drought is expected, leaving boat docks and ramps high and dry for longer periods. Periodic wetter periods also have happened as predicted. In addition to the 2011 flood, another record flood occurred in 2019 (on the middle Missouri River, downstream from the Dakotas).

Even without climate-induced flow reductions, water demands from the Missouri are expected to intensify. Currently very little Missouri River water is used for cropland irrigation, compared to most western rivers. A warmer and drier climate would change that if farmers become dependent on irrigation. Also, the demand for air conditioning during warmer summers will increase demand for hydroelectric power. At the same time, the population of the Dakotas is increasing, with more cities tapping into Missouri River water. Piping that water to other regions experiencing severe droughts is likely. With less water, greater variability in inflows, and changing demands, the Army Corps of Engineers will need to revise the Master Manual—an enormously complicated and politically charged process involving seven states and numerous public interest groups. The historic "surplus" of unallocated water that now flows through the Dakotas is likely to be short lived. Because of climate change, the past century is no longer a template for future water management.[52]

In summary, the Missouri River has a rich and fascinating history. Native Americans relied heavily on the river's natural resources. When the first explorers of European descent arrived thousands of years later, the upper Missouri River ecosystem was described in glowing terms for its panoramic beauty and incredible wildlife. European-style agriculture, industry, and water resource management led to great transformations. Natural ecosystem benefits were diminished. Dam construction has been the most transformational event on the river since the glaciers. The extent to which the long-term impacts of the dams can be mitigated depends on the effects brought about by climate change and a willingness to reclaim and preserve some free-flowing sections of North America's longest and most iconic river.

Chapter 11 Prairie Wetlands and Lakes

Some airline passengers are surprised by the numerous lakes and wetlands that are visible as they fly over the Dakotas. Large or small, whether lakes, marshes, fens, or wet meadows, they contribute significantly to the economy of the region because of the abundant fish and wildlife found in such places, their importance for livestock watering, and their role in recharging groundwater. Most were formed in millions of depressions left by receding glaciers, especially on the Missouri Coteau and Prairie Coteau (fig. 11.1). Lakes with clear, deep water are crowded with homes and cabins on the shore, some of the most highly valued real estate in the region. Upslope wetlands affect the lakes by capturing and filtering surface runoff, and fish move back and forth between the two.

Thus, lakes and wetlands are highly interactive and are considered together in this chapter. Wetlands are discussed first, beginning with an overview of the benefits they provide. This is followed by a description of the various kinds of wetlands and their biodiversity, how climate and other factors cause wetlands to change through time, and whether wetland reconstruction can mitigate the problems resulting from wetland drainage. The chapter ends with a discussion of how lakes are formed, factors affecting water quality, and lakes' propensity for winterkill. The surprisingly rapid expansions of Waubay and Devils Lakes are described, as are the probable effects of climate change in the future.

Prairie Wetlands

Fish and Wildlife

Some of the best fishing and waterfowl hunting in North America is found in the Prairie Pothole Region (PPR) of North America, which includes the glaciated portions of the Dakotas (fig. 11.2). All kinds of animals can be found, but birds are the most conspicuous. Best known are various kinds of ducks. Also common are grebes, coots, bitterns, herons, and rails, some of which are secretive and seldom seen. Medium-sized mammals are year-round residents, such as mink, muskrat, and beaver. Many other mammals forage in wetlands but spend most of their time on the uplands, such as the red fox, coyote, racoon, and striped skunk. Some species, large and small, live mostly in wetland margins or on nearby uplands. White-tailed deer survive bitter winters in areas with scarce woodland by seeking shelter in frozen but tall cattail marshes, thereby minimizing exposure to the cold night sky. In the spring some of the smallest but most vocal vertebrates are amphibians, especially the leopard frog, chorus frog, and spring

Fig. 11.1. Wetlands and lakes on the Prairie Coteau were created by continental glaciers, as described in chapter 2. Photo by Craig Novotny.

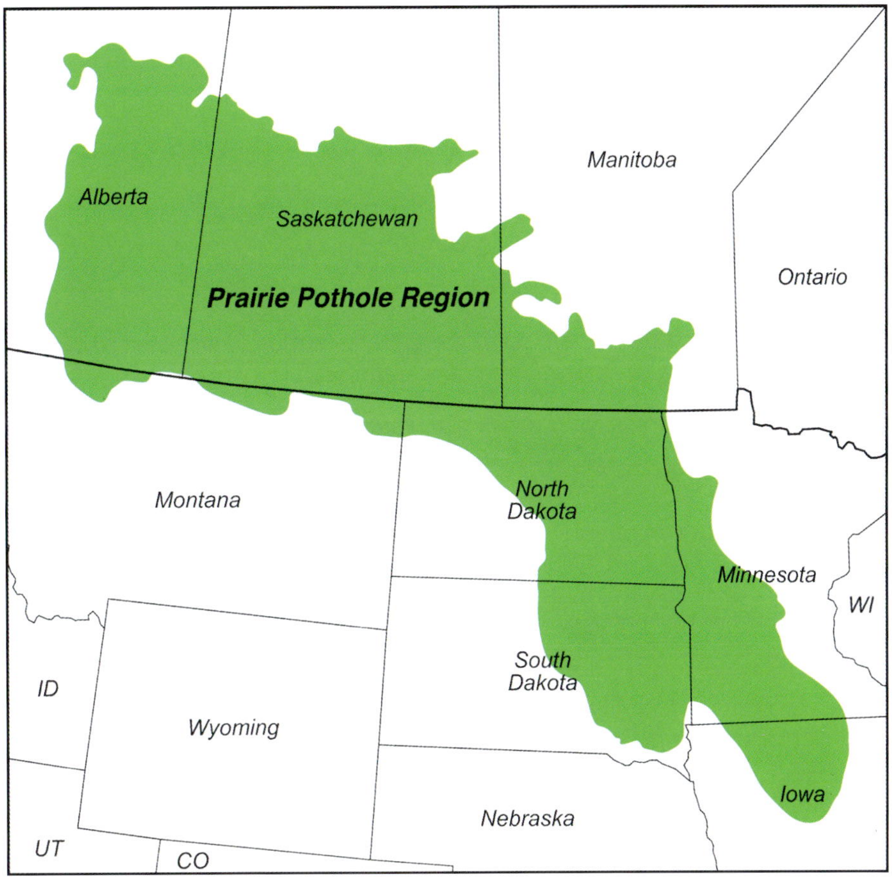

Fig. 11.2. The Prairie Pothole Region includes thousands of lakes, ponds, marshes, fens, and wet meadows. Some wetlands in the Dakotas are commonly referred to as potholes or sloughs. The region is described by waterfowl biologists as the nursery of North American ducks. Adapted from Renton et al. (2015).

peeper. Algae and numerous wetland plants form the energy base for the entire ecosystem. Often mentioned as critical for water birds are invertebrates such as amphipods, fairy shrimp, brine shrimp, and mussels. Popular game fish in nearby lakes also depend on wetlands. For example, yellow perch and northern pike swim to accessible marshes, where they attach their sticky eggs to submerged plants that later provide escape cover for young fish (known as fry).[1] Many, if not most, of these fish—those not stocked by game and fish departments—began life in these shallow waters. Clearly, the loss of wetlands through drainage is a catastrophic loss for outdoor enthusiasts and biodiversity. Nearly half the birds found throughout the Dakotas use wetlands for breeding or foraging.[2]

The wildlife of wetlands can be conspicuous and raucous. Well-known examples are yellow-headed and red-winged blackbirds that nest in cattails. Others, though rarely seen, have loud spring courtship calls that are varied and hard to forget. Favorite examples are frogs, toads, sedge wrens, rails, and bitterns. With the arrival of millions of ducks and geese in the spring and fall, an iconic natural wonder unfolds that has inspired many naturalists, young and old.[3] The blue-winged teal arrive from wintering grounds as far away as South America, the canvasbacks from Chesapeake Bay, and the northern pintails from wetlands along the Gulf of Mexico. The PPR, though comprising only 10 percent of North America's duck nesting habitat, produces from 50 percent to 80 percent of the ducks hatched each year on the continent (figs. 11.3, 11.4, 11.5, table 11.1, box 11.1).[4]

Away from the PPR, south and west of the Missouri River, natural wetlands are less common.

Fig. 11.3. Waterfowl during spring migration on the Prairie Coteau. Different kinds of ducks and geese migrate together. The seven species visible in this photo are the mallard, gadwall, green-wing teal, lesser scaup, ring-necked duck, northern pintail, and white-fronted goose. Photo by Dennis J. Larson.

Blue-winged teal

Pintail

Canvasback

Fig. 11.4. Blue-winged teal, canvasbacks, pintails, and other kinds of ducks (see table 12.1) nest in the Prairie Pothole Region after migrating long distances from their winter ranges. Photos by Kent C. Jensen.

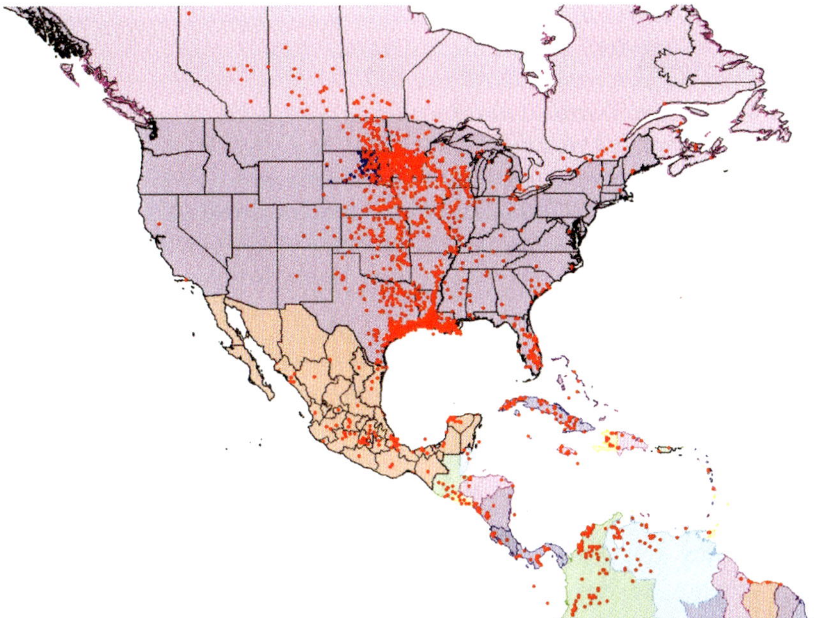

Fig. 11.5. Each red dot is a recovery location of a blue-winged teal that was numbered and leg-banded in South Dakota. Most bands were collected in the Prairie Pothole Region but others were recovered in South America, Canada, and Alaska, confirming the importance of the Prairie Pothole Region to Western Hemisphere waterfowl populations. Courtesy of the South Dakota Department of Game, Fish, and Parks (Jensen et al. 2014).

Table 11.1. Some common ducks and other birds that nest in or near the wetlands of the Prairie Pothole Region.

WATERFOWL	
American wigeon	*Anas Americana*
Blue-winged teal	*Anas discors*
Canada goose	*Branta canadensis*
Canvasback	*Aythya valisineria*
Gadwall	*Anas strepera*
Green-winged teal	*Anas crecca*
Hooded merganser	*Lopodytes cucullatus*
Lesser scaup	*Aythya affinis*
Mallard	*Anas platyrhynchos*
Northern pintail	*Anas acuta*
Northern shoveler	*Anas clypeata*
Redhead	*Aythya americana*
Ring-necked duck	*Aythya collaris*
Ruddy duck	*Oxyura jamaicensis*
Wood duck	*Aix sponsa*
OTHER BIRDS	
Avocet, American	*Recurvirostra americana*
Bittern, American	*Botaurus lentiginosus*
Blackbird, red-winged	*Agelaius phoeniceus*
Blackbird, yellow-headed	*Xanthocephalus xanthocephalus*
Coot, American	*Fulica americana*
OTHER BIRDS, continued	
Godwit, marbled	*Limosa fedoa*
Grebe, eared	*Podiceps nigricollis*
Grebe, horned	*Podiceps auritus*
Grebe, pied-billed	*Podilymbus podiceps*
Grebe, western	*Aechmophorus accidentalis*
Harrier, northern	*Circus cyaneus*
Heron, black-crowned night	*Nycticorax nycticorax*
Heron, great blue	*Ardea herodius*
Killdeer	*Charadrius vociferus*
Phalarope, Wilson's	*Steganopus tricolor*
Plover, piping	*Charadrius melodus*
Rail, sora	*Porzana carolina*
Rail, Virginia	*Rallus limicola*
Sparrow, savannah	*Passerculus sanwichensis*
Tern, black	*Chlidonias niger*
Willet	*Catoptrophorus semipalmatus*
Wren, marsh	*Cistothorus palustris*
Yellowthroat, common	*Geothlypis trichas*

Source: Kantrud et al. 1989. Knopf and Samson (1997) list the endemic vertebrates of the Great Plains, including fish, amphibians, reptiles, birds, and mammals.

The climate is drier and most depressions have eroded away on the much-older land surfaces. Shallow, temporary wetlands are found here and there where runoff accumulates during wet periods, providing resting habitat for migrating birds. To provide a more reliable source of water for thirsty cattle, ranchers have dug thousands of stock ponds to expose shallow groundwater, and small dams are often constructed on intermittent streams. The grasslands surrounding such places are favorable for waterfowl nesting if they are not overly grazed.

The numbers of water birds in western South Dakota and southwestern North Dakota during migration may be lower than in the PPR,[5] but many birds use them when those in the east are dry. At such times, the more widespread western grasslands are used for nesting. Notably, nest success may be higher in the western Dakotas because of less predation, such as by red fox, raccoon, and skunk. While constructed wetlands do not provide all the benefits of those that are natural, they are of considerable value for wildlife.

In addition to waterfowl, 35 species of shorebirds

Box 11.1. Summer Nesters

Semipermanent wetlands occupy a small portion of Dakota landscapes, but they are conspicuous in the spring and early summer because of colorful birds and a cacophony of sounds. The highly vocal yellow-headed and red-winged blackbirds defend their territories from all intruders, including humans. Others are seen less frequently, because of their secretive behavior and camouflaged feathers, but are heard from long distances—such as the laughing sora, chatty sedge wren, and resonant bittern. Ducks will quack vociferously early in courtship, but they are relatively quiet later when hens are swimming across open water with ducklings close behind. There could be diving ducks such as the regal canvasback, the perky lesser scaup, and the stiff-tailed ruddy duck. Dabbling ducks like the familiar mallard, or the swift blue-winged teal and streamlined pintail, will nest in grassy uplands, walking their broods to water immediately after fledging, sometimes a mile away. Other open-water birds could be the white-billed American coot, chick-toting western grebes, and the diminutive pied-billed grebe. Flying overhead, quiet but conspicuous, are likely to be a few black terns and a solitary northern harrier. Foraging along the shore may be various herons, egrets, sandpipers, and phalaropes, though they nest elsewhere. Even louder than the birds on some warm evenings will be choruses of frogs and toads. During extended droughts, or with drainage, sedimentation, water pollution, and invasive plants, such places become disturbingly quiet.

either breed or migrate through the Northern Great Plains, mostly using the shallow parts and mudflats of natural and constructed wetlands. The favorable mix of grasslands and wetlands provides essential habitat for many species of waterbirds that are conservation priorities, such as the whooping crane, snowy egret, marbled godwit, American avocet, green heron, Wilson's phalarope, pied-billed grebe, and American bittern. Also, 11 species of amphibians and reptiles live in prairie wetlands (fig. 11.6, table 11.2). A noteworthy example is the tiger salamander, which prospers in seasonal and semipermanent wetlands. Larval salamanders sometimes exceed 2,000 per acre and are often seen on roads when moving en masse between wetlands in the fall.[6] Larval salamanders grow rapidly, reaching an average weight of 12 ounces by June (fig. 11.7). Because of their numbers and appetite, they reduce the density of macroinvertebrates, an important food for waterbirds. But tiger salamanders have predators, too, such as fish and herons. One means of escape is to occupy brackish wetlands, as they are more tolerant of saline water than fish.

Lakes and wetlands provide habitat for a number of commercially valuable plants and animals. Baitfish, frogs, and crayfish are caught and sold to anglers throughout the region and beyond. For example, fathead minnows, abundant in deeper wetlands and shallow lakes, have been shipped to bait shops as far away as Lake Erie.[7] The wholesale value of the minnow harvest is millions of dollars. Also, income from trapping muskrats and mink for their fur can be significant, and profits from the sale of native wetland plant seed, such as prairie cordgrass, wedgegrass, and sedges, can exceed that from domestic grains.[8] Proceeds from license fees and federal excise taxes paid by hunters, anglers, recreational boaters, and even target shooters help fund fish and wildlife conservation and recreation projects. Other funds come to the region for the maintenance of National Wildlife Refuges, National Grasslands, and Waterfowl Production Areas, among others.[9] State governments and nonprofit organizations also fund wildlife conservation. It is not difficult to calculate the economic value of wetland wildlife; they are worth the investment.

Water Filtration, Retention, and Water Quality

Less widely appreciated and more difficult to evaluate monetarily are the ways that wetlands filter

Yellow-headed blackbird

Red-winged blackbird

Sedge wren

American bittern

Green-backed heron

Virginia rail

Avocet

Long-billed curlew

Fig. 11.7. The aquatic larval form of the western tiger salamander, a common resident in prairie wetlands, especially those that are slightly or moderately brackish. Photo by Dave Mushet, U.S. Geological Survey.

sediments and nutrients from the water, and how they reduce undesired flooding, detoxify some pesticides, and sequester carbon.[10] Filtration is accomplished with the help of gravity by particles of silt and clay that accumulate on the bottom. Water that passes through these sediments is cleaner. Sedimentation has accelerated with plowing and cultivation, but the filtering capacity of wetlands continues until the basin is filled. Filtering of nutrients occurs when nitrogen, phosphorus and other elements are absorbed by wetland plants, including algae. While nitrates are beneficial for crop production, they are a pollutant in water. This filtration process is especially valuable in removing the high levels of nutrients applied to croplands that otherwise would wash into lakes, causing algal blooms and poor conditions for swimming

Fig. 11.6. Eight nongame birds found in Dakota wetlands. Photos by Kent C. Jensen.

Table 11.2. Amphibians and reptiles found in wetlands of the Prairie Pothole Region

Tiger salamander	*Ambystoma tigrinum*
Leopard frog	*Rana pipiens*
Chorus frog	*Pseudacris nigrita*
Wood frog	*Rana sylvatica*
Spring peeper	*Pseudacris crucifer*
American toad	*Bufo americanus*
Great Plains toad	*Bufo cognatus*
Dakota toad	*Bufo hemiophrys*
Painted turtle	*Chrysemys picta*
Snapping turtle	*Chelydra serpentine*
Plains garter snake	*Thamnophis radix*
Red-sided garter snake	*Thamnophis sirtalis*

Source: Kantrud et al. (1989) and other reports. Knopf and Samson (1997) list the endemic vertebrates of the Great Plains, including fish, amphibians, reptiles, birds, and mammals.

and boating. Anaerobic bottom sediments enable another mechanism for improving water quality, namely, denitrification, a process that converts nitrate (NO_3) into nitrite (NO_2) and finally into molecular nitrogen, a harmless gas (N_2).

By their very nature, wetland basins collect water. Some *fill and spill*, whereas others have no outlet and are drawn down by evaporation and subsurface leakage. Obviously, a landscape with many wetlands will retain much of the water and sediment that otherwise would become part of a creek, river, lake, or reservoir. As a result, wetlands function to lessen flood magnitude, frequency, and duration.[11] Research shows that having between 3 percent and 7 percent of a watershed in wetlands can minimize floods and improve water quality. The severity of flooding in the Dakotas in recent decades has been attributed in part to the loss of wetlands by drainage. These places include Waubay and Devils Lakes and the Big Sioux, James, Souris, and Red Rivers. Restoration of drained wetlands is one means proposed to reduce both the flooding in the Mississippi River and Red River watersheds and the size of the nutrient-enriched "dead zones" in the Gulf of Mexico and Lake Winnipeg.[12]

Carbon storage and detoxification are benefits that are less well known. Wetlands accumulate large amounts of carbon-rich plant detritus that decomposes slowly in anaerobic sediments underwater.[13] Restoring wetlands that have been drained increases the amount of carbon sequestered in a landscape. Wetlands also release methane, a more effective greenhouse gas than carbon dioxide, but methane has a lower concentration and shorter residence time in the atmosphere.[14] Knowing the net effect of carbon storage and methane production on climate change is challenging, as is confirming the potential of wetlands to detoxify pesticides.[15]

Defining Wetlands and the Prairie Wetland Complex

Wetlands are transitional between terrestrial and aquatic ecosystems, places where the water table is usually at or near the land surface or the basin is covered at times by shallow water.[16] But more than water is required. The U.S. Department of Agriculture specifies that, for legal protection, a wetland must have three attributes:[17]

- A predominance of hydric soils, those known to develop only after many years of submergence
- Inundation or saturation with water at a frequency and duration sufficient to support a prevalence of hydrophytic plants adapted for saturated soil in the area
- A prevalence of hydrophytes under normal circumstances

Defining wetlands in this way minimizes inevitable controversies when commercial developments are proposed where wetlands are protected.

Many factors contribute to the various kinds of wetlands that are now recognized. They include the duration, depth, and source of water along with geologic substrate and climate. To make sense of this variation, ecologists Robert Stewart and Harold Kantrud identified five categories based on hydroperiod (length of time standing water is normally present each year) and water quality (fresh or brackish):[18]

- Temporary, usually dry by early summer
- Seasonal, usually dry by midsummer
- Semipermanent, with water throughout most growing seasons
- Alkaline, variable hydroperiods but with brackish water
- Fens, consistently wet from springs but with little surface water

An obvious feature of prairie wetlands of all types are concentric vegetation zones. The zones differ in water depth, hydroperiod, and the hydrophytes that are present (figs. 11.8, 11.9, 11.10, table 11.3). Plants in semipermanent wetlands are tallest in deepwater marshes, including the well-known cattails and bulrushes rooted in the bottom but with emergent stems and leaves. Floating and submerged aquatic plants also are found, such as duckweed and pondweed. Shorter plants become more common in shallow water near the shore,

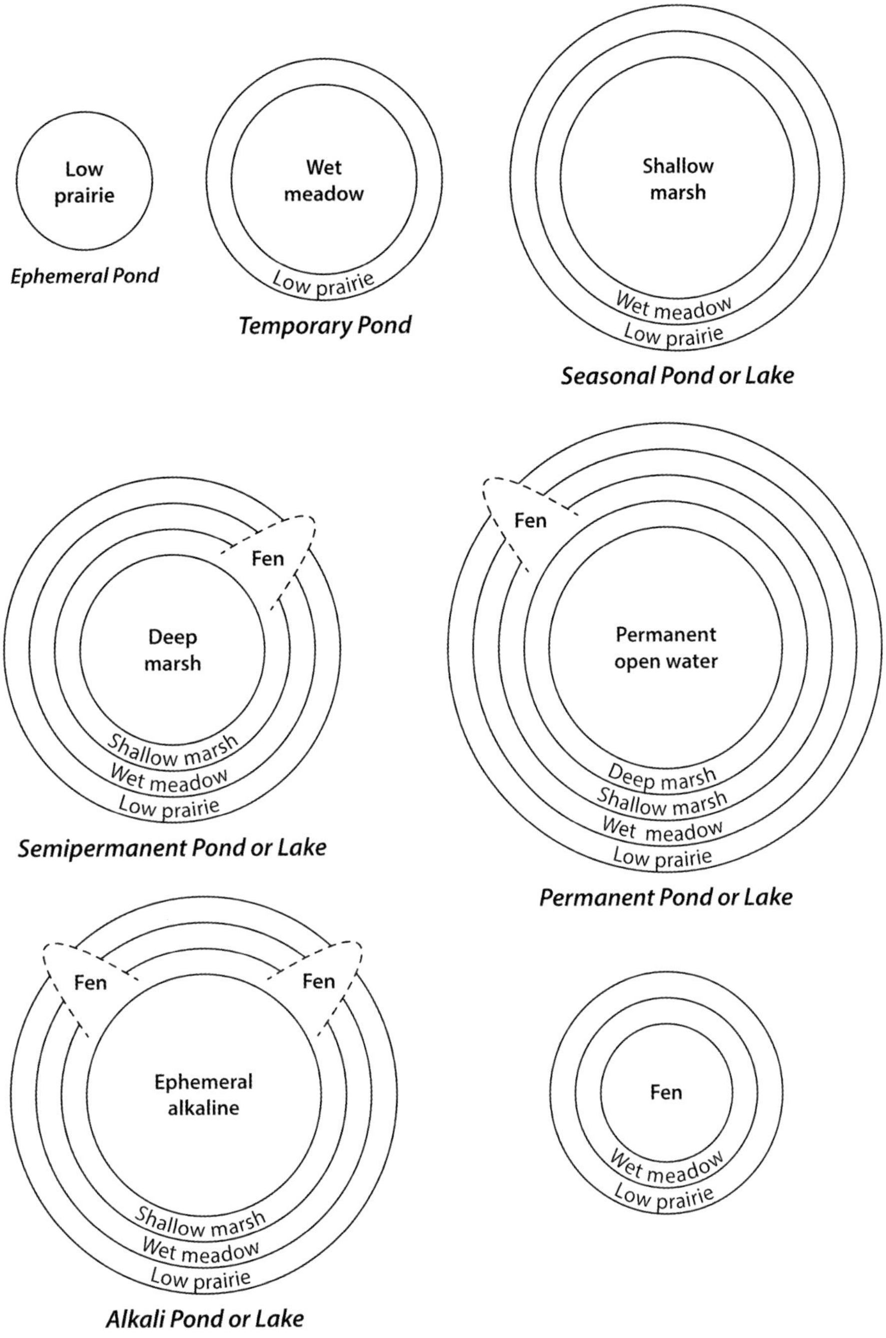

Fig. 11.8. Arrangement of wetland zones across a range of wetland types defined by hydroperiod (the typical duration of surface water). The most complex zonation is found in semipermanent wetlands; temporary wetlands have only one zone. Lakes are distinctive by having water that is too deep for most aquatic vascular plants (hydrophytes), although seasonal and semipermanent wetlands with emergent plants are common in shallow bays. Adapted from Stewart and Kantrud (1971).

Fig. 11.9. A small semipermanent wetland, commonly referred to as a prairie pothole, with four concentric zones defined by different plants: submerged coontail and pondweeds in the center, narrow-leaved cattail on water's edge, a zone with bur reed, and a zone adjacent to the upland prairie with woolly sedge.

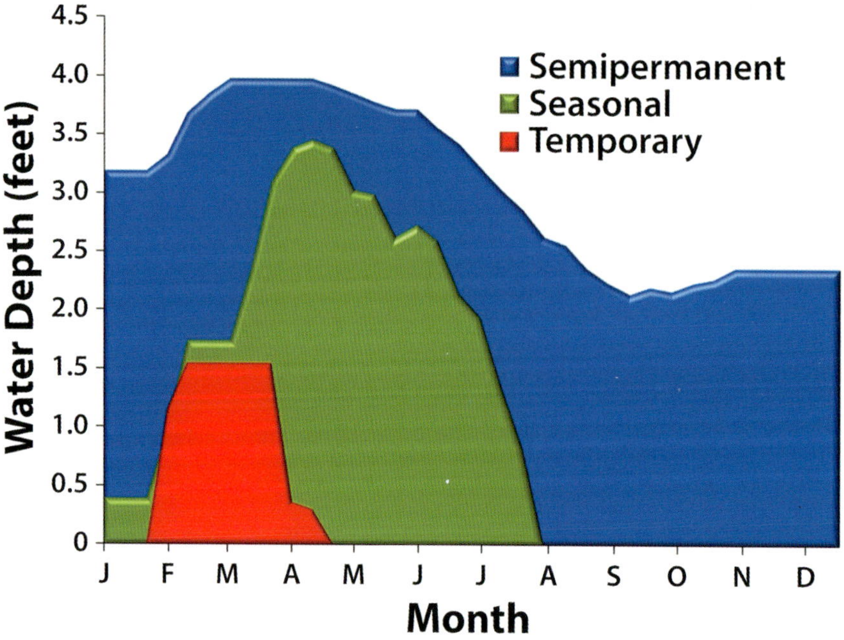

Fig. 11.10. Comparison of the length of time surface water is present for the three main wetland types in a typical year. Many animals move from temporary wetlands to seasonal and semipermanent wetlands during the course of the summer. Adapted from Johnson et al. (2005).

Table 11.3. Some characteristic plants in different kinds of wetlands

TEMPORARY WETLANDS & WET MEADOWS	
Aster, panicled	*Symphyotrichum lanceolatum*
Bluegrass, fowl	*Poa palustris*
Canarygrass, reed	*Phalaris arundinacea*
Cordgrass, prairie	*Spartina pectinata*
Daisy, white doll	*Boltonia asteroides*
Reedgrass, northern	*Calamagrostis stricta*
Rush, Baltic	*Juncus arcticus ssp. littoralis*
Sedge, clustered field	*Carex praegracilis*
Sedge, Sartwell's	*Carex sartwellii*
Sedge, woolly	*Carex pellita*
SEASONAL WETLANDS & SHALLOW MARSHES	
Bur reed	*Sparganium eurycarpum*
Mannagrass, American	*Glyceria grandis*
Sedge, wheat	*Carex atherodes*
Sloughgrass, American	*Beckmannia syzigachne*
Smartweed	*Polygonum coccineum*
Spikerush, common	*Eleocharis palustris*
Spikerush, pale	*Eleocharis macrostachya*
Water plantain, northern	*Alisma trivale*
Whitetop	*Scolochloa festucacea*
SEMIPERMANENT DEEP-MARSH EMERGENTS AND SUBMERGENTS	
Bulrush, hardstem	*Schoenoplectus tabernaemontani*
Bulrush, slender	*Schoenoplectus heterochaetus*
Cattail, broad-leaved	*Typha latifolia*
Cattail, hybrid	*Typha xglauca*
Cattail, narrow-leaved	*Typha angustifolia*
Pondweed, sago	*Stuckenia pectinata*

Source: Stewart and Kantrud 1971; Johnson et al. 2004. See Marriott and Faber-Langendoen (2000a) for wetland plant communities in the Black Hills, and Smeins (1967) for wetland vegetation of the Red River Valley and adjacent drift prairie.

such as river bulrush, bur reed, prairie cordgrass, and several kinds of sedges.

Water salinity is affected by various factors, namely, the chemistry of the glacial substrate and whether or not a wetland loses water and dissolved salts by spillage and leakage into groundwater, or alternatively, retains salts by losing water primarily through evaporation. Saline wetlands are more common in drier parts of the PPR. In general, biodiversity is lower in saline wetlands because much of the flora and fauna is intolerant of significant salt concentrations (fig. 11.11).

The most frequent disagreements about classification pertain to temporary wetlands. They are the most numerous, have the shortest hydroperiod (fig. 11.12), and are usually the shallowest and smallest in size, all suggesting that they might be less important. But the term *temporary* betrays their ecological value. Waterfowl biologists have discovered that, because they thaw earlier in the spring than semipermanent and some seasonal wetlands, invertebrates such as fairy shrimp, clam shrimp, and tadpole shrimp emerge in time for migrating waterfowl and shorebirds. Not surprisingly, the food web of temporary wetlands is populated by organisms that can complete their life cycle in a few weeks, move to nearby wetlands, or become dormant.[19] Mobile species fly or walk; others tolerate extended periods of desiccation or produce drought-tolerant eggs. Some animals routinely rely on a complex of wetland types, such as waterfowl. Similarly, the leopard frog, the most common amphibian in prairie wetlands, requires about 90 days of ponded water to develop from egg to adult, but courtship occurs in temporary wetlands. Even when dry, temporary wetlands provide nesting habitat for ducks, prairie grouse, various songbirds, and pheasants.

Temporary wetlands often occur at higher elevations in the landscape, from where water seeps into groundwater and other wetlands at lower elevations.[20] Salts are exported along with the leaked water, leaving fresher water in temporary wetlands. Groundwater below temporary wetland basins in dry years may drop by as much as

Fig. 11.11. Brackish semipermanent wetland with typically sparse emergent vegetation. Bulrushes are common in brackish wetlands, cattails in freshwater wetlands.

Fig. 11.12. A temporary wetland, on the left, and a temporary pond. The wetland is dominated by smooth-cone sedge, wooly sedge, spike rush, smartweed, and reed canary grass. The taller, light-colored patches are reed canary grass, an invasive plant in the wet-meadow zone of prairie wetlands. The pond in the middle of a cornfield is in a swale that had been cultivated. Swales and temporary wetlands are often used by migrating waterfowl.

10 feet by the end of the growing season. This loss of groundwater must be recharged each spring for surface water to form and persist.

Farmers have found that shallow wetlands are relatively easy to drain with ditches or tile. Also, they cultivate some of these wetlands in dry years without any penalties. The hydrophytes are lost, especially when herbicides are used, but seeds, invertebrate eggs, and waterfowl respond in years when wetlands are too wet for crops—or when a grain crop is lost due to preharvest flooding.

Seasonal wetlands, like temporary wetlands, also warm up early. Their deeper water and longer hydroperiod makes it possible for some vertebrates in some years to complete their life cycles there. Nesting ducks can number as high as 750 pairs per square mile in wet years.[21] Seasonal wetlands increase in abundance to the west and northwest as aridity increases.

Wetlands of different types usually occur in groups or complexes and are connected by groundwater, surface water, and exchanges of organisms and their propagules (fig. 11.13, box 11.2). To illustrate, temporary and seasonal wetlands are especially important for waterfowl in the spring; but nearby semipermanent wetlands are critical later in the year, when the taller emergent plants provide hiding cover for vulnerable young birds and molting adults that are flightless (fig. 11.14).[22] In autumn, semipermanent wetlands and lakes provide important feeding and resting areas for migrating birds. The wildlife value of the wetland complex is diminished if one or more of the wetland types is degraded or destroyed.

The Special Case of Fens

Fens are defined as wetlands with quaking organic soils that usually are saturated throughout the year by groundwater emanating from springs, often on the fringes of semipermanent wetlands (see fig. 11.8).[23] Small pools sometimes form but they cover a small proportion of these wetlands. There is no conspicuous wet-dry cycle, and there is little variation in plant and animal life from year to year. Most Dakota fens are calcareous with a neutral or alkaline pH because most of their water passes through glacial till that includes limestone, dolomite, or sandstone particles high in calcium. This distinguishes them from bogs that have an acidic pH and high peat content. The only true sphagnum bog known to exist in the Dakotas is

Box 11.2. Watchful Water Birds

Once arriving on breeding grounds, migratory water birds instinctively evaluate where to settle and build nests. Their survival and that of their progeny is on the line. The characteristics of the potential nest site and surrounding landscape are scrutinized. The number, size, and water levels of nearby wetlands is important, as is food availability, the amount of grassland and cropland, and the risk of disturbances, predation, and competition from other nesters. This approach to evaluating the landscape mosaic was verified by wildlife biologist David Naugle and his associates in 1999. Black terns are more numerous in landscapes with high wetland density and where grasslands have not been tilled for agriculture. They forage widely, as much as several miles from their nests. Larger water birds such as ducks fly more rapidly and can scout much larger areas in a short time. Areas experiencing drought or with scarce grassland are flown over to find more favorable conditions elsewhere. Later in the breeding season, some ducks walk their broods a mile or more from shallow wetlands that are drying up to those with water. Choices made by breeding ducks exposed to highly variable field conditions from year to year produce frequent geographic shifts in nesting populations. Large numbers of ducks may breed one year in the Prairie Pothole Region of the Dakotas, while in the next year a drought forces most birds to continue north to the cooler prairies and parklands of Canada. Where they settle each year remains an intriguing part of the mystery of migration. How climate change in the future will affect historical patterns is difficult to predict.

Fig. 11.13. Prairie wetlands occur naturally in complexes consisting of different wetland types, from temporary to semipermanent. Rugged, rocky terrain has left most of the grassland intact in this example from the northern part of the Prairie Coteau. Photo by Kurt Forman, U.S. Fish and Wildlife Service.

Fig. 11.14. Two examples of semipermanent wetlands with approximately equal proportions of open water and emergent plant cover. Except in drought years, these wetlands have relatively deep, continuous surface water. They are slowest to warm up in spring because of thick ice during the winter. Commonly referred to as marshes or sloughs, semipermanent wetlands are the most diverse biologically because they vary greatly in water depth and salinity from one wetland to another.

on Turtle Mountain. It has nine rare plants, some of which have been found on fens in the Prairie Coteau.[24] The plant diversity of fens is higher than all other kinds of wetlands.

Most fens are only a few acres in size and rarely do they make up more than one percent of a wetland complex.[25] In the Dakotas, they are found on the Missouri and Prairie Coteau, situated on the slopes of moraines (fig. 11.15) and on slopes with exposed sand and gravel. Some occur in isolated pockets along the margins of saline ponds and lakes. Fens occur in the Black Hills and along the ancient beaches of Glacial Lake Agassiz but are less common in drier parts of the Dakotas.

Another characteristic of fens is the low concentration of nitrogen and phosphorus associated with natural spring water.[26] The distinctive plants of many fens may be attributable to their adaptations for tolerating nutrient deficiencies. Indeed, nutrient enrichment from agricultural runoff or atmospheric deposition probably enables other wetland plants to displace them, another threat to the persistence of fens and the rare species they harbor. Fens also are lost when the volume of spring water declines due to wetland drainage or groundwater pumping.

Water Fluctuations, Muskrats, Drought, and Fire

Early in the study of prairie wetlands, scientists discovered that some species, such as terns, ducks, and muskrats, are most abundant when the proportion of emergent plant cover in semipermanent wetlands was roughly equal to that of open water. This ratio changes in response to wide-ranging variation in weather and muskrat population sizes, a process now known as the cover cycle or the wet-dry cycle of a wetland (fig.11.16).[27] The four stages of the cycle are *dry marsh* with dense emergent plant cover and little or no surface water (though the soils may be wet), *regenerating marsh* when reflooding promotes vegetative reproduction and seed germination, *degenerating marsh* with more emergents and often larger muskrat populations, and *lake marsh* with more open water and relatively little emergent cover. Cycling through this sequence usually requires

Fig. 11.15. Fens receive their water from springs on slopes, which often causes slumping and the formation of small wetland terraces, such as here on the Prairie Coteau. The soil is high in organic matter and often springy underfoot. Very little open water is present. The trees in the background are hackberry.

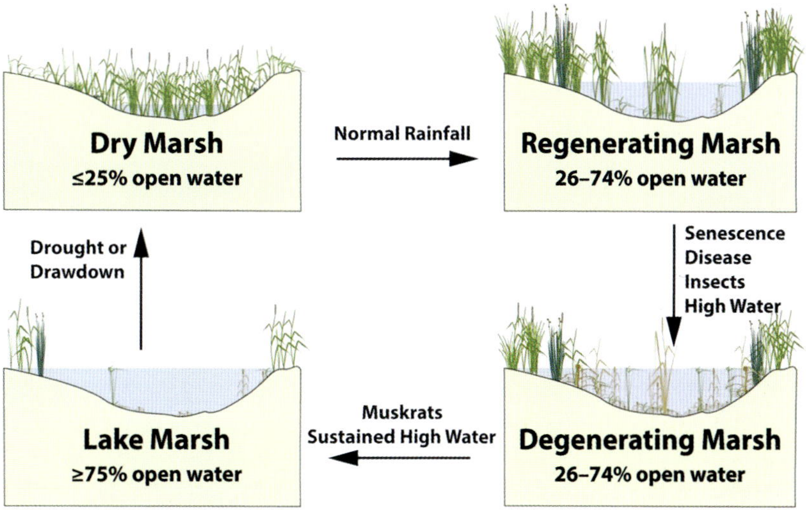

Fig. 11.16. Semipermanent wetlands often shift through four phases during a period of 10–20 years. Different water levels favor different plant and animal species. Muskrats can cause significant changes. Intermediate phases occur and some wetlands shift from one type to another during protracted wet or dry periods. Semipermanent wetlands have about equal proportions of open water and plant cover during the regenerating and degenerating marsh stages. Adapted from van der Valk and Davis (1978).

10–20 years, although the dry or lake marsh stages may persist during extended periods of drought or unusually wet weather.

Wetlands benefit from periods of drought, but why would it not be beneficial for wetlands to be consistently wet—as often is desired by hunters, bird watchers, and others? The dry stage is essential because decomposition of detritus is accelerated when water levels are low, soil temperature is warmer, and oxygen more abundant, which frees up nutrients in dead plant material and sediment for new plant growth when flooding reoccurs. The prodigious plant growth that follows is beneficial to other members of the food web, leading to a flush of algae, zooplankton, and invertebrates needed by fish, amphibians, and birds. The eventual, seemingly inevitable lake stage eliminates plants and animals that cannot tolerate deep water. But with time, water levels decline. Fluctuations in water level benefit species diversity, as different species become more abundant during different stages of the wet-dry cycle. Each plant group contributes seed, many of which remain dormant until new opportunities for germination return.[28] Thus, the biological diversity of the wetlands depends on such fluctuations.

Another factor affecting the wet-dry cycle and wetland habitat is the muskrat, a rodent known for rapid population growth.[29] Muskrats can have four litters per year, each with up to six or more young. Even their young sometimes bear a litter in their first year. Because they require an abundance of emergent vegetation for food and shelter, which varies greatly during the wet-dry cycle, their effect on wetlands is episodic. After long dry periods it may take several years for muskrats to recover. Moreover, muskrat survival is low when the water is shallow enough to freeze to the bottom, and predation rates are higher when muskrats are forced to emigrate across open ground to find deep water. In addition to cold winters and drought, muskrat population size is affected by disease and predation by mink, foxes, and fur trappers. Rapid changes in muskrat populations over just a year or two are legendary.

Notably, muskrats eat the lower stems and roots of deepwater emergents such as bulrushes and cattail, and they use the upper stems and leaves to construct dome-shaped lodges (fig. 11.17), some that are 6 feet high and 15 feet across. Large muskrat populations with as many as 20 lodges per acre remove dense plant cover to produce patches of

Fig. 11.17. Muskrat lodge constructed with bulrushes and cattails in a semipermanent wetland. Muskrat cutting of emergent plants creates open water and the lodges are used by numerous birds for nesting, resting, and foraging.

open water, which is beneficial to some birds. In fact, muskrat activity affects most wetland processes, including plant growth, decomposition, vegetation structure, and, indirectly, the dynamics of invertebrate and vertebrate populations. Closely related species use different parts of the muskrat-engineered marsh. For example, marsh watchers are familiar with the aerial, territorial battles between yellow-headed and red-winged blackbirds. The former strongly defend the more robust emergent cover in deeper water, while the vanquished redwings accept vegetation closer to shore, closer to terrestrial predators. Similarly, Forster's terns feed on fish and they select large, dry muskrat lodges for their nests, while black terns feed on invertebrates but usually choose smaller, wet lodges on which to nest. Muskrat lodges also are nest sites for waterfowl, particularly Canada geese. When muskrat populations are high, their "eat outs" can hasten the shift from the degenerating marsh stage to the lake marsh stage.

Unexpectedly, perhaps, fire is another important factor affecting wetlands. Wetlands are especially flammable whenever stems and leaves are dry. Fires were common in prehistoric times, having been started by people intent on driving or attracting game or by lightning strikes in the surrounding prairie.[30] When deeply flooded, prairie wetlands would not burn, providing safe places for humans and animals alike, but at other times flame intensity would have been frightening. Wetland fires are now much less frequent as a result of fire suppression and the conversion of grassland to farmland.

Fire affects plant growth, wildlife, and the wetland complex.[31] Burning increases growth by releasing nutrients such as phosphorus and potassium, and fall burns affect water levels by removing tall, dense cover that would otherwise trap drifting snow. That could be beneficial. However, winter cover for pheasants and other wildlife is less after such fires. Burning dense patches of cattails and common reed can increase biodiversity by creating habitat for other species, but managers know that prescribed burns should be scheduled so that nesting or winter cover are not destroyed in large portions of the wetland complex.

Learning from research and experience, wildlife biologists have adjusted the way National Wildlife Refuges and Waterfowl Production Areas are managed.[32] Established after the drought of the 1930s, water control structures were constructed to reduce the potential adverse effect of future droughts on ducks and geese.[33] The focus was on keeping wetlands wet. That changed after research found that waterfowl populations depended on the condition of surrounding grasslands as much as water levels. For a time, the focus was broadened to provide dense nesting cover, often through prairie reconstruction with introduced plants. Usually no livestock grazing or haying were allowed, and fires were suppressed whenever possible. But that favored invasive plants. Plant and animal diversity declined (see box 4.3). Today, prescribed fires and livestock grazing are employed to tip the balance in favor of native grasses and forbs, and some water control structures are being removed. Managers are optimistic that such initiatives will benefit waterfowl as well as other wildlife.

Wetland Drainage, Restoration, and Construction

Corn Belt farmers learned early in the twentieth century that shallow wetlands were productive for crops when dry enough to till.[34] This observation led to extensive drainage using open ditches and buried drainpipes (tile drainage). The speed and scale of drainage accelerated when large machines (fig. 11.18) were built to dig wider and deeper ditches. Tiling and ditching were encouraged with government funds. By the mid-1980s, the percentage of prairie wetlands that had been drained was astonishing: 90 percent in Iowa, 97 percent in Minnesota, and 50 percent and 35 percent in North Dakota and South Dakota, respectively.[35] In 1985 a concerned Congress passed laws intended to curb this practice, but drainage continues. The loss of wetlands in the PPR of the Dakotas has averaged 12,857 acres per year since the 1980s and 15,377 acres per year since 2001.

Ongoing drainage has been facilitated by *pattern field tiling*, a cheaper and more efficient method that uses plastic pipe buried in trenches. Thousands of drainage permit applications have been filed during the past decade.[36] These permits allow wetlands to be drained if they are not enrolled in certain USDA farm programs. Increased crop productivity has been demonstrated, but concerns were raised that protected wetlands might be inadvertently drained if tile was placed too close to a wetland or below the elevation of the wetland bottom.[37] Research on the environmental consequences of tiling has not caught up with the practice, but well-known effects are water pollution from fertilizers and downstream flooding.[38]

Scientists and many landowners now place great value on the ecosystem benefits provided by the wetlands that remain. One means to slow or even reverse the decline is to restore those that have been drained or otherwise degraded.[39] Restoration commonly begins by plugging the tile or an open ditch to create a hydroperiod similar to the original wetland. Once the natural water regime has been reestablished, many of the native plants regenerate from dormant seeds without human intervention; others require seeding or transplanting. Similarly, some animal species

Fig. 11.18. Ditching machine of the type used to drain wetlands in the early twentieth century in the Prairie Pothole Region. Courtesy of the Minnesota Historical Society.

arrive soon after restoration begins, bringing to mind the adage "if you build it they will come." Early colonists include mallards, teal, pheasants, coot, killdeer, blackbirds, and frogs (fig. 11.19). Many invertebrates fly independently to the new wetland, and others arrive attached to the feet or feathers of waterfowl. Successful restoration depends on whether the land surrounding wetlands is grassland or farmland and whether other natural wetlands are nearby to supply plant and animal colonists. For this reason, managers often prefer to restore wetlands located near others.

A reconstructed wetland can appear quite natural, but the diversity of uncommon plant and animal species rarely matches that of one that has not been drained.[40] Achieving a satisfactory restoration can be hindered by the availability and affordability of desired seeds from commercial suppliers. Also, some invertebrates with limited mobility are slow to recover. Another factor limiting biodiversity is the highly competitive nature of invasive plants, such as common reed, hybrid cattail, reed canarygrass, quack grass, Garrison foxtail, and Kentucky bluegrass—all of which suppress native species. Control measures are costly and often ineffective. For this and other reasons, prairie wetlands cannot be fully restored or replicated, a strong argument for preserving those that remain. Still, reconstructed wetlands are valuable for attenuating floods, supporting the food web, converting nitrates to atmospheric nitrogen, providing water and habitat for wildlife and livestock,

Fig. 11.19. Rapid repopulation by spring peeper and chorus frog tadpoles in a one-year-old wetland restoration project. Photo by Craig Novotny.

filtering and recharging groundwater, storing runoff, and sequestering carbon.

Aside from agriculture, wetlands often are destroyed if they are in the way of suburban expansion and the widening and rerouting of roads and highways. Wetland trading is a commercial, market-based method to mitigate such losses. Developers purchase newly restored or constructed wetlands to compensate, which provides funding for the restoration or construction of new wetlands that will be sold in the future—a form of mitigation known as conservation finance. Rules and regulations involving third-party oversight govern such transactions.[41] In addition, various governmental and nongovernmental agencies pay landowners to protect their wetlands. Some agreements prohibit wetland drainage in perpetuity, while others allow farming during dry years. Wetland losses have been slowed by such programs, but they continue.

Prairie Lakes

North Dakota professor Robert T. Young wrote about the life of Devils Lake,[42] commenting, "We can little comprehend the vast flood of water which passed this way from the southern and western front of the great ice sheet." As the last glaciers retreated, vast lakes formed in what is now the northern and eastern Dakotas. Some, known as Agassiz, Dakota, Minnewaken, Mose, and Souris, dried up thousands of years ago but their flat bottoms and shoreline markings are still visible today. Another legacy of glaciation is the many hundreds of smaller lakes that formed when huge chunks of ice buried in glacial till melted and the overburden collapsed to form a basin (till lakes). Others were created by torrents of meltwater flushing away the till, forming water-collecting basins (outwash lakes). These natural lakes are concentrated in the Prairie Coteau, Missouri Coteau, and on Turtle Mountain (fig. 11.20).

Lakes are defined generally as water bodies that average more than six or seven feet deep and about 20 acres in surface area.[43] Their depth drops during droughts, and some wetlands become lakes during extended wet periods. All but a few Dakota lakes dried up during the 1930s. Many lake bottoms were farmed until the water returned. Artificial lakes formed by earthen and concrete dams also occur, both in glaciated and in unglaciated terrain, but most are less than a few hundred acres in size. Exceptions are the five reservoirs on the Missouri River—the Great Lakes of the Dakotas (see chapter 10)—and moderate-sized reservoirs like Lake Ashtabula on the Sheyenne River, Angostura Reservoir on the Cheyenne River, and Shadehill Reservoir on the Grand River.

Lakes range from being infertile (nutrient poor,

Fig. 11.20. Shallow lakes and wetlands on the Missouri Coteau in central North Dakota. Photo by David Mushet, U.S. Geological Survey.

oligotrophic) to fertile (nutrient rich, eutrophic). Infertile lakes are relatively deep with clear water and are preferred for water sports, especially boating and swimming. Their deep basins, shallow bays with emergent and submerged aquatic vegetation, and rocky points and gravel bars explain their high diversity of popular game fish: northern pike, bass, walleye, and panfish. In contrast, fertile lakes are generally shallower and more turbid from the suspension of bottom sediments. They also are more likely to have toxic cyanobacteria blooms and are more vulnerable to winterkill. Fish production peaks in those that are intermediate in fertility, known as mesotrophic lakes. Over time, lakes change from oligotrophic to mesotrophic to eutrophic as a consequence of natural and human-caused inflow of sediment and nutrients from the surrounding landscape. This process, known as *eutrophication*, has been accelerated by cultivating prairie watersheds and by shoreline developments, with the result that the qualities of oligotrophic lakes are declining.

Many of the ecological characteristics of lakes are determined by whether they are part of a flow-through or a closed-basin system. Flow-through lakes have streams entering above and leaving below. The elevation of the outflow, known as the *pour point*, fixes the maximum elevation of the surface water, thereby narrowing the range of water depth during wet and dry cycles. Flow-through lakes also export nutrients, salts, and pollutants continuously. Oligotrophic flow-through lakes with clear water and picturesque, wooded shorelines are preferred for cabins, year-round homes, state parks, and resorts (fig. 11.21).

Less preferred for recreation are closed-basin lakes that can become saline as the water evaporates and salts accumulate.[44] They also respond to periods of deluge and drought by expanding and contracting in area and depth. Such lakes are the first ecosystems in a landscape to indicate weather extremes, climate change, wetland drainage, and upland cultivation. As described below, the two largest lakes have closed basins, Devils Lake in North Dakota and Waubay Lake in South Dakota. They illustrate the fascinating ways, albeit catastrophic for some, that closed-basin lakes can change rapidly. Freshening of these lakes and large wetlands in recent decades from wet weather has improved conditions for waterfowl and game fish. Good examples are Bitter Lake and Devils Lake.[45]

Winter Kill and Brackish Water

Anglers are well aware that fish populations can plummet from one year to the next. This occurs when deep snow on lake ice casts a shadow on algae and other submerged plants, thereby greatly reducing the photosynthesis required to produce well-oxygenated water throughout the winter. Oxygen levels can drop to one or two parts per million.[46] Shallow eutrophic lakes are most vulnerable to winterkill because much of the oxygen is depleted, deep snow or not, as the large amount

Fig. 11.21. Summer cabins and year-round homes are common on Lake Poinsett, a lake on the Prairie Coteau named by the explorer John C. Frémont to honor the American physician and secretary of war Joel Poinsett. This natural lake, one of the largest in the Dakotas, has a campground, restaurants, boat ramps, and tackle shops. Photo by Greg Latza, PeopleScapes Photography.

of organic matter in sediments decomposes. Deep oligotrophic lakes are least vulnerable. Lake managers sometimes try to maintain adequate oxygen levels by plowing snow from ice or by adding aeration systems, which work in some years. Thin ice near aerators can be dangerous for anglers.

As noted previously, shallow lakes can be brackish, a fact implied by names such as Bitter Lake and Medicine Lake. The number of saline lakes and the degree of salinity are much higher on the Missouri Coteau, due primarily to the large number of closed lake basins, a relatively dry climate, and the salt content of the glacial deposits. A study of 178 lakes in south-central North Dakota in the mid-1960s and mid-1970s found that the specific conductance of lake water had an extraordinarily wide range (365 to 70,000 microsiemens per centimeter).[47] The wide variation in lake water chemistry depends on physiographic setting, position relative to regional and local groundwater, basin composition (outwash or till), and rate of evaporation. Sampled again in 2012–2013, scientists found that wetter weather had diluted the previously saline lakes and wetlands, causing major changes in invertebrate and vertebrate populations.[48] Game fish rarely occur in water greater than 4,000 microsiemens per centimeter; salt-tolerant fish such as fathead minnows and brook sticklebacks cannot survive in most water exceeding 12,000 microsiemens, although the threshold depends on the kind of salt.

Notably, with fresh and saline lakes in close proximity, a buffet of food types is available for mobile organisms such as waterfowl.[49] Some saline lakes are more attractive to ducks because preferred foods such as fennel-leaf pondweed and widgeon grass often are common. Invertebrates also change with salinity. Snails, amphipods, and insects are abundant in fresh and slightly brackish lakes but are replaced by plump fairy shrimp in saline lakes. Interestingly, although ducks frequently use brackish and saline wetlands, they are not physiologically adapted for long stays. A common pattern is for mobile waterbirds to forage in saline lakes before returning to fresher water for drinking. Using saline lakes for breeding can be successful if fresh water is available near the shore, such as near springs. Duck broods and molting adults cannot survive on saline lakes without springs or seeps.

Human disruption of surface and groundwater flows among wetlands and lakes alters water quality. For example, construction of a road that bisected a lake complex in North Dakota greatly altered the salinity, plants, and animals on both sides of the road. One newly isolated pond supported game fish, another fathead minnows and brook sticklebacks, and a third brine shrimp and widgeon grass.[50]

In general, lakes very widely in size, depth, water quality, permanence, biodiversity, and how people use them. Clear, deep lakes with relatively stable water levels and game fish are most popular. Shallower and more brackish lakes are visited less often, except when reflooded with fresh water. But all wetlands and lakes are important habitat for waterfowl and other wildlife. Some argue that Dakotans "love their lakes to death," given that the water quality of lakes has deteriorated over time because of nutrients from human wastes and runoff from crop fields and well-manicured lawns. Accelerated eutrophication increases the chances of winterkill and diminishes other values that attracted people to lakes in the first place.

Waubay Lake Basin

The two largest natural lakes in the Dakotas, Waubay Lake in northeastern South Dakota and Devils Lake in northeastern North Dakota, illustrate the ways that lakes can change. Waubay Lake is one of 10 lakes in a 409-square-mile watershed. Between 1993 and 1999, seven of these lakes coalesced when the water level rose 20 feet, forming one closed-basin lake (figs. 11.22 and 11.23). Two flow-through lakes spilled water into the coalescing

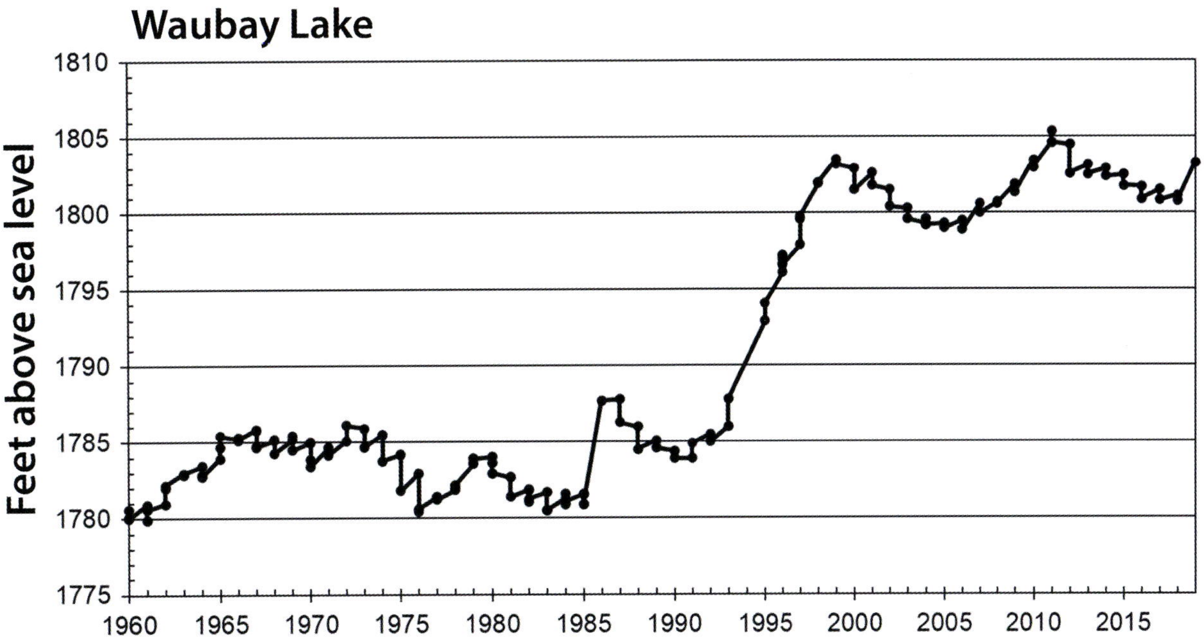

Fig. 11.22. Depth of Waubay Lake since 1960, the period of continuous record. Note the rapid rise in water level starting about 1990. Source: U.S. Geological Survey.

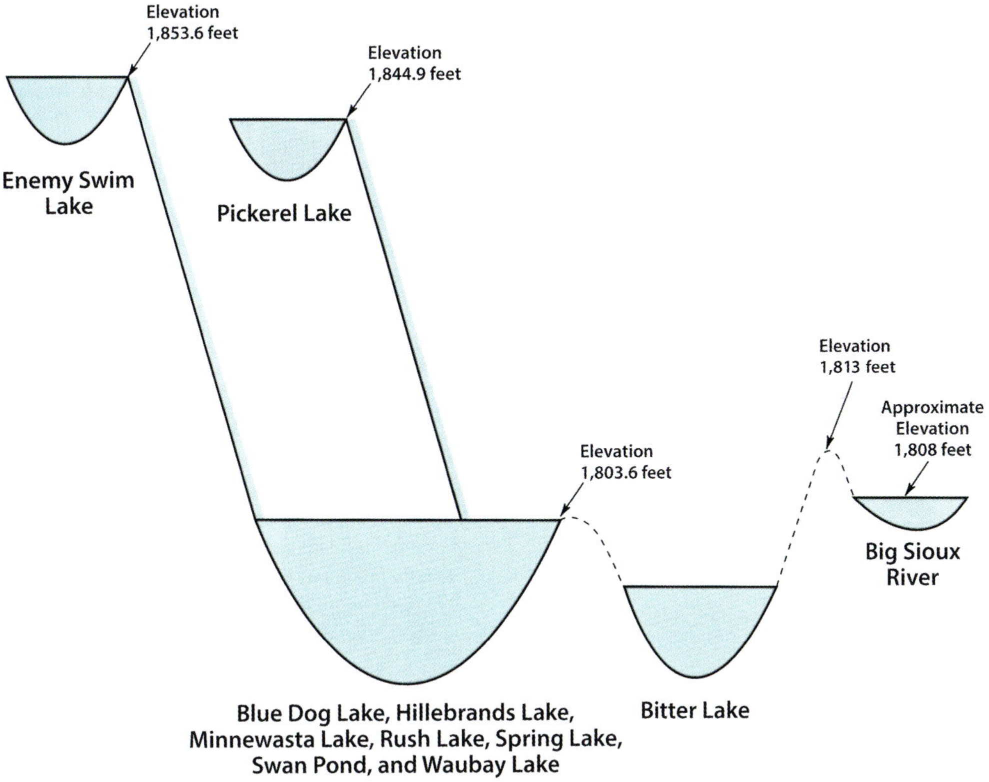

Fig. 11.23. Flow pathways among lakes of the Waubay Lake Complex. The closed-basin Bitter Lake is lowest in the topography and receives water from other lakes during high-water periods. Elevation is feet above sea level. There is no evidence that Bitter Lake has ever overflowed to the Big Sioux River. Its water quality ranges widely from brackish during low water periods to fresh during high water. Adapted from: Niehus et al. (1999).

lakes, doubling the surface area and flooding about 25 square miles (16,000 acres) of wetlands, grassland, farmland, buildings. and roads.[51] Thousands of acres of cottonwood, willow, and oak woodland also were submerged. Standing-dead, bleached-white trees mark former shorelines.

Cooler and wetter weather was the primary cause. Annual precipitation at Webster increased from an average of 21.1 inches in 1960–1990 to 27.4 inches in 1991–1997, an average increase of 6.3 inches each year. Evaporation was reduced because of cooler summers. A second cause of flooding was the drainage of wetlands and increased runoff from tilled land. Hydrologists concluded that a 10-year drought as severe as that in the Dust Bowl of the 1930s would be required to return the lake back to pre-1992 levels.[52] Water levels decreased by about five feet after the record high in 2011 but began to rise again in 2019.

Paleoecologists asked whether this kind of flooding had occurred in the past. Historical data show that water levels were lowest in the 1930s and highest after about 1995 (see fig. 11.22). Lake area when the land was first surveyed in 1868–1877 was about two-thirds of that in the 1990s. Going back beyond written records of flooding is possible using three lines of evidence: elevation of still identifiable wave marks, width of growth rings in nearby unflooded trees, and the salinity of lake water.[53] Trained observers could not find physical evidence of water levels higher than those in the latest flooding, although there were some shoreline wave marks nearly as high. The lowest and last lake in the chain, Bitter Lake, is a large, closed-basin lake with high evaporative rates. It has no obvious outflow channel, indicating no significant outflow since glaciation.

Tree-ring evidence for flooding was found in a 161-year-old bur oak growing on an island in Waubay Lake National Wildlife Refuge.[54] A core from this tree was cross-dated with another core collected earlier from a preserved bur oak beam from the Officers' Quarters at nearby Fort Sisseton, built in about 1864.[55] The two overlapping cores yielded a chronology from 1674 to 1998 of ring widths that corresponds well to known precipitation records—narrow rings in the Dust Bowl years and wide rings following heavy rain in 1977 and again in the 1990s when massive flooding occurred. The ring-width series revealed five wet periods during 325 years (fig. 11.24). Each peak was wet enough to produce lake levels that would have been well above average.

Another method for determining the historical periodicity of floods is to examine pollen and invertebrates found in bottom sediments. Spring Lake, one of the 10 lakes in the Waubay Lake complex, was chosen for coring because it reportedly did not dry out completely during the 1930s. Hence, the bottom remained anaerobic and the materials of interest did not decompose. One invertebrate, an ostracod, leaves shells behind with highly variable ratios of magnesium and calcium (figs. 11.25 and 11.26). High ratios indicate a deep lake, when the water was dilute and fresher. Results from the ostracod data match up well, though not perfectly, with those from tree rings: five high-water years and two droughts since 1675, with intervals of 140 to 160 years (ostracods and tree rings, respectively) between high lake levels (see fig. 11.24).

Devils Lake Basin

Changes to Devils Lake were similar to those observed at Waubay Lake during the same time period. This semiclosed lake basin reached its highest recorded level in 2011, nearly 32 feet above 1993 levels and high enough to spill into Stump Lake (fig. 11.27).[56] If the lake had risen six feet higher, it would have overflowed into the Sheyenne River, a tributary of the Red River. Historically, from 1880 to 1905, Devils Lake was relatively fresh and had a prodigious commercial fishery; thousands of northern pike from the lake were shipped to fish markets. But during the dry years that followed, the low water became highly

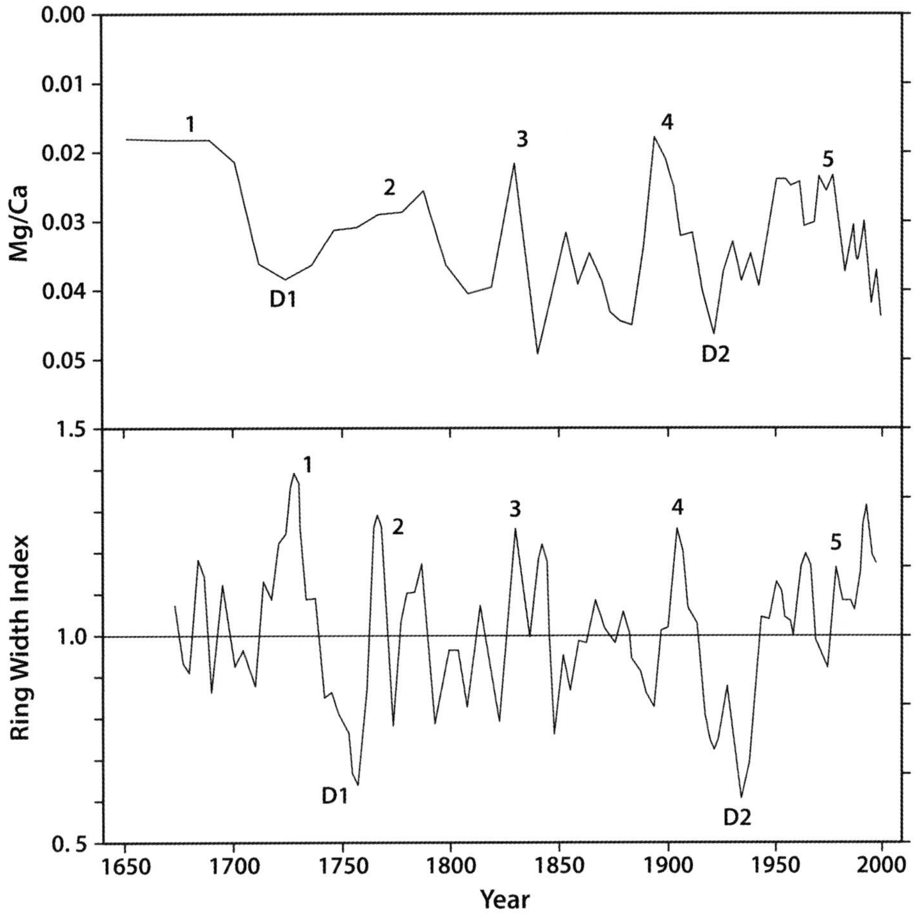

Fig. 11.24. Evidence for past water levels in the Waubay Lake complex using tree-ring analysis and the magnesium-to-calcium ratio in ostracod shells preserved in lake sediments. Five wet periods (1–5) and two periods of drought (D1, D2) are indicated using both methods. See text for explanation. A ring index of one is the average for the 325-year period. Ring index values were smoothed to clarify growth trends. Adapted from Shapley et al. (2005).

saline and the salt-tolerant brook stickleback was essentially the only fish present.[57] Total dissolved solids dropped dramatically when flooding started in the early 1990s. Evidence from ancient shorelines and diatoms and ostracods in lake sediments revealed similar changes prior to recorded history.[58] During the past 4,000 years the lake went dry four times, spilled into nearby Stump Lake five times, and discharged into the Sheyenne River three times.

By 2013, Devils Lake had flooded about 250 square miles of land (fig. 11.28). The beds of highways were raised and many less traveled roads were closed. When farmsteads and shore towns like Devils Lake and Minnewaukan were flooded, an artificial outlet and pumping station were proposed to pump water into the Sheyenne River, a proposal that became surprisingly controversial.[59] Downstream residents did not want more flooding with what they thought would be salty water. Also, because the pumped water would transport unwanted organisms and brackish water into Canada via the Red River, there was the possibility of violating international water quality

Fig. 11.25. Mark Shapley (left) is shown with the apparatus used to extract a core of bottom sediments in Spring Lake, part of the Waubay Lake complex. Above, Mark and Daniel Engstrom are wrapping the core for transport to their Universty of Minnesota laboratory.

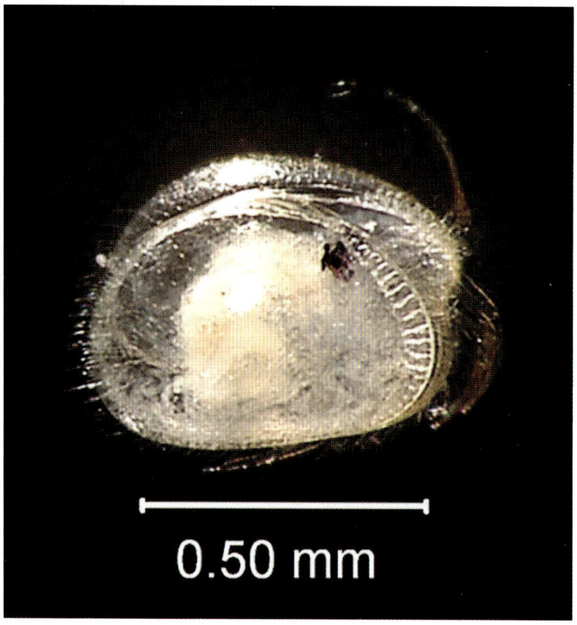

Fig. 11.26. Microphotographs of a live ostracod (a crustacean) on the left and two fossilized ostracod shells on the right. Chemical analysis of ostracod shells in lake-bottom sediments can be used to determine previous water levels (Tressler 1957, Griffiths 1995). Photos by Alison J. Smith, Kent State University.

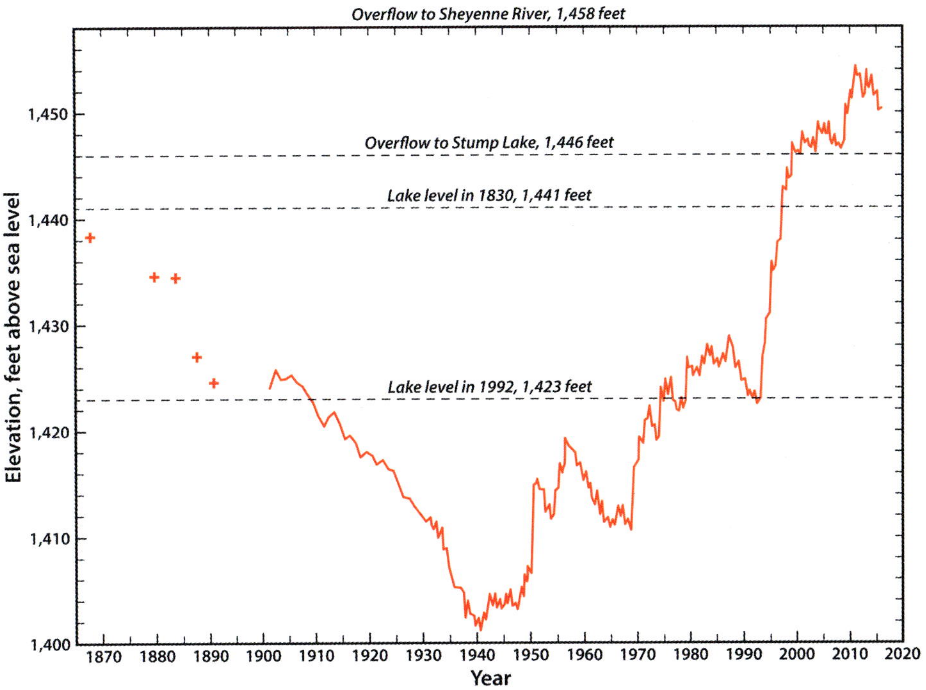

Fig. 11.27. Depth of Devils Lake during the past 150 years. The lake did not spill into the Sheyenne River during this time, as it did several times during the Holocene. Adapted from Haskell et al. (1996) and Larson (2012).

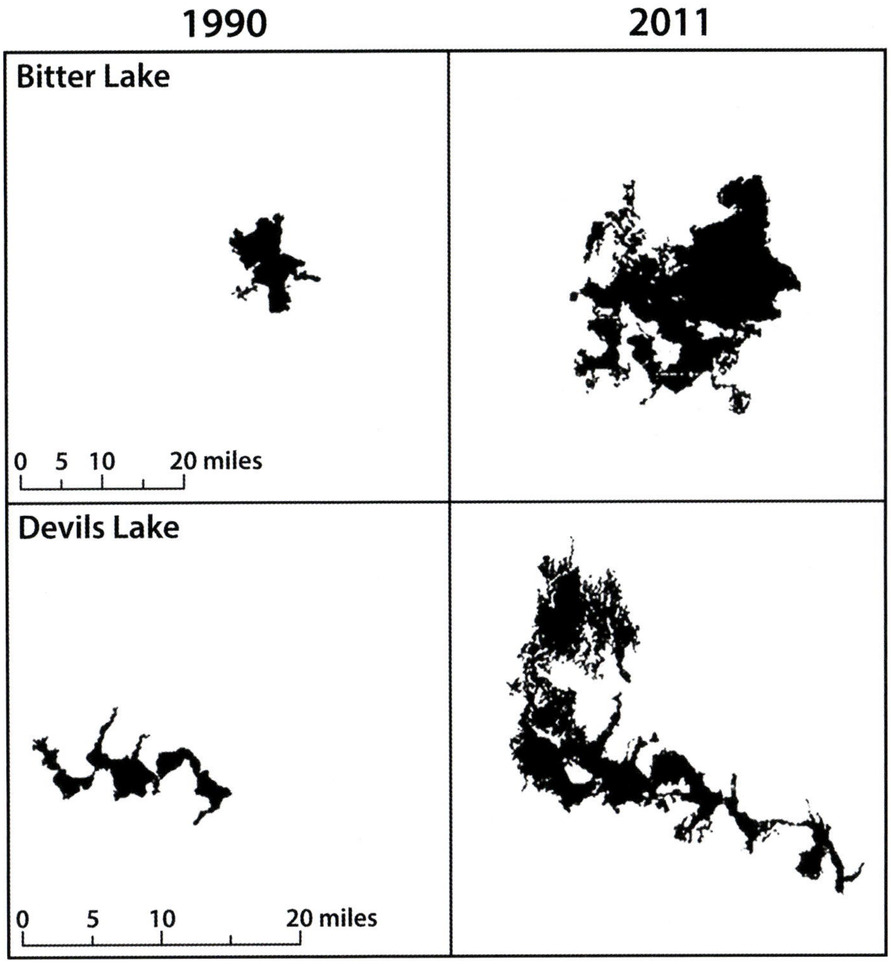

Fig. 11.28. Expansion of Bitter Lake (part of the Waubay Basin) and Devils Lake from low water levels in 1990 to high water levels in 2011. Adapted from Vanderhoof and Alexander (2016).

agreements. Other groups argued that pumped water could degrade channels and cut a lower outlet level for Devils and Stump Lake, causing more frequent catastrophic flooding downstream in the future.

As a partial solution, some groups recommended restoration of drained wetlands in the Devils Lake watershed, where about half of the original wetlands had been drained and cultivated. Hydrologists estimated that restoring 60,000 acres of wetlands would reduce runoff and stabilize or lower lake water levels. Restoration would be costly and take more time but would be less expensive than the construction, maintenance, and operation of the outlet pump stations, especially when factoring in other amenities, including major improvements in ecosystem services and more sustainable income from outdoor recreational activities. Government efforts to control flooding and protect communities exceeded $1 billion from the mid-1990s to 2013. After much debate, the outlet and pumping station were chosen rather than wetland restoration.

Overall, studies of Waubay and Devils Lakes found that recent high-water levels exceeded, or at least matched, past water levels. Land use changes in the past century, such as wetland drainage, conversion of prairie to cropland, blockage of flow patterns by roads, and a wetter climate are all factors that contributed to the most recent high water levels that changed the landscape so dramatically—and probably will again in the future. The ecological and socioeconomic consequences of overflowing lakes are complex and viewed differently by neighboring stakeholders. Unquestionably, the flooding has been catastrophic for taxpayers and the affected farmers and town residents, but there have been benefits as well. These lakes are more popular than ever for boating, swimming, and fishing. Dilution and expansion of historically brackish lakes have favored the growth and reproduction of desired game fish, like perch, walleye, northern pike, and smallmouth bass. The sale of licenses supports the game and fish departments of both states, and enhanced recreational opportunities have dramatically improved local economies. Thousands of anglers participate year-round in what is commonly described as a world-class fishery.[60] Local businesses have benefited, including bait and tackle shops, motels, restaurants, and local fishing guides—a relatively new profession in the Dakotas. Guides have escorted more than 15,000 clients to Devils Lake in open-water seasons and thousands more during the ice fishing season.

The Implications of Climate Change for Wetlands and Lakes

Clearly, wetlands and lakes are sensitive to changes in weather and climate.[61] With higher temperatures, low humidity, and strong winds, evaporation can lower water levels by as much as two inches per week in midsummer. A warmer, drier climate will lead to more frequent, more prolonged drawdowns—fewer small wetlands and shallower lakes. Nearly all lakes in the Dakotas, and all the prairie wetlands except those supported by larger springs, dried up at times during the 1930s. Conversely, as described, relatively cool, wet periods have led to flooding.

Climate scientists project that the Prairie Pothole Region will be warmer in the future with the same or a little more annual precipitation (3–15 percent more).[62] Most likely this climate will be effectively drier because the expected increase in precipitation would be insufficient to offset the higher evaporative demand of average air temperatures that are 6–8° F warmer by the end of the century. Future precipitation is projected to vary by season—wetter in fall, winter, and spring but no change in summer.[63] A rise of only a few degrees could produce more dry wetland basins, slow down or even stop the wetland cover cycle illustrated in fig. 11.16, and affect salinity and invertebrate communities. Precipitation since 1993 in the eastern Dakotas, especially in North Dakota,

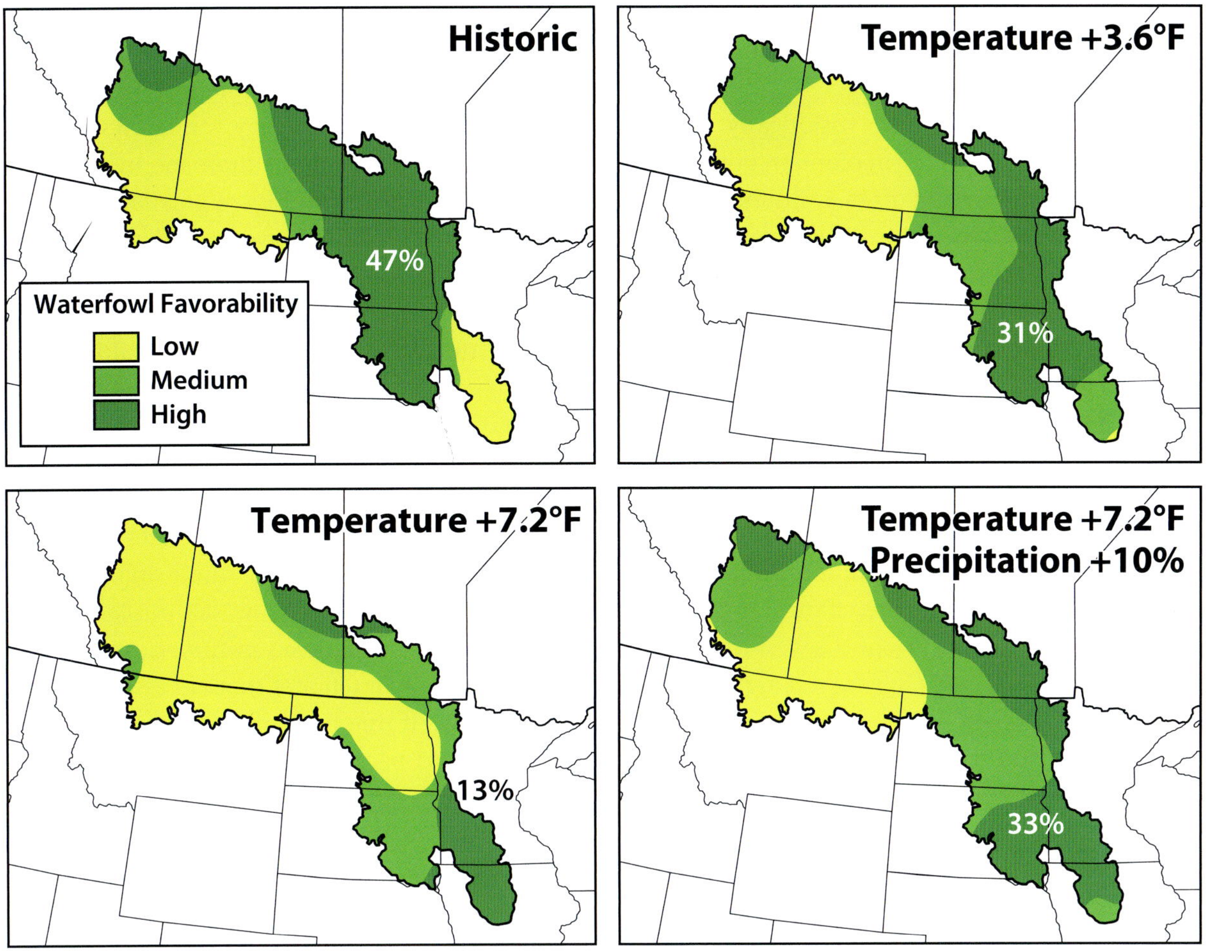

Fig. 11.29. Favorability of Prairie Pothole Region wetlands for waterfowl production based on historical weather data and three climate scenarios. Dark green indicates where the climate has been or potentially will be most favorable for waterfowl production. Climate favorability shifts away from the Dakotas under warmer scenarios to Minnesota and Iowa, which have less grassland nesting habitat and fewer wetlands. Warming combined with increased precipitation (lower right) would most likely ameliorate the temperature effect. Adapted from Johnson et al. (2010).

has substantially exceeded these long-term projections, although this period also has been punctuated with severe single year droughts in 2012 and 2016 and a 2-year drought in 2020–2021. Will this wet weather trend continue as projected by climatologists? North Dakota has warmed more than any of the contiguous states.

The projected climate would not affect all parts of the Dakotas uniformly (fig. 11.29). To illustrate, waterfowl production in the drier climate of the Missouri Coteau is highly episodic because of weather variability that includes droughts.[64] A drier climate would increase the number of low-production years, even though favorable grassland habitat may persist. The eastern, more humid Dakotas, including the Prairie Coteau, could benefit from some warming that could increase wetland cycling between wet and dry extremes, compared to current conditions. Unfortunately, more of the surrounding prairie has been cultivated in the east, most wetland drainage has occurred there, and the cost of wetland restoration is higher because land is more expensive.

Total duck numbers have historically tracked weather patterns, with a maximum population during the past 60 years of 33.6 million in 1956 following several wet years. During a drought, three years later, there were only 17.6 million ducks.[65] Not all species are affected to the same degree. For example, ruddy ducks and redheads are expected to tolerate future climate change better than others, such as black terns and mallards.[66] Nor will impacts due to climate change be isolated from other factors such as continued wetland drainage and loss of grassland nesting habitat. Strategic spending of conservation dollars could soften the impact of climate change and loss of habitat. Shifting support from the west to the east could pay off if wetlands in the west are too often dry.[67] As of 2010, about 35 percent of remaining PPR wetlands had been protected with long-term and perpetual easements from government agencies and nonprofit conservation groups. This is an impressive number, but the majority of wetlands are still unprotected.[68] Given the studies and analyses so far, it is hard to be sanguine about the future of lakes, prairie wetlands, and waterfowl if agriculture intensification continues.

Thus, the conservation challenges are formidable. To use a baseball analogy, strike one for wetlands was the high rate of drainage in the past, strike two was historical loss of grassland nesting habitat in wetland watersheds, and strike three may be the negative effect of climate change on the shrinking number of wetlands that remain. If the future is wetter, the effects could be less severe.[69] The ecologist Nancy McIntyre and her team has projected increased numbers of six water bird species on the Missouri Coteau during this century, assuming minimal wetland drainage.[70] In contrast, Kyle McLean and his associates cautioned that a wetter climate could dilute many saline wetlands, which would allow salt-intolerant fish such as perch and fathead minnows to colonize and consume invertebrates that are essential food for waterfowl. If that happens, they suggested that the PPR could become less favorable for ducks.[71] All projections agree that the future climate will be warmer, but they disagree at this time about whether the future climate will be effectively wetter or drier. There is much to learn about lakes and wetlands on the Northern Great Plains.

Chapter 12 **Buttes, Badlands, and Sandhills**

The buttes and badlands of the Missouri Plateau, south and west of the Missouri River, are the result of millions of years of erosion. Conspicuous from a distance, buttes rise above the surrounding plains and have flat tops. In contrast, badlands are mostly hidden in river breaks below the plains. Sandhills are much different in origin and location, having developed on the other side of the Missouri after the last glaciers retreated. All three landforms constitute a small portion of the Dakotas, but they add diversity to the landscape and are legacies of the region's geologic history.

Buttes

Stark reminders of the astonishing power of water erosion over incomprehensible time periods, the flat tops of dozens of buttes are the remnants of broad land surfaces that once were several hundred feet higher than they are today.[1] To understand their origin, imagine meandering rivers flowing eastward from the Rocky Mountains. Sediments were slowly deposited in layers (strata) on floodplains and in the bottoms of seas and lakes, eventually becoming thousands of feet thick. By chance, some of the sediments were cemented more firmly than others. Then imagine periods of regional uplifting that raised the mountains and western Great Plains still higher above sea level. Water flowed more rapidly, erosion increased. A network of shallow river valleys gradually coalesced into canyons, leaving large mesas that continued to erode, eventually becoming small enough to be labeled buttes (figs. 12.1, 12.2, and 12.3).

The buttes persisted because the sedimentary rocks at the top, dubbed *caprock*, were relatively more resistant to erosion. Of course, the caprock continues to disintegrate, eventually reducing buttes to little more than rounded hills or ridges. A notable example is the Killdeer Mountains, which extend about eight miles between the now obscure South and North Killdeer Buttes in western North Dakota (see chapter 7).[2]

The mere presence of a butte creates habitat diversity. North slopes are cooler much of the year, receiving less intense sunlight than elsewhere, and leeward sites provide protection from wind, allowing for the accumulation of drifted snow—which enhances plant water supply. Large rocks and boulders on the slopes provide microenvironments more favorable for some plants and animals that otherwise would be less common in the area. For example, rainwater falling on rocks flows quickly to the soil below, providing a reservoir of water for nearby plants with roots under the rock. Moreover, the water from rain and

Fig. 12.1. The sedimentary strata of this butte near the Black Hills are remnants of once-widespread rocks that were created when the Hills and surrounding land were covered with water during the Permian and Triassic periods. The reddish rock is iron-rich sandstone, and the white caprock is gypsum. Both are part of the Spearfish Formation. Ponderosa pine, yucca, and little bluestem are common in this area (see also fig. 1.9).

Fig. 12.2. Thunder Butte at sunrise in north-central South Dakota, surrounded by mixed-grass prairie and cropland. Millions of years ago, essentially all the land was at the same elevation as the top of this butte. Photo by Fr. Tony Grossenburg CC BY-SA 3.0.

Fig. 12.3. Chimney Butte, left, and Table Butte in western North Dakota have the same caprock, a sandstone of the Eocene Golden Valley Formation (55 million years old). Note the large boulders of broken caprock on the slopes. The shrub in the foreground is silver buffaloberry. Woodlands on the slopes have green ash, Rocky Mountain juniper, chokecherry, and silver buffaloberry. Photo by John Bluemle.

melting snow infiltrates rocky soils more deeply than where the soil has more silt and clay, such as on the grasslands below. Subsoil water is less likely to be evaporated before plants can use it. The net effect is that, compared to the surrounding plains and butte tops, the rocky slopes provide a moister environment. Shrubs and sometimes trees are common, for example, snowberry (buckbrush), skunkbush sumac, chokecherry, wild rose, buffaloberry, serviceberry, rubber rabbitbrush, Rocky Mountain juniper, green ash, and ponderosa pine (table 12.1). Common grasses include bluebunch wheatgrass and Indian ricegrass. Coyotes and foxes excavate dens, chipmunks scurry among the rocks, and raptors perch on vantage points above.[3]

Where soil is still present on the tops of buttes, the plants can be similar to those growing in the mixed-grass prairie below (see table 4.1). This seems counterintuitive because the soils are quite different. The tops have shallow, coarse-textured soils, the result of fine particles having been eroded by wind more rapidly than they are replaced by caprock weathering. In contrast, the surrounding plains have deep soils because of accumulation from above. Many people assume that the shallow soils on butte tops are drier, but in fact both environments are dry. On top the soils are dry because of low water-holding capacity and because very little snow accumulates there. The surrounding plains are dry because, even though the soils

Table 12.1. Some characteristic trees and shrubs associated with buttes and badlands in western North Dakota and South Dakota

Common name[a]	Latin name	Buttes	Little Missouri River badlands	White River badlands
TREES				
Ash, green	*Fraxinus pennsylvanica*	x	X	X
Aspen	*Populus tremuloides*	–	X	–
Cottonwood, plains[R]	*Populus deltoides*	x	X	X
Elm, American	*Ulmus americana*	–	X	–
Elm, slippery	*Ulmus rubra*	–	–	X
Hackberry	*Celtis occidentalis*	–	–	X
Juniper, Rocky Mountain	*Juniperus scopulorum*	X	X	X
Pine, limber	*Pinus flexilis*	rare[b]	–	–
Pine, ponderosa	*Pinus ponderosa*	X	X	X
Willow, peachleaf[R]	*Salix amygdaloides*	x	X	X
SHRUBS				
Buffaloberry, silver	*Shepherdia argentea*	x	X	X
Chokecherry	*Prunus virginiana*	x	X	X
Cinquefoil, shrubby	*Dasiphora fruticosa*	x	–	–
Greasewood[H]	*Sarcobatus vermiculatus*	–	X	X
Juniper, common	*Juniperus communis*	–	X	–
Juniper, creeping	*Juniperus horizontalis*	x	X	X
Rabbitbrush, rubber	*Ericameria nauseosa*	x	X	X
Sagebrush, big	*Artemisia tridentata*	x	X	–
Sagebrush, silver	*Artemisia cana*	–	X	–
Saltbush, spiny[H] (shadscale)	*Atriplex confertifolia*	x	X	–
Saltbush, four-wing[H]	*Atriplex canescens*	–	X	X
Saltbush, silverscale[H]	*Atriplex argentea*	–	X	X
Snowberry, western (buckbrush)	*Symphoricarpos occidentalis*	x	X	X
Sumac, skunkbush	*Rhus trilobata*	x	X	X
Serviceberry (juneberry)	*Amelanchier alnifolia*	x	X	X
Winterfat[H]	*Krascheninnikovia lanata*	–	X	X
Wormwood, longleaf	*Artemisia longifolia*	–	X	–

[a]Grasses and forbs are not listed here but include some of the same species found in the adjacent mixed-grass prairie (see table 4.1). X = common, x =occasional. A dash indicates that the species is absent or not common. Superscripts indicate the habitat of some species: R = riparian or wet habitats, H = halophyte.

[b]Limber pine has been found in two places in the Black Hills and on at least one butte in southwestern North Dakota.

receive drainage from above and have a relatively high water-holding capacity, water infiltration into fine-textured soils is slow. Much of the water held near the surface evaporates before the plants can use it. Consequently, the plants of both habitats must be drought tolerant, more so than those growing on slopes with deep, coarse-textured soils.

Another curious observation about steep-sided buttes is that livestock, bison, and other large herbivores usually cannot climb to the top, leading scientists and managers to wonder whether a comparison of butte-top grasslands with the prairie below could serve as a basis for evaluating the effects of grazing by large mammals. The problem with this approach is that observed differences could be due to differences in other environmental factors, such as soil characteristics, microclimate, or time since the last fire. A better comparison would be the vegetation on the top of steep-sided buttes not accessible by livestock to buttes that are accessible because of more gentle slopes—and have been grazed. This would provide a reasonable comparison, assuming the buttes have comparable caprocks and that good estimates of grazing intensity over a period of time are available. Assessing the effects of livestock grazing on grasslands can be difficult (see chapter 4).

Notably, the tops of some buttes produce good crops of harvestable hay. One rancher managed to lead two horses to the top of a butte in southwestern North Dakota during the drought of the 1930s when forage was scarce. He then used the horses to pull a mower up the steep slope with a rope and pulley.[4] The hay was dropped over the rim. The absence of heavy grazing combined with long periods between fires might have led to sufficient accumulation of plant biomass to justify this extraordinary effort.

Badlands

Some river breaks on the Missouri Plateau are veritable badlands—steep, sharply eroded escarpments impassable to early explorers traveling with horses and wagons (figs. 12.4–12.7). Erosion in such places is accelerated by cycles of freezing and thawing or wetting and drying that loosen fine particles from relatively soft, fine-textured mudstones, claystones, and siltstones. Cloudbursts then produce erosive flash floods. Landslides and rockfalls are common.[5]

Though sometimes thought of as barren wastelands, the resources of the largest tracts of badlands have led to the establishment of two national parks—Theodore Roosevelt National Park along the Little Missouri River and Badlands National Park along the White, Bad, and Cheyenne Rivers (see figs. 1.9 and 1.10). Among several motivations for establishing both parks were the astonishing fossil beds that had been exposed as the sedimentary strata eroded. As Joy Hauk wrote in 1969, "The first adventurers into the badlands must have found fossilized bones almost everywhere." The badlands also are appreciated for their unique landforms, wildlife, natural wildlands, and recreational opportunities.

Badlands can be observed along various rivers in the western Dakotas, but most people experience them in the national parks. Unknown to many visitors, the bedrock exposed along the Little Missouri River is much older than along the South Dakota rivers, a difference that is reflected by the fossil record. The Paleocene Bullion Creek and Sentinel Butte formations in North Dakota have fossils of turtles, crocodiles, alligators, fish, mollusks, and snails—all cold-blooded species. Fossils of small mammals are found in the youngest rocks, but they are rare. In contrast, fossils found in the younger White River badlands are mostly mammals, including Eocene and Oligocene ancestors of the horse, camel, rhinoceros, pig, and deer—and some with no living descendants anywhere on Earth, such as titanotheres the size of an elephant and a diversity of oreodonts. Fossils of semiaquatic turtles and alligators are found as well.[6]

Aside from fossils, living animals are part of the appeal of badlands in both states. Many are

Fig. 12.4. The Little Missouri River Badlands flank the meandering Little Missouri River in Theodore Roosevelt National Park. Riparian woodlands are dominated by plains cottonwood in the North Unit of the park, where this photo was taken. Some adjacent slopes have groves of Rocky Mountain juniper, green ash, chokecherry, and various other shrubs (see table 12.1). Aspen is found in relatively moist, north-facing ravines.

Fig. 12.5. The Little Missouri River Badlands in the South Unit of Theodore Roosevelt National Park. With fire suppression, Rocky Mountain juniper is expanding into mixed-grass prairie and shrublands. Big sagebrush and silver sagebrush are common. Ponderosa pine is not present in the park, but it occurs a short distance south along the river (see fig. 1.3).

Fig. 12.6. The White River Badlands in Badlands National Park are more barren and steeply eroded than the Little Missouri River Badlands. Note the eroding "tablelands" on the right with mixed-grass prairie.

Fig. 12.7. Flat, alluvial sediments develop in the bottoms of lower ravines of the White River Badlands, where a white crust of salts forms on the soil surface as water evaporates. Salt-tolerant plants growing in this meadow include foxtail barley, thickspike wheatgrass, meadow brome (introduced), saltbush, gumweed, and milkvetch. The shrubs on the slope in the foreground are greasewood. Trees on the ridge in the distance are Rocky Mountain juniper and ponderosa pine.

attracted by food and shelter not usually found on the surrounding prairie. Burrowing is relatively easy and raptors glide on updrafts created by the rugged topography. Common birds include the rock wren, cliff swallow, ferruginous hawk, prairie sharp-tailed grouse, great horned owl, golden eagle, turkey vulture, lark bunting, and grasshopper sparrow. Mammals include elk, bison, and bighorn sheep in the parks, along with mule deer, white-tailed deer, pronghorn, coyote, bobcat, swift fox (now rare), porcupine, beaver, badger, black-tailed prairie dog, desert cottontail, bushy-tailed woodrat, least chipmunk, and numerous species of bats, mice, shrews, and voles. The prairie rattlesnake, bullsnake, and eastern short-horned lizard (horned toad) are common, and the rare black-footed ferret has been successfully reintroduced to a prairie dog town in Badlands National Park (see chapter 4).

The Badland Vegetation Mosaic

More than steep escarpments, the badlands of both states have extensive mixed-grass prairie and semiarid shrublands on the surrounding flat lands. There are also riparian meadows, shrublands, and woodlands on the well-watered floodplains of the rivers below—all part of the badland mosaic.[7] The plant life of the grasslands varies greatly: Well-developed sandy-loam soils have needle-and-thread, blue grama, junegrass, western wheatgrass, and other mixed-grass prairie species; fine-textured soils have some of the same plants, but more western wheatgrass; coarse-textured sandy soils support prairie sandreed and Indian ricegrass. Hillcrests and slopes carved from sandstone or shale have little bluestem, stiff sunflower, and narrow-leaved purple coneflower (echinacea). Shrublands are dominated by silver sagebrush, big sagebrush (in far-western counties), western snowberry (buckbrush), skunkbush sumac, creeping juniper, and, on saline soils, greasewood and spiny saltbush (see table 12.1). Seemingly barren slopes have gumbo lily and the rare Dakota buckwheat.

Woody draws with trees are conspicuous, usually on north slopes and dominated most often by Rocky Mountain juniper. Occasionally American elm, boxelder, and ponderosa pine are found. The juniper is now expanding into the prairie in some areas, probably because of the suppression of prairie fires that otherwise would have killed young trees.[8] Various birds consume juniper "berries" (actually fleshy cones), thereby dispersing the seed. Ponderosa pine grows with juniper on some ridges and knolls south of Theodore Roosevelt National Park and on Sheep Mountain Table in Badlands National Park. Aspen is found in the North Dakota badlands on north slopes were snow accumulates, but not in the South Dakota badlands.

The riparian zone is dominated by plains cottonwood, but green ash and Rocky Mountain juniper often grow in the understory and may become dominant as the cottonwoods die (see chapter 10). Understory plants include western wheatgrass, silver sagebrush, western snowberry, skunkbush sumac, woods rose, Virginia creeper, and clematis. Relatively coarse soils on the floodplain and some upland sites have silver sagebrush; black sagebrush is found on some terraces above the floodplain.

A noticeable difference between the Little Missouri and White River badlands, visible from satellites (see fig. 1.4), is less plant cover in South Dakota. More of the land is barren and woody draws and shrublands are less common. The best explanation seems to be a drier climate. The mean annual temperature in the South Dakota badlands is 7.5° F warmer than in western North Dakota (51° F and 43.5° F, respectively), which could lead to higher levels of water stress for plants, less plant cover, and more erosion.[9] Geologic differences could be another factor.

Conserving Valued Resources in Badland Parks

Asking people what they appreciate most about the badlands elicits various replies: the unusual topography, tranquil prairies that support herds of bison and prairie dog towns, areas of Wilderness designated by Congress, and fossil exhibits featuring incredible, now-extinct reptiles and mammals. To maintain these features, park managers are faced with challenges that include fire management, the culling of bison herds to avoid excess grazing, curbing the spread of invasive plants, and reducing the impact of diseases impacting prairie dogs and other wildlife—all in the context of continued erosion and a changing climate.[10]

Erosion is a natural process, but managers everywhere do what they can to avoid accelerating the rate at which soil is lost. The White and Little Missouri Rivers are already laden with silt; intact soil has become precious. With regard to grazing, the perimeters of both national parks are fenced to exclude domestic livestock, partly because national park policy is to favor native plants and animals and partly because, without livestock, it is easier to regulate the number of large herbivores and minimize the spread of exotic plants. The fences also help prevent close intermingling of livestock with bison and elk, both of which are known to carry brucellosis, a disease that, it is feared, could decimate the local livestock industry for long periods of time. Fencing combined with vaccination programs has been effective in preventing the spread of brucellosis.

Having bison in both national parks is a high priority for managers and visitors alike. However, without predation by wolves and grizzly bears and with movement restricted by fences, herd size can easily exceed the goal set by rangeland scientists. For example, the 50 bison reintroduced to Badlands National Park in 1963 and 1964 led to a herd of 1,250 animals by 2015, approximately double the size recommended by managers. The herd is culled opportunistically, with some live animals trapped and transferred to ranches and parks.[11] The available forage has been adequate to sustain larger herds, but water for the bison is scarce, and there are concerns about the animals moving onto private land. Harvested animals are made available to food pantries, but animals that die from other causes are left on the grassland to support native species that are part of the scavenger food web (see chapter 4).[12]

Notably, fences are of mixed value for bighorn sheep (fig. 12.8). On the one hand, they separate the bighorns from domestic sheep, which carry

Fig. 12.8. Bighorn sheep are native to the badlands of the Northern Great Plains and have been reintroduced to Theodore Roosevelt National Park and Badlands National Park. Photo by Justin Meissen.

diseases for which the bighorns have little tolerance, such as pneumonia. On the other hand, the fences also confine the native sheep, often to suboptimal habitat. Research suggests that bighorns benefit from open range that includes extensive prairies adjacent to rugged terrain in which they can escape predators.[13]

Most prairie fires in the badland parks are now prescribed, that is, ignited intentionally to accomplish a management objective. Unfortunately, the conditions considered safe for prescribed fires are infrequent and fire control can be expensive, with the result that some of the vegetation is not burned as often as managers would like. Prolonged fire intervals often lead to the spread of Rocky Mountain juniper into grasslands. Indeed, this native juniper is sometimes viewed as an undesired invasive species,[14] along with numerous other weedy plants that become problems when the fire-free interval is too long. Of greatest concern in Theodore Roosevelt National Park grasslands are leafy spurge, Canada thistle, Kentucky bluegrass, and smooth brome. In Badlands National Park the problematic invasives are Japanese brome, Kentucky bluegrass, yellow sweetclover, and yellow salsify. In the riparian zone, invasive plants include saltcedar, Russian olive, Russian knapweed, spotted knapweed, cheatgrass, white top, musk thistle, Canada thistle, bull thistle, leafy spurge, black henbane, and common mullein. Prescribed fires in the spring can reduce the abundance of some exotic species, although herbicides are sometimes required. Using herbicides leads to difficult decisions in national parks because native plants often are killed as well.[15]

In addition to invasive plants, new insects and pathogens are a challenge. The emerald ash borer could spread westward, killing many green ash—a common tree in badlands and the most widespread and abundant native tree on the Northern Great Plains (see chapter 7). Another worrisome disease is sylvatic plague (*Yersinia pestis*), a lethal, introduced bacterium spread by fleas that has caused rapid declines in the black-tailed prairie dog populations since 2008.[16] Reintroduced black-footed ferrets also have declined, due to their own susceptibility to the plague and to reductions in prairie dogs, on which they prey almost exclusively (see chapter 4). The ferret is already protected by the Endangered Species Act, and now some observers are convinced that the black-tailed prairie dog should be listed as well. To combat plague, prairie dog burrows are dusted with insecticides that kill the fleas, and research is under way to develop an oral vaccine for the rodents. Ferrets are vaccinated as well.

And then there is the concern of rapid, human-caused climate change, over which land managers have no control. The best-available projections indicate that the climate will become warmer in the western Dakotas, with droughts and flash floods that are more frequent and more intense. The anticipated droughts could lead to less plant cover, which would increase the rate of erosion as well as reduce the carrying capacity for bison and other large herbivores. Some invasive species may be favored, depending on how fire frequency is affected. A period of wetter-than-average weather could lead to more forage, but if not grazed adequately, the forage could become highly flammable during the inevitable dry periods that follow. Erosion in badlands is a natural process, but if extreme rainfall events occur after a long dry spell, the rate of erosion could be accelerated because of less plant cover.[17] The most adverse effect of such erosion from badlands may be extraordinary amounts of silt deposition in reservoirs downstream, reducing their longevity.

Can the effects of climate change be mitigated? Studies in both national parks concluded that severe flooding events could be moderated by upstream withdrawals for irrigation or the construction of stock dams on tributaries, but there is still debate about the effectiveness or wisdom of such measures (see chapter 9).[18] Stopping erosion in naturally eroding landscapes is not a priority, but greatly accelerated erosion—for whatever reason—is likely to cause undesired consequences. Climate change further complicates resource management decisions.

Sandhills and Sand Plains

Ten thousand years ago, a prominent feature of Dakota landscapes was very large lakes fed by rivers flowing from melting glaciers (see fig. 2.8). The climate was warming but cold winters occurred then as they do now. Rivers would flood during the spring, depositing lightweight silt and clay on floodplains as floodwaters receded. Heavier sand particles were deposited in sandbars, which gradually moved downstream, shifting from place to place as the channel changed. Large sand deposits accumulated in deltas where the rivers flowed into lakes, a process that continued for hundreds of years. As the climate became more arid, shorelines receded and the loose sand was blown downwind, sometimes accumulating in *dunes* that were 100 feet high. Elsewhere the sand was spread more evenly, forming *sand plains*. Whether hilly or flat, these sandy soils are the legacy of large rivers that existed long ago. Now almost entirely stabilized by plant growth, the dunes are commonly referred to as *sandhills* (fig. 12.9).

The most extensive sandhills and sand plains in the Dakotas are found downwind from the former deltas of four rivers: the Souris River delta in glacial Lake Souris, the Sheyenne and Pembina River deltas in Glacial Lake Agassiz, and the James

Fig. 12.9. Sand dunes often are easily identified by their abrupt, "choppy" topography, such as in this area in the Sheyenne National Grassland in southeastern North Dakota. Nearly all dunes and sand plains in the Dakotas and Nebraska are now stabilized because of a climate that has been favorable for plant growth and because grazing intensity usually is moderate. Stabilized dunes are commonly referred to as sandhills. Grasses and forbs include prairie sandreed, sand bluestem, sand dropseed, silky prairie clover, and blazingstar (see table 12.2). The trees are plains cottonwood; the tall shrub (lower right) is buckthorn, an invasive shrub. Photo by Cory Enger.

River delta in Glacial Lake Dakota (see figs. 1.2 and 2.8). Similarly, the Nebraska Sandhills—the largest tract of sandhills in North America, which extends into South Dakota—is composed largely of sand transported by the North and South Platte Rivers.[19]

Then and now, sand deposits initially are active, that is, they continue to be windblown because of little or no plant cover. But gradually the seedlings of some plants become established, probably during a relatively wet spring that coincides with fortuitous seed burial at a suitable depth. If the seeds are too close to the surface, the delicate seedlings wilt or are blown away; if covered by too much sand, there is insufficient energy for shoot growth to the surface. Even if new leaves emerge, blowing sand often buries the young plants. However, mature plants eventually stabilize the dune. Most produce horizontal underground stems known as rhizomes, from which new sprouts develop. The plants commonly live for decades, but if heavy grazing, burrowing, vehicular traffic, or fire coincide, especially during a drought, plant mortality increases and the dune could become active again. Accompanying winds create *blowouts* (fig. 12.10). Paleoecologists have determined that Dakota sand dunes were reactivated during long dry periods that occurred a few thousand years ago, as discussed in chapter 2, and that some of the dunes were active much more recently, during the drought of the 1930s. One study found that all of the Souris sandhills were active over large areas at

Fig. 12.10. Blowouts, such as this one, are currently rare in the Dakotas, but they become more common during extended droughts or when livestock grazing is excessive. Some native plants grow better in blowouts than on stabilized sandhills. Photo by Jeffrey Printz.

some time during the past 1,200 years, though not everywhere at the same time.[20]

Plant Adaptations

Only a few species of plants are sand pioneers, that is, adapted for becoming established on loose sand (table 12.2). Their preference for such habitat is reflected by their names: sand bluestem, prairie sandreed, blowout grass, sand lovegrass, sand bur, and sand sunflower. And there are others: Indian grass, field sagewort, lemon scurfpea, hairy prairie clover, and rush skeletonplant. Some species become part of the plant community only after the dune is stabilized, possibly because organic matter from the decomposing roots of pioneer species makes the soil more suitable for their physiological requirements. Such plants include many species found in the surrounding mixed-grass prairie. With sufficient time between disturbances, trees and shrubs further stabilize the sand, plants such as aspen, bur oak, green ash, hackberry, American plum, snowberry, chokecherry, buffaloberry, wild rose, and poison ivy.[21]

Each plant species has adaptations that enable survival in the shifting dune environment. Nearly all are perennials, an adaptation for habitats where seedling mortality is high. Also, the roots and rhizomes of dune plants are long and grow rapidly. Often, a plant may be completely covered with blowing sand, but the shoots and rhizomes grow upward using the energy of stored carbohydrates until new green leaves emerge on the surface. And new roots, known as adventitious roots, may grow from buried stems. If the roots are exposed by the loss of sand from around the base, they rapidly grow deeper. Nitrogen is often a limiting nutrient, but a few species have the ability to fix nitrogen from the atmosphere. As the roots and leaves of nitrogen-fixing plants die and decompose, their nitrogen and other nutrients become available for other plants. Also, the decomposing plant material further stabilizes the developing soil. The

Table 12.2. Some plants found on sandhills and sandplains in North Dakota and South Dakota[a]

GRASSES & SEDGES	
Bluestem, sand	*Andropogon hallii*
Dropseed, sand	*Sporobolus cryptandrus*
Flatsedge, Schweinitz's	*Cyperus schweinitzii*
Grass, blowout [Rare]	*Redfieldia flexuosa*
Indiangrass	*Sorghastrum nutans*
Lovegrass, sand	*Eragrostis trichodes*
Needle-and-thread grass	*Hesperostipa comata*
Panicgrass, Leiberg's	*Dicanthelium leibergii*
Ricegrass, Indian	*Achnatherum hymenoides*
Sandbur	*Cenchrus longispinus*
Sandreed, prairie	*Calamovilfa longifolia*
FORBS	
Blazingstar	*Liatris spicata*
Clover, silky prairie	*Dalea villosa*
Sagewort, cudweed	*Artemisia ludoviciana*
Scurfpea, lemon	*Psoralidium lanceolatum*
Spiderwort, prairie	*Tradescantia occidentalis*
Sunflower, perennial	*Helianthus nutallii*
SHRUBS	
Buckthorn	*Rhamnus cathartica*
Buffaloberry, silver	*Shepherdia argentea*
Cherry, sand	*Prunus pumila*
Chokecherry	*Prunus virginiana*
Ivy, poison	*Toxicodendron rydbergii*
Leadplant	*Amorpha canescens*
Plum, American	*Prunus americana*
Rose, prairie	*Rosa arkansana*
Silverberry	*Elaeagnus commutata*
Snowberry, western	*Symphoricarpos occidentalis*
Sumac, smooth	*Rhus glabra*
Willow, prairie	*Salix humilis*
TREES	
Ash, green	*Fraxinus pennsylvanica*
Aspen	*Populus tremuloides*
Cottonwood, plains	*Populus deltoides*
Hackberry	*Celtis occidentalis*
Oak, bur	*Quercus macrocarpa*

[a]Not listed are various mixed-grass prairie species that become more common as soil organic matter increases in sandy soils (see table 4.1).

seeds of some sand-loving species (psammophiles) are large, with enough energy in the endosperm to enable rapid and sustained seedling growth for a longer time; and soon after establishment, these plants produce durable, sometimes waxy stems and leaves that minimize the abrasion of blowing sand.

As is true for grassland plants in general, sandhill plants must be adapted to tolerate periods of desiccation. Their rigid stems and leaves resist wilting and they use water efficiently, which means the amount of carbon fixed by photosynthesis per gram of water is relatively high. Herbaceous dune plants produce more belowground biomass than aboveground biomass (having a high root-to-shoot ratio), which helps ensure that the shoots acquire enough water and nutrients to survive. Augmenting the resource-acquiring function of the roots are *mycorrhizae*, a mutualistic relationship between roots and certain fungi. The fungi derive their energy from carbohydrates stored in the roots while fungal filaments known as hyphae extend from the roots into the soil. Like root hairs, the hyphae greatly increase the amount of water and nutrients transported to the plant. To conserve moisture, the stems and leaves of dune plants are often angled upward to reduce direct sunlight interception and their gray-green color reflects much of the light. Such adaptations moderate leaf temperature and the amount of water evaporated from leaves.

But sandy soils often are not as dry as they appear on the surface. Water from rain and melting snow soaks rapidly into the coarse-textured soil and, once the surface sand dries to a depth of only a few inches, it forms a barrier that greatly reduces further evaporation. Digging into the soil late in the summer commonly reveals moist sand, often wetter than on surrounding fine-textured soils. In fact, plant growth rates on sandy soils can be higher than on adjacent fine-textured soils, where much of the water remains on the surface and is evaporated before the plants can use it. The presence of aspen, bur oak, and other trees away from rivers can be an indicator of relatively moist sandy soils (figs. 12.11 and 12.12).

Benefiting from Sandhill Ecosystems

Sandhills are widely appreciated for livestock grazing, wildlife habitat, outdoor recreation, and as a source of water. Leeward slopes provide shelter from the wind on cold winter days, and the combination of plants and animals in such places is different because the sandy soils provide a different habitat than the surrounding prairie. Also, most of the sandhill plants are native. Attempts were made to plow and cultivate sandy soils, but the outcome was less profitable than raising livestock. The native sandhill forage is nutritious, and the dunes often sustain plant growth at times when surrounding grasslands and croplands are suffering from drought, as during the 1930s. Surface water is usually available in the ponds that develop between some dunes, even during drought years, and the creeks and rivers fed by sandhill aquifers continue to flow.[22] Ranchers have learned how to manage their herds without causing blowouts. Thus, the landscape as a whole is similar to that which existed for thousands of years previously. Cattle have replaced bison, but sustainable land management is relatively easy when managers are able to work with native plants rather than trying to replace them.

In addition to providing a diverse habitat, the sandhills and associated lakes and ponds are commonly used for outdoor recreation. The Sheyenne National Grassland and Fort Ransom State Park in North Dakota are popular destinations, largely because of the Sheyenne River sandhills. The same is true for the Souris National Wildlife Refuge and Denbigh Experimental Forest on the Souris River sandhills, Icelandic State Park on the sandhills of the Pembina River delta, and the Hankinson sandhills south of the Wild Rice River.[23] Lacreek National Wildlife Refuge in South Dakota is on the

Fig. 12.11. Aspen groves and open-grown bur oak trees are often found on stabilized dunes and sand plains, such as here on the now rounded hills of the Sheyenne National Grassland.

northern fringe of the Nebraska Sandhills and is especially important for waterfowl. The popularity of sandhills is further illustrated by the Dakota Dunes development sandwiched between the Big Sioux and Missouri Rivers, which is now the site of a casino, hotel, many homes, and a place for all-terrain vehicle enthusiasts.

The intriguing mosaic of prairie, riparian, and sandhill habitats has led to research designed to protect them. A top priority is to determine the actual abundance and habitat preferences of rare species and different kinds of ecosystems.[24] The status of the relatively rare greater prairie chicken and other birds and mammals is monitored, as are populations of rare plants, including the western prairie fringed orchid and white lady's slipper. The Sheyenne Sandhills are widely known as one of the largest remaining tracts of tallgrass prairie. Wildflowers provide food and habitat for numerous pollinating insects, including different kinds of bees (see box 4.2).

To help conserve these species and increase profitability, some sandhill ranchers are collaborating with conservation biologists to coordinate livestock grazing with prescribed fires. That can be good for beef production, as cattle are attracted to recently burned grassland, but if the burned area is small, or animal numbers too high, blowouts

Fig. 12.12. A bur oak forest with straight forest-grown trees in the Hankinson Sandhills on the Sheyenne National Grassland, south of the Wild Rice River. Understory plants include various grasses and sedges along with Virginia creeper, poison ivy, sweet cicely, bedstraw, and buckthorn (an invasive shrub). Chokecherry and smooth sumac are common on the forest edge.

can greatly reduce forage availability for a decade or more. Notably, some native plants and animals benefit from shifting sands, which occasionally calls for an unusual decision: should one produce more forage for livestock or create habitat for blowout species? Complicating the decision is a desire to constrain the spread of undesirable invasive species. As noted, pioneer dune species are by nature colonizers and some species, introduced from Eurasia, are even better adapted—but less desirable—than the natives. Sandhill weeds include leafy spurge, Russian thistle, kochia, Japanese and downy bromes, and rough pigweed. Fortunately, most of these plants currently occur primarily along roadsides, in adjacent cropland, and on heavily grazed rangeland. Leafy spurge is an exception.

With regard to climate change, the most reliable projections are for a warmer and wetter climate in the eastern Dakotas where most of the sandhills occur. Warmer temperatures will lead to higher rates of evapotranspiration, but that may be balanced by added precipitation. Still, extended droughts may occur. If that happens, blowouts would be formed, as in the past.[25] The sandhill landscape could become quite different, with active dunes and less forage for livestock. However, the dunes have been stabilized for many decades, due to fire suppression and moderate levels of grazing, perhaps lower than when unfenced herds of bison were abundant in the area. Consequently, the soil could be less subject to wind erosion because of accumulated organic matter.

In general, the sandhill ecosystem is fragile, subject to easily triggered episodes of rapid erosion. With more refined management and soils that now have higher levels of organic matter, and more available water for plant growth, it is possible that most sandhills will remain more productive than the surrounding grasslands. Much depends on the duration of anticipated droughts. Sandhills, along with buttes and badlands, help sustain a much-appreciated diversity of native plants and animals.

Chapter 13 Working toward Sustainability

Native ecosystems in the Dakotas have been developing for thousands of years, with each kind of plant, animal, and microbe evolving adaptations for tolerating fluctuating environmental conditions and coexisting with its neighbors, large and small. When one part of an ecosystem changes, such as with the arrival of a new species or during an unusually long drought, all other parts of the ecosystem are affected, sometimes in ways that are not easily predicted. Understanding how such places have survived for so long and why they change is a fascinating endeavor that is pertinent to developing sound land management practices. This final chapter focuses on several initiatives that are under way and have the potential for curbing some of the undesired trends identified in previous chapters. All depend on the collaboration of landowners, conservation biologists, policy makers, and society at large.

Responding to disturbing trends has been the focus of land stewards for many years. Soil erosion was slowed by contour farming, minimum-till or no-till agriculture, and the containment of sediments during road repair and construction. Also, some erodible croplands were withdrawn from cultivation through programs that provide economic incentives to do so, such as USDA's Conservation Reserve Program. Strategic crop rotations and cover crops are now used to reduce the need for expensive and sometimes problematic fertilization or weed control. Biological or integrated pest management is practiced where practical and advisable. Downward trends of some wildlife species were slowed or reversed through the protection or restoration of critical habitat, such as in National Wildlife Refuges and Waterfowl Production Areas. The restoration of some wetlands contributed to the filtration required for clean water.

Human influences on the land are nothing new.[1] The first people likely contributed to the extinction of mammoths and other large mammals about 10,000 years ago, as discussed in chapter 2. Later, the size of bison herds would have been reduced locally by indigenous hunters, but at that time, still without horses, the number harvested was a small fraction of those available.[2] Gardens on floodplains became larger after the Mandan, Hidatsa, and Arikara arrived 500 to 1,000 years ago. Air quality would have been pristine—except when smoke from inevitable prairie fires obscured the horizon. There were no or few troublesome weeds from other continents. Hazards existed, but people seeking to live near rivers soon learned how to avoid most floods. Other precautions would have been required to avoid large predators, such as grizzly bears, although an abundance of less formidable prey may have minimized that problem. Rattlesnake bites might have been a more serious threat than bears.

Immigrants of European descent brought new ambitions, guns, and other tools. Beaver were trapped almost to extinction for their fur; bison were extirpated from the Northern Great Plains; and elk, pronghorn, deer, and other animals were hunted without restrictions. Two primary lifestyles soon were adopted—livestock husbandry and farming. With a little care, erosion caused by livestock could be avoided, but that was not true for farming. Profitable crops did not seem possible unless the protective prairie sod was plowed (see chapter 5).

Today there are powerful market incentives to cultivate large tracts of land more intensively than ever before. Herbicides and pesticides are used more often, and some rangelands previously reserved for livestock grazing are being converted to corn and soybeans.[3] The magnitude of these agricultural impacts over the Dakotas is compounded by the continued expansion of invasive plants, gradual loss of soil organic matter in many areas, and the consequences of fire suppression, streamflow management, and further fragmentation of landscapes by oil and gas developments, wind farms, feed lots, rural subdivisions, and still more roads. The most heavily affected ecosystems in the Dakotas are tallgrass prairies, wetlands, and rivers. Potential options for their restoration, where that is logical and feasible, have been discussed in previous chapters. But as the climate continues to change, what can be done to curb these trends while producing the food and other commodities that sustain so many people?

Providing Habitat for Native Species and Ecosystem Services

In this era of genetic engineering and rapid climate change, conserving multimillion-year-old gene pools is the right thing to do. Unfortunately, many native species have been squeezed into small tracts of habitat that still persist for one reason or another. Such species include the black-footed ferret, swift fox, prairie dog, ferruginous hawk, mountain plover, Dakota skipper butterfly, western prairie fringed orchid, and uncounted others that are rarely seen or thought about. Some of the habitat required by these species is protected by wildlife refuges managed by the U.S. Fish and Wildlife Service. More habitat has been created by landowners, often in collaboration with nongovernmental organizations such as The Nature Conservancy, Ducks Unlimited, Pheasants Forever, and various land trusts. The motivations for protecting habitat include viable populations of fish and game and a desire to retain native species for their inherent values. It can be exhilarating to walk through a native prairie with numerous plants in bloom, their busy pollinators flitting from flower to flower, and the calls of meadowlarks and other birds. The values of wildlife habitat were recognized by state legislatures with the establishment of game and fish departments. For similar reasons, departments of agriculture now promote the restoration of pollinator habitat. Children learn about biodiversity in schools.

Small patches of native plants and animals often are referred to as *natural areas*—genuine refugia for some species. Unfortunately, these remnants of once-widespread ecosystems often are not large enough to sustain other species. Also, they may be too small to enable the prairie fires or floods that help maintain biological diversity and slow the invasion of undesired plants. For these reasons, some would say they are not natural at all. Nevertheless, such areas provide elements of an ecosystem's diversity that can be enjoyed, studied, and used for the restoration of degraded ecosystems in the future. Enlarging them is possible by reconstructing prairie habitat on adjacent croplands and elsewhere (box 13.1), or by adopting the approach of regenerative agriculture discussed in chapter 5. Larger refugia are more valuable

Box 13.1. Prairie Restoration

Restoration is a promising approach to counterbalance losses of native prairie. The goal is to assist in the recovery of ecosystems that have been degraded, damaged, or destroyed. Prospective sites in the Dakotas include mines, weedy prairies, heavily grazed pasture, and retired cropland. These sites form a range in the degree to which historical conditions of flora, fauna, and environment can be reached. Experience has shown that complete restoration, analogous to restoring a faded Renaissance painting, is not attainable because of cost, limited availability of native plant seed, scarce information on historical conditions, and the degree of damage.

Most challenging is to reestablish prairie on large areas of land subject to various kinds of mining activities, where the natural soil profile and much of the soil biota are destroyed. Low-diversity mixtures of plant seed are planted routinely, sometimes including introduced species (mostly grasses). If the topsoil is stored and put back properly, some ecosystem services are regained, such as livestock grazing, wildlife, and carbon sequestraton. However, biodiversity remains low for many years.

A higher level of restoration can be achieved by repairing overgrazed pastures and retired cropland. Damaged pastures need rest from grazing, prescribed fire, and reduction of invasive plants. The seed in pasture soils, known as the seed bank, routinely includes invasive species that slow progress towards domination by native species. Planting certain native species among the pasture plants, known as interseeding, is sometimes recommended. Restoration of abandoned cropland involves purchasing essentially all of the seed, which can be costly, but the cost of weed control is modest because the seed bank usually is dominated, not by pernicious perennial grasses such as smooth brome but by more easily controlled annual weeds. Restoration work on pastures and farmed land can recover most ecosystem services. One limiting factor appears to be the presence of enough forb diversity to provide nectar for pollinators and numerous other insects. Their recovery promotes higher avifauna diversity.

The land with highest potential for approximating historical conditions is native, unplowed prairies with few invasive plants but which have been degraded due to inattentive management. In tallgrass prairies, which are most in need of improvement and expansion, prescribed burning usually is recommended to stimulate germination of the native seed bank and to reduce weed pressure, especially by cool-season, invasive grasses such as Kentucky bluegrass and smooth brome (see box 4.3). Livestock grazing can play a role as well.

In general, prairie reconstruction is a challenge but scientists and land managers have learned much about how to proceed. Useful references include Wedin and Fales (2009), Smith et al. (2010), Galatowitsch (2012), Kurtz (2013), Norland (2015), Dixon et al. (2017), Yurkonis (2013), Yurkonis et al. (2019), and websites of the Natural Resources Conservation Service and Fish and Wildlife Service.

because the population sizes of uncommon species become larger and less vulnerable, in part because they are genetically more diverse.[4] They also provide more ecosystem services, as discussed in previous chapters.

Fortunately, the Dakotas have large areas on portions of numerous public lands that can be considered "natural" (fig. 1.11): Black Hills National Forest, Wind Cave National Park, Custer State Park, Badlands National Park, Theodore Roosevelt National Park, Custer National Forest, and the six national grasslands (Fort Pierre, Sheyenne, Little Missouri, Buffalo Gap, Grand River, and Cedar River).[5] Most of the national grasslands are in the western half of the two states. Also, large portions of tribal lands are essentially natural habitat. The conservation of wildlife and biological diversity is an important objective in each one, a reflection of priorities set by Congress, state legislatures, and tribal councils. Often such places are now cared for with an approach known as *ecosystem management*, which includes the following guidelines:[6]

- Conduct environmental impact assessments before proceeding with new initiatives, providing opportunities for adjacent landowners and the public to participate in a collaborative way.
- Protect or restore air and water quality.
- Maintain native species diversity and monitor the population sizes of vulnerable species so that trends are documented and used to improve management practices.
- Where life, property, and cultural resources can be protected, promote beneficial fires and floods.
- Discourage the construction of buildings in places where prescribed fires and floods are deemed important for achieving management goals.
- Seek collaborative ways of minimizing developments on adjacent land that cause adverse impacts inside the management unit, including the various effects of oil and gas development on water resources and scenic vistas.[7]
- Considering the absence of wolf and grizzly bear predation, evaluate hunting options that might maintain genetically diverse bison and elk herds—where they exist—at population sizes the habitat can support.
- Restore degraded habitats and close roads wherever feasible to reduce habitat fragmentation and the spread of invasive plants.
- Adopt management practices that minimize the adverse effects of periodic droughts, exceptionally wet years, and climate change.

National forests and national grasslands are highly valued because of the timber and forage they provide while also protecting watersheds and providing wildlife habitat and opportunities for outdoor recreation. Accomplishing these multiple goals is possible but often leads to heated debates because it pits anxious users who depend on public lands for their livelihood with those who make their living elsewhere. The Black Hills National Forest is a notable example because it is the largest tract of public land in the two-state area. The climate is more moderate than the surrounding plains and supports extensive forests, various kinds of wildlife not found elsewhere, and a thriving tourism industry. Unfortunately, the natural inclination to suppress fires over the years has led to the potential for fires in some areas that cannot be suppressed, because of fuel accumulation. Also, the public land is fragmented with private residences, resorts, and a variety of inholdings that make effective prescribed fires difficult to arrange, as discussed in chapter 8. Efforts are under way to resolve this problem with a combination of timber harvesting, forest thinning, and prescribed fire. The same kind of challenges can occur on private land, such as on much of Turtle Mountain.

The Potential for Large Prairie Preserves

There are now frequent discussions about the need for conservation of large prairie landscapes. The value of such places has been demonstrated in Badlands National Park, Wind Cave National Park, Theodore Roosevelt National Park, and the sovereign lands of some tribes—all large enough for prescribed fires, bison herds, and prairie dog colonies sufficient to support black-footed ferrets. In all cases the landscape has a diversity of habitats favorable for different kinds of wildlife, including rocky crags above and floodplains below. Such places also are popular destinations for tourists, photographers, biologists, and ecologists alike. Wind Cave National Park has the added benefit of being adjacent to three other large tracts of public land: Custer State Park, Black Hills National Forest, and Buffalo Gap National Grassland. Similarly, 50 miles to the east, Badlands National Park is adjacent to Buffalo Gap National Grassland and the Pine Ridge Indian Reservation. In North Dakota, Theodore Roosevelt National Park is adjacent to the Little Missouri National Grassland (see fig. 1.11).

The National Grasslands are a notable example of repurposing land after an undesired outcome. Privately owned for 50 years or more and with marginal crop production, these semiarid lands reverted to the public domain when many homesteaders abandoned farming because of drought in the 1930s, when soil was eroding at an alarming rate. The small land parcels were consolidated into national grasslands managed by the U.S. Department of Agriculture. Deliberately planted with perennial grasses and with the passive invasion of native species, the abandoned croplands are becoming prairie again. Introduced grasses such as smooth brome and Kentucky bluegrass are still more abundant than managers would like (see box 4.3), but the national grasslands illustrate the potential for reestablishing prairie ecosystems after years of cultivation and the advantages of managing these lands for livestock and wildlife.

Although many farmers abandoned their land during the Dust Bowl, others persisted.[8] Some farms and ranches were eventually enlarged by acquiring land from neighbors. Their livelihood depended on raising several crops along with a few cows, chickens, and pigs. Today, most farms are specialized and much larger. Also, many wetlands have been drained and some pastures plowed, further fragmenting wildlife habitat. Fertilization, weed and pest control, and sometimes irrigation became routine, which, combined with the costs of purchasing and operating new machinery, greatly increased the costs of cultivating "the breadbasket of the world." Recognizing the importance of keeping farmers in business, Congress approved subsidies to assist with recovering from inevitable droughts, hailstorms, floods, and other causes of crop failures.

Today, allocating public funds in this way has led to debates about the sustainability of some farms. At a time when more people are concerned about climate change and continuing losses of soil and wildlife habitat, probing questions have been raised:

- Do federal subsidies enable or encourage the plowing of native rangeland or draining of remaining wetlands?
- Are there sufficient federal or state incentives for implementing conservation practices on private land?
- Would public investments in farming be more compelling or appreciated if some marginal croplands were converted back to grasslands dominated by native plants?
- When and where are the profits from raising perennial grasses, producing protein in the form of meat, and providing wildlife habitat likely to be as great or greater than producing grain, seeds, silage, vegetable oil, and root crops?
- Recognizing the high level of carbon sequestration in perennial grasslands compared to annual croplands, can the market for carbon sequestration or other ecosystem services be expanded?
- Can the adverse effects of industrial development in one place in the Dakotas be mitigated to a significant degree by habitat improvements in another place, creating a market for ecosystem services, as discussed below?
- Have state and federal laws been passed that mandate mitigation, so that the sale of ecosystem services can be profitable to those wanting to change the way their land is used?

Such questions, and the limitations of small prairie preserves, led to discussions about how to establish natural areas that are much larger, perhaps as large as an average county in the Dakotas. One such endeavor in eastern Montana, known as the American Prairie Preserve, is to consolidate the management of portions of the Charles M. Russell National Wildlife Refuge, Upper Missouri River Breaks National Monument, the Fort Belknap Indian Reservation, and an estimated 500,000 acres of private land to be purchased by the American Prairie Foundation as it becomes available. The goal is to dedicate 2 million–3 million acres to native plains wildlife. Some landowners throughout the Northern Great Plains see opportunities for using their land differently than before, whether for tourism or by selling the protein of bison, elk, and pronghorn, all living among prairie dogs, mountain lions, coyotes, swift foxes, black-footed ferrets, and other species—possibly including wolves (figs. 13.1 and 13.2).[9]

Fig. 13.1. Plains bison (*Bison bison*) is the largest mammal in North America. Nearly extinct in the late 1800s, the survival of this icon is a testament to the success of conservation work, as reviewed most recently by O'Brien (2017) and Aune and Plumb (2019). There are now thousands of bison in herds that are variously labeled as commercial, domestic, conservation, legacy, and wild. Truly wild bison are rare (box 13.2). Photo by Chamois Andersen, Defenders of Wildlife.

Fig. 13.2. A recent challenge for conservation biologists has been saving the black-footed ferret from extinction. This one, raised in captivity, is becoming one of several hundred now living in the wild—all descendants of 18 individuals captured in a Wyoming prairie dog town and bred in captivity (see chapter 4). Success cannot yet be claimed. Source: Defenders of Wildlife.

Funding such large preserves is a formidable challenge, but private foundations, land trusts, and conservation groups are interested.[10] Moreover, some landowners, even if not willing to sell their land, are potential collaborators if there is a way for them to benefit. Research has shown that "selling nature" can be as profitable or more so than traditional ranching. The wildlife biologist Daniel Licht wrote that "permanently retiring portions of the Great Plains [in large prairie preserves] benefits taxpayers, the failing producers, the remaining producers, and future generations." He concluded that the costs of some government subsidies are greater than the benefits provided. Similarly, the economists Willard W. Cochrane and C. Ford Runge wrote that "permanent conversion of excess farmland [land that produces surplus grain] to natural areas could lessen the need for expensive farm programs, improve the farm economy, and enhance recreational and ecological benefits."[11]

One avenue for funding large prairie preserves is *conservation finance*. With this approach, land trusts and *conservation banks*, also known as *mitigation banks*, pay landowners to make habitat improvements that are subsequently certified by a third party, such as the U.S. Fish and Wildlife Service. The bank then recovers the cost of the improvements from an industry or other entity seeking to at least partially mitigate environmental degradation that is deemed necessary for a desired and otherwise approved development. The payments that landowners receive could be applied to restoring and selling additional conservation credits, or for enlarging a prairie reserve.[12]

An example of conservation finance has recently been announced between the Microsoft Corporation and two nongovernmental organizations, the Southern Plains Land Trust and the Environmental Defense Fund. Microsoft, after adopting a policy of becoming carbon neutral, purchased grassland carbon credits offered for sale by the land trust, which had paid the owners of two large ranches in southeastern Colorado to restore their cultivated land and degraded rangeland and to not plow up rangeland for crops in the future. These ranches, covering 28 square miles, now sequester 8,000 metric tons of carbon in the soil annually, which partially offsets Microsoft's carbon emissions and probably would not have been achieved without Microsoft's involvement. As noted, restored grasslands sequester much more carbon than most croplands and degraded rangelands do. In addition, wildlife habitat and other conservation values have been restored in grasslands, wetlands, and 20 miles of prairie streams. The land trust keeps daily records on the carbon footprint of the two ranches, including the carbon costs of using and applying nitrogen-based fertilizers and fueling vehicles and other machinery. A third-party verifier monitors rangeland condition. Other landowners have expressed an interest in participating in the future, and similar projects are under way elsewhere.[13]

What are the possibilities for restoring large tracts of native prairie in the Dakotas? One starting point could be to use Badland National Park and Buffalo Gap National Grassland as the core, with the potential involvement of the Oglala Sioux and other neighbors. The Rosebud Reservation already has started the Wolakota Buffalo Range project. One of its goals is "regenerating indigenous ecosystems."[14] A similar possibility is in the vicinity of Theodore Roosevelt National Park and Little Missouri National Grassland. Bison herds have not been reintroduced to the national grasslands, but they do exist in both national parks. Nearby landowners and a large corporation might choose to participate. Introducing the gray wolf, if that ever happened, would be debated. The costs of such measures could be a reasonable investment considering the benefits, reduced environmental costs, and reduced federal subsidies. Strategic fair-chase hunting of bison would be an alternative to wolves, with the goal of maintaining the wild bison gene pool (box 13.2).[15]

Box 13.2. Conservation of the American Bison: Success or Failure?

Bison bison is the largest living mammal in North America (fig. 13.1), and there used to be millions of them. Gray wolves and often grizzly bears would kill individuals that were not sufficiently vigilant, had poor vision, or could not run fast enough. Frigid winters killed those with insufficient fat reserves or that were not adept at pawing through deep snow for the food they needed. Diseases culled susceptible animals; smaller or less competitive bulls were less successful at breeding. These and other selection pressures led to the evolution of wild bison, too wild to raise like cattle. They survived until a new form of predation arrived for which they had no defense—indiscriminate hunters with firearms. By the late 1800s only about 200 of the plains bison remained, raising concerns about extinction.

Today it is common to see herds of bison in the Dakotas. Some are known as *conservation herds* and are kept in state parks, national parks, and tribal lands, all surrounded with stout fences. To avoid excess grazing by too many animals due to insufficient predation, the herds are culled periodically. Some are transferred to ranchers interested in starting *commercial herds*, for the primary purpose of selling grass-fed bison meat. It would appear that the American bison has been saved from extinction. However, except in the Greater Yellowstone Area, where the animals are unfenced and predation by large carnivores continues, some wildlife biologists have declared that bison are ecologically extinct on the Great Plains. Essentially, they have been domesticated—fenced, protected from natural predators, fed during harsh winters or droughts, and bred for ease of handling. Indeed, bison are not considered wildlife in most states, including North Dakota. In South Dakota bison are classified as wildlife only when penned in parks.[a] There are practical reasons bison and their natural predators do not range freely in the Northern Great Plains, but restoring wild bison on some large tracts of mixed-grass prairie is feasible. Adjacent towns and landowners could profit from ecotourists attracted by herds of an iconic animal and most of its associated flora and fauna—living together more or less as they did for millennia.

[a]Bailey 2013; O'Brien 2017; Aune and Plumb 2019.

Establishing large prairie preserves where ranches are common is more feasible than in farmlands where the soils have been tilled and human population density is higher. The best strategy in the eastern Dakotas is to provide incentives that reduce erosion, conserve remaining wetlands, restore wetlands and adjacent prairie where feasible, and maintain habitat for fish and wildlife where that is possible. Still, the case can be made for large prairie preserves in the Prairie Coteau, Missouri Coteau, and sandhills—places where much of the land is too steep, rocky, or sandy for cultivation and where native plants still dominate large areas. Bison could be reintroduced if stouter fencing can be provided, though some level of conservation management can be achieved using cattle instead of bison. Some marginal cropland already has been restored to perennial grass cover, such as in Waterfowl Production Areas, CRP lands, and various conservation easement programs. Such areas would be more effective if they could be enlarged.

Anticipating the Future

Many farmers surely noticed undesirable trends long before the 1930s. Clouds of dust in the air and flushes of sediments in creeks must have become more frequent as more of the prairie sod was broken. Wildlife populations declined due to habitat loss and hunting pressure. Some concerned ranchers and farmers would have taken steps to adjust their herd sizes or methods of cultivation, knowing that the sustainability of their livelihood depended on conserving the soil. But to stop plowing was not an option at the time; they were dependent on exotic species. Moreover, their attempts to curb soil erosion were often overwhelmed by crop-damaging weather and low market prices.

Major improvements have been made in farming practices, but the condition of rivers, lakes, wildlife habitat, and soils is often disappointing and now there is climate change to consider—warming, longer droughts, and more frequent storms that cause flooding. To prepare, there are two general approaches for everyone to consider.[16] One is to minimize further carbon emissions from the burning of fossil fuels. The other is to minimize the adverse effects of inevitable climate changes by modifying land management practices.[17]

Reducing carbon emissions requires a global effort, but land managers on the Great Plains can contribute in significant ways:

- Sequester carbon in the soil and provide other benefits by conserving prairies, wetlands, and other tracts of perennial vegetation that remain (box 13.3); agricultural intensification may be recommended in some areas.[18]
- Restore wetlands and perennial grassland ecosystems on land that is marginal for crop production, or in strips adjacent to cropland, thereby benefiting from the carbon sequestration provided by such places (see box 13.1).
- Incorporate regenerative farming methods that include cover crops and which contribute to carbon sequestration; minimize the hours that machinery is needed to produce and harvest crops.
- Use strategic crop rotations that include nitrogen-fixing plants, thereby minimizing the need for fertilizers that require large quantities of fossil fuel for their manufacture and application.
- Consider the costs associated with feedlots and the corn used to feed cattle; raise grass-fed beef where practical.
- Purchase fuel-efficient machinery; use solar panels when possible, such as for water heaters and grain drying.
- Minimize the costs of heating and cooling buildings by planting shelterbelts; use solar heating when feasible.
- Manage forests in the Black Hills, on Turtle Mountain, and elsewhere to maximize carbon sequestration in the wood of trees.

Of course, all consumers affect carbon emissions by their food choices. Buying food produced locally reduces emissions associated with food transport; consuming the meat of grass-fed livestock produces lower emissions than meat produced in feedlots and shipped long distances. Properly managed, grasslands on the Great Plains are profitable sources of protein. Some large food manufacturers and retailers now buy their grain from producers committed to minimizing their use of fertilizers and other agrochemicals, thereby encouraging sustainable practices and protecting water quality.

The second challenge is adjusting to projected climate changes: an increase of up to 9° F in mean annual temperature by 2085, a frost-free season 20 to 30 days longer by 2050, and higher mean annual precipitation but more frequent droughts and floods (see chapter 3). What is likely to happen? In a book titled *The New Normal: The Canadian Prairies in a Changing Climate,* the following climate-induced changes were forecast:[19]

- Forests, woodlands, and aspen parklands will decline due to reduced tree regeneration caused by drought, more frequent fires, and more frequent outbreaks of insects and pathogens.
- There will be increases in crop productivity per acre due to warmer conditions and more carbon dioxide—but only when water or nutrients are not limiting.
- Despite increases in mean annual precipitation, native tallgrass and mixed-grass prairie will become drier due to more evapotranspiration,

Box 13.3. Prairie as Biofuel and Soil Amendments

An innovative approach to expanding grassland and reducing the carbon footprint of farms is to grow perennial plant biomass and sell it to future biofuel refineries. This low-input practice stores more carbon in the soil, reduces soil erosion and stream flooding, increases wildlife diversity and abundance, and diversifies farm income, as discussed in chapter 5.

Presently, ethanol is produced at large refineries using specialized microbes to ferment corn seed and stover (leaves and stalks). Nearly half of the U.S. corn crop is converted to ethanol (USDA 2020). Fermentation using different microbes can produce ethanol from a monoculture of switch grass, an especially productive species of the tallgrass prairie. Alternatively, a thermal process using extreme heat, for example from a commercial-grade microwave oven, produces three by-products from any mixture of grassland plants: bio-oil, syngas, and biochar. Along with ethanol, the less familiar bio-oil and syngas have been used as fuel in internal combustion engines. Biochar is a form of charcoal used to improve soil structure and sequester carbon. The chemical procedures to produce biofuel from prairie biomass have been tested successfully in the laboratory, but they are not yet scaled up to commercial refineries.

Farm-scale experiments have shown that biomass grown for biofuel and soil amendments on retired farmland can reach five tons per acre for mixtures of species and six tons per acre for switchgrass monocultures. Given this volume, and the advantages of farming perennial plants, why are prairie biofuels not being produced? The answer is availability and price. Most grasslands were cultivated long before the technology to produce liquid biofuel was known, and now, an oversupply of subsidized corn has made biofuel from corn profitable. Moreover, massive amounts of corn and soybeans are produced each year while the area of tallgrass prairie continues to shrink. Biofuels would be used more often and more carbon could be sequestered in the soil if federal supports became available for reconstructing prairies for this purpose.[a]

[a]Zilverberg et al. 2014a, 2014b; Lark et al. 2015.

greater seasonality of rainfall, and extended droughts; warm-season grasses may extend farther north.[20]

- Many wetlands will be dry for longer periods, leading to declines in waterfowl and aquatic species.

Farming, livestock husbandry, and forest management practices surely will change in the future, as will conservation practices. Traditional priorities will continue, such as minimizing soil erosion, selecting profitable crops, providing habitat for highly desired species, and preventing rare and threatened species from becoming extinct or locally extirpated. But climate change complicates everything. New agro-business strategies will be adopted more widely to minimize the adverse effects of extreme weather events. Conservation biologists already are asking if wildlife preserves established long ago to provide habitat for certain species will remain suitable for those species. Should South Dakota continue to provide habitat for a species if new evidence suggests that comparable habitat in North Dakota and Saskatchewan will be better? Similarly, should South Dakota think more about providing habitat for new arrivals from Nebraska? Answering such questions will involve discussions about single species, but is single species management a good approach?

The realities of rapid climate change have led to the following conclusions and guidelines for conservation:[21]

- A network of connected native habitat over a large area usually is better than isolated reserves; create or conserve access routes (corridors) for sensitive species.
- Heterogeneous habitats with multiple sensitive species are more consequential than homoge-

neous, representative habitats set aside for a single species.

- Habitats set aside for the conservation of multiple species will change; manage for the delivery of ecosystem services rather than for specific species.
- Restoration of native habitats will become more important in the future, to conserve the genes of those native species that have survived climate change in the past; identify and conserve climate change refugia.[22]
- Assessments of the costs and benefits of conservation practices are improved when based on peer-reviewed research, the experience of land managers, and public involvement.
- Controversial conservation strategies should be considered, such as learning to live with some exotics when the environmental costs of controlling them seems high, or saving endangered animals by capturing and moving them to places that may have more favorable habitat in the future (assisted colonization).[23]

In general, if widespread undesired trends are to be curbed, it is clear that economic incentives often will be required. Accordingly, the public already has enabled federal and state programs that pay landowners for protecting or improving the soil that remains and providing or restoring wildlife habitat. In addition, conservation finance has emerged as a new form of banking, and non-governmental organizations now work collaboratively with landowners, providing opportunities for compensation in exchange for adopting new methods. Farmers also can now be more precise in the ways their fields are cultivated and fertilized; regenerative agriculture enables profits while maximizing environmental protection.[24] Life on the Northern Great Plains has changed dramatically since the advent of horse-powered machinery—a mere five generations ago. Change will continue; there are reasons and viable options for doing better. However, as the agrarian philosopher Wendell Berry asserts, "It won't happen unless a lot of people—consumers and producers, city people and country people, conservationists and land users—get together deliberately to make it happen."[25]

Afterword

There are limits to how much ecosystems can be abused without costly consequences to human health and long-term well-being. This principle gained traction when alarming rates of soil erosion prompted Congress to establish the Soil Conservation Service in 1935. Additional legislation followed for the purpose of curbing air and water pollution, providing wildlife habitat, and conserving species threatened with extinction. But soil organic matter continues to decline on most farmland, riparian ecosystems are still severely affected, water quality is a growing concern, and debates about forest management on public lands are often contentious. Worldwide, the human population has grown exponentially, which has had dramatic effects everywhere. In an instant of geologic time, the widespread effects of the Industrial Revolution have initiated a new era in earth's history—the Anthropocene, a time when human activity is even changing the climate. This is the most stunning revelation of our lifetime.

Turning the tide requires the involvement of producers and consumers. All of us are land stewards, part of the same system. Change will not happen unless all sectors of our economy and society participate, whether they manage private or public land, grow food or conserve wildlife habitat, care for land in towns and cities, or establish policies that reduce carbon emissions and ensure the availability of clean air and water. This is true everywhere. In the Northern Great Plains, farming now has the most widespread impacts. Changing directions is difficult, but the various forms of regenerative agriculture offer the most promise for improving the overall quality of farms, grasslands, woodlands, lakes, rivers, and wetlands while still being profitable. We have enjoyed learning and writing about the ingenuity of Dakotans as they continue to experiment with new methods in the twenty-first century. The challenges are great, time seems short, the ongoing quest to make a sustainable living continues.

Appendix: Latin Names of Animals Mentioned in Text

BIRDS

Avocet, American	*Recurvirostra americana*
Bittern, American	*Botaurus lentiginosus*
Blackbird, Red-winged	*Agelaius phoeniceus*
Blackbird, Yellow-headed	*Xanthocephalus xanthocephalus*
Bobolink	*Dolichonyx oryzivorus*
Bunting, Lark	*Calamospiza melanocorys*
Coot, American	*Fulica americana*
Cowbird, Brown-headed	*Molothrus ater*
Crane, Sandhill	*Antigone canadensis*
Crane, Whooping	*Grus americana*
Crossbill, Red	*Loxia curvirostra*
Crow, American	*Corvus brachyrhynchos*
Cuckoo, Black-billed	*Coccyzus erythropthalmus*
Cuckoo, Yellow-billed	*Coccyzus americanus*
Curlew, Long-billed	*Numenius americanus*
Dickcissel	*Spiza americana*
Dipper, American	*Cinclus mexicanus*
Eagle, Golden	*Aquila chrysaetos*
Egret, Snowy	*Egretta thula*
Flycatcher, Great Crested	*Myiarchus crinitus*
Godwit, Marbled	*Limosa fedoa*
Goshawk, Northern	*Accipiter gentilis*
Grackle, Common	*Quiscalus quiscula*
Grebe, Pied-billed	*Podilymbus podiceps*
Grosbeak, Black-headed	*Pheucticus melanocephalus*
Grosbeak, Blue	*Passerina caerulea*
Grouse, Ruffed	*Bonasa umbellus*
Grouse, Sharp-tailed	*Tympanuchus phasianellus*
Hawk, Ferruginous	*Buteo regalis*
Hawk, Swainson's	*Buteo swainsoni*
Henslow's sparrow	*Centronyx henslowii*
Heron, Great Blue	*Ardea herodias*
Heron, Green	*Butorides virescens*
Jay, Blue	*Cyanocitta cristata*
Kestrel, American	*Falco sparverius*
Killdeer	*Charadrius vociferus*
Lark, Horned	*Eremophila alpestris*
Longspur, Chestnut-collared	*Calcarius ornatus*
Longspur, McCown's	*Rhynchophanes mccownii*
Magpie, Black-billed	*Pica hudsonia*
Meadowlark, Eastern	*Sturnella magna*
Meadowlark, Western	*Sturnella neglecta*
Nighthawk, Common	*Chordeiles minor*
Oriole, Baltimore	*Icterus galbula*
Oriole, Bullock's	*Icterus bullockii*
Ovenbird	*Seiurus aurocapilla*
Owl, Burrowing	*Athene cunicularia*
Owl, Great horned	*Bubo virginianus*
Owl, Short-eared	*Asio flammeus*
Pelican, American White	*Pelecanus erythrorhynchos*
Phalarope, Wilson's	*Phalaropus tricolor*
Pheasant, Ring-necked	*Phasianus colchicus*
Plover, Mountain	*Charadrius montanus*
Plover, Piping	*Charadrius melodus*
Prairie Chicken, Greater	*Tympanuchus cupido*
Raven, Common	*Corvus corax*
Sandpiper, Upland	*Bartramia longicauda*
Sparrow, Clay-colored	*Spizella pallida*
Sparrow, Grasshopper	*Ammodramus savannarum*
Sparrow, Lark	*Chondestes grammacus*
Sparrow, Savannah	*Passerculus sandwichensis*
Swallow, Cliff	*Petrochelidon pyrrhonota*
Swan, Tundra	*Cygnus columbianus*
Tanager, Scarlet	*Piranga olivacea*
Tern, Black	*Chlidonias niger*
Tern, Forster's	*Sterna forsteri*
Tern, Least	*Sternula antillarum*
Towhee, Eastern	*Pipilo erythrophthalmus*

Towhee, Spotted	*Pipilo maculates*
Turkey, Common	*Meleagris gallopavo*
Vulture, Turkey	*Cathartes aura*
Warbler, Yellow	*Setophaga petechia*
Waxwing, Cedar	*Bombycilla cedrorum*
Woodpecker, Lewis's	*Melanerpes lewis*
Woodpecker, Red-headed	*Melanerpes erythrocephalus*
Wren, Rock	*Salpinctes obsoletus*
Wren, Sedge	*Cistothorus platensis*
Yellowthroat, Common	*Geothlypis trichas*

FISH

Bass, Largemouth	*Micropterus salmoides*
Bass, Smallmouth	*Micropterus dolomieui*
Bass, White	*Morone chrysops*
Bluegill	*Lepomis macrochirus*
Buffalo, Smallmouth	*Ictiobus bubalus*
Bullhead, Brown	*Ictalurus nebulosus*
Burbot	*Lota lota*
Carp, Bighead	*Hypophthalmichthys nobilis*
Carp, Common	*Cyprinus Carpio*
Carp, Silver	*Hypophthalmichthys molitrix*
Carpsucker, River	*Carpiodes carpio*
Catfish, Blue	*Ictalurus furcatus*
Catfish, Channel	*Ictalurus punctatus*
Catfish, Flathead	*Pylodictis olivaris*
Chub, Flathead	*Platygobio gracilis*
Chub, Sickelfin	*Macrhybopsis meeki*
Chub, Silver	*Macrhybopsis storeriana*
Chub, Sturgeon	*Macrhybopsis gelida*
Crappie, Black (Calico Bass)	*Pomoxis nigromaculatus*
Crappie, White (Calico Bass)	*Pomoxis annularis*
Drum, Freshwater	*Aplodinotus grunniens*
Gar, Longnose	*Lepisosteus osseus*
Gar, Shortnose	*Lepisosteus platostomus*
Goldeye (Mooneye)	*Hiodon alosoides*
Herring, Skipjack	*Alosa chrysochloris*
Minnow, Brassy	*Hybognathus hankinsoni*
Minnow, Fathead	*Pimephales promelas*
Minnow, Plains	*Hybognathus placitus*
Minnow, Western Silvery	*Hybognathus argyritis*
Paddlefish	*Polyodon spathula*
Panfish (several species)	
Perch, Yellow	*Perca flavescens*
Pike, Northern	*Esox lucius*
Salmon, Chinook	*Oncorhynchus tshawytscha*
Salmon, Kokanee	*Oncorhynchus nerka*
Sauger	*Sander canadensis*
Shad, Gizzard	*Dorosoma cepedianum*
Shiner, Emerald	*Notropis atherinoides*
Shiner, Red	*Cyprinella lutrensis*
Shiner, Spottail	*Notropis hudsonius*
Sicklefin Chub	*Macrhybopsis meeki*
Smelt, Rainbow	*Osmerus mordax*
Stickleback, Brook	*Culaea inconstans*
Sturgeon, Pallid	*Scaphirhynchus albus*
Sturgeon, Shovelnose	*Scaphirhynchus platorynchus*
Sucker, Blue	*Cycleptus elongatus*
Sucker, Quillback	*Carpiodes cyprinus*
Sucker, Shorthead Redhorse	*Moxostoma macrolepidotum*
Sucker, White	*Catastomus comersonii*
Trout, Brown	*Salmo trutta*
Trout, Lake	*Salvelinus namaycush*
Trout, Rainbow (Steelhead)	*Oncorhynchus mykiss*
Walleye	*Stizostedion vitreum*

INSECTS

Tortrix, Aspen	*Choristoneura conflictana*
Butterfly, Dakota Skipper	*Hesperia dacotae*
Beetle, Mountain Pine	*Dendroctonus ponderosae*
Beetle, Emerald Ash Borer	*Agrilus planipennis*
Beetle, Flea (several species)	
Moth, Forest Tent Caterpillar	*Malacosoma disstria*
Hawkmoth, Five-spotted	*Manduca quinquemaculatus*
Hawkmoth, Pink-spotted	*Agrius cingulata*
Butterfly, Monarch	*Danaus plexippus*
Locust, Rocky Mountain (extinct)	*Melanoplus spretus*
Beetle, Soybean Stemborer	*Dectes texanus texanus*

MAMMALS

Bear, Black	*Ursus americanus*
Bear, Grizzly	*Ursus arctos horribilis*
Beaver, North American	*Castor canadensis*
Bighorn Sheep	*Ovis canadensis*
Bison, American	*Bison bison*
Bobcat	*Lynx rufus*
Chipmunk, Eastern	*Tamias striatus*
Chipmunk, Least	*Neotamias minimus*
Cougar (see mountain lion)	
Coyote	*Canis latrans*
Deer, Mule	*Odocoileus hemionus*
Deer, White-tailed	*Odocoileus virginianus*
Elk	*Cervus elaphus*
Ferret, Black-footed	*Mustela nigripes*
Fox, Red	*Vulpes vulpes*
Fox, Swift	*Vulpes velox*
Ground Squirrel, Richardson's	*Urocitellus richardsonii*
Hare, Snowshoe	*Lepus americanus*
Jackrabbit, White-tailed	*Lepus townsendii*
Mink	*Mustela vison*
Moose	*Alces alces*
Mountain Lion	*Puma concolor*
Mouse, Deer	*Peromyscus maniculatus*
Mouse, Bear Lodge Meadow Jumping	*Zapus hudsonius campestris*
Mouse, White-footed	*Peromyscus leucopus*
Muskrat	*Ondatra zibethicus*
Opossum	*Didelphis virginiana*

Otter, River	*Lontra canadensis*
Pocket Gopher, Northern	*Thomomys talpoides*
Prairie Dog, Black-tailed	*Erethizon dorsatum*
Porcupine	*Cynomys ludovicianus*
Pronghorn	*Antilocapra americana*
Rabbit, Desert Cottontail	*Sylvilagus audubonii*
Rabbit, Eastern Cottontail	*Lepus sylvaticus*
Raccoon	*Procyon lotor*
Shrew, Least	*Cryptotis parva*
Shrew, Northern Short-tailed	*Blarina brevicauda*
Skunk, Striped	*Mephitis mephitis*
Squirrel, Fox	*Sciurus niger*
Squirrel, Eastern Gray	*Sciurus carolinensis*
Vole, Meadow	*Microtus pennsylvanicus*
Weasel, Long-tailed	*Mustela frenata*
Wolf, Gray	*Canis lupis*
Woodrat, Bushy-tailed	*Neotoma cinerea*

Notes

Chapter 1: The Land, Ecosystems, and Ecology

1. Handy-Marchello and Swenson 2018; see chapter 2.
2. Unlike South Dakota, about 30 percent of western North Dakota is north of the Missouri River.
3. Costanza et al. 1997, 2014; Zedler 2003; Wall 2012; Johnson 2019.
4. Brussard 2012.
5. Sagoff 2002; Blanchard and Briefer 2015. Salt Lake City, Utah used regulations and zoning to protect the quality of drinking water flowing from the Wasatch Mountains.
6. *Des Moines Register*, April 11, 2017, and January 23 and June 22, 2018, https://www.desmoinesregister.com/story/money/agriculture/2017/03/17/judge-dismisses-water-works-nitrates-lawsuit/99327928/.

Chapter 2: Landscape History

1. Handy-Marchello and Swenson (2018) and Hoover (2009) review ancient human history; Bluemle (2016) and Gries (1996) review geologic history. Geologic and topographic maps are accessible on the websites of geological survey offices in each state.
2. Martin and Wright 1967; Alroy 2001; Sandom et al. 2014; Widga et al. 2017; see chapter 5.
3. Cretaceous seas in central North America are known as the Western Interior Seaway. Mountain ranges were found in the Dakotas a billion years ago or more, but all were leveled by erosion long before the time our historical account begins.
4. John Bluemle, personal communication.
5. This period of mountain building west of the Dakotas is known as the Laramide Orogeny. Mountains had formed in approximately the same area before, but they were worn down by erosion (Knight et al. 2014).
6. Robertson et al. 2004; DePalma et al. 2019, reviewed by Douglas Preston in the *New Yorker* (April 8, 2019) and by Colin Barras in *Science* (2019). The Cretaceous-Paleogene (KPg) boundary is also known as the Cretaceous-Tertiary (KT) boundary.
7. Axelrod 1985.
8. Calendar years and radiocarbon years differ slightly because the amount of carbon 14 in the atmosphere during the past 40,000 years has not been constant.
9. Bluemle 2000, 2016. The Red River Valley was a prominent feature several million years before glaciation. There is some evidence (erratics and patterned ground) that pre-Wisconsin glaciers extended to the Killdeer Mountains, Dickinson, and Bowman in southwestern North Dakota, but most of that evidence has eroded away.
10. *Eskers* may be 100 feet high, a few hundred feet wide, and several miles long. The most prominent esker in the Dakotas is the Dahlen Esker, near Dahlen, North Dakota. *Kames* are mounds or conical hills that formed as the ice melted. *Drift* is a general term for material transported by a glacier and deposited by or from the ice, or by running water emanating from a glacier; *till* is the unconsolidated material deposited by glaciers without subsequent reworking by meltwater.
11. Union Grove State Park, Newton Hills State Park, and Blood Run State Park are located in the loess hills of South Dakota.
12. Bluemle 2000, p. 43; good examples of ice thrusting in central North Dakota are the Prophets Mountains in Sheridan County and the hills adjacent to Devils Lake.
13. Chapman et al. 1998; Bluemle 2000, p. 13.
14. Teller and Leverington 2004.
15. Bluemle 2016.
16. Hoover (2009) reports that people first arrived in the land that would become South Dakota about 12,000 years ago; Handy-Marchello and Swenson (2018) report about 13,500 years ago for North Dakota. See also Ahler et al. (1991), Hoffman and Graham (1998), Raventon (2003), and Risjord (2012). There is evidence that the first Native Americans, known as paleo-Indians, camped along the banks of major rivers and glacial lakes. They

were nomadic and were assisted by dogs that pulled small sleds (*travois*), but for thousands of years they had no horses. For centuries the Mandan and Hidatsa were centered in North Dakota, the Arikara (Ree) in South Dakota.

17. Hoover 2009; Handy-Marchello and Swenson 2018; the Dakotas came to be known as the Sioux and included the following tribes: Mdewakanton, Sisseton, Teton, Lakota, Oglala, Brulé, Wahpekute, Wahpeton, Yankton, and Yanktonai. Scientists now are benefiting from the traditional knowledge of indigenous people (see pages 1249–1340 in volume 10 of the journal *Ecological Applications*, published in 2000).
18. Martin and Wright 1967.
19. Pielou 1991; Grimm 2001; Yansa and Ashworth 2005; Bradbury 1980. Severson and Sieg (2006) provide dates for the presence of spruce in North Dakota. Watts and Bright (1968) studied Pickerel Lake in northeastern South Dakota. Baker et al. (1996) studied northeastern Iowa.
20. Grimm 2001; Watts and Bright 1968; McAndrews 1966; Shay 1967. Data from Wordworth Pond in North Dakota revealed that spruce, tamarack, poplar, and black ash disappeared from that area about 9,000 years ago. Burgess (1964) used land survey records in eastern North Dakota to document vegetation changes from 1871 to 1961, and Fredlund and Tieszen (1997) used phytoliths and carbon isotopes to study grassland expansion in the southern Black Hills. Grimm et al. (2011) studied Kettle Lake in North Dakota.
21. Muhs et al. 1997; Mangan et al. 2004; Mason et al. 2011.
22. Bluemle 2000.
23. Raventon 2003; Severson and Sieg 2006; Thompson 2009; Risjord 2012.
24. Lavin et al. 2011, p. 50.
25. Severson and Sieg 2006; Myers 2009. Other explorers in the late 1700s and early 1800s were David Thompson, Alexander Henry (the younger), Francois Larocque, Jean Baptiste Truteau, John Evans, Manuel Lisa, Major Stephen Long, Joseph Nicollet, John Frémont, James Audubon, and George Catlin.
26. Cutright (1969), Botkin (1995), Isenberg (2000), Johnsgard (2003), and others summarized the natural history of the plants and animals mentioned in the journals of the Lewis and Clark expedition.
27. Severson and Sieg 2006.
28. Roe 1972; see also Isenberg 2000, Bailey 2013, O'Brien 2017; Aune and Plumb 2019.
29. Raventon 2003; Danbom 2014.

Chapter 3: Climate

1. Cook et al. 2004; Matthew Bunkers, quoted by Mark Meierhenry in the March–April 2008 issue of *South Dakota Magazine* (https://www.southdakotamagazine.com/old-growth-pines). Lauenroth et al. (1999) describe the climate and grasslands of the Northern Great Plains.
2. Winter and Rosenberry (1998) describe floods when an intense drought in the central Dakotas (1988–1992) was followed by a period of heavy rains; see also Johnson et al. (2004) and Shapley et al. (2005).
3. Keating 1824; Ross 1856.
4. See online reports of the Intergovernmental Panel on Climate Change (IPCC) and the Wikipedia page on the scientific consensus on climate change (https://en.wikipedia.org/wiki/Scientific_consensus_on_climate_change).
5. More CO_2 in the atmosphere leads to the formation of more carbonic acid in water; atmospheric CO_2 concentrations continue to rise and were 414 ppm in early 2020.
6. Laura Roti, "No One Turned Off the Tap," *Morning Ag Clips*, December 8, 2019, https://www.morningagclips.com/no-one-turned-off-the-tap/.
7. WeatherTrends360 provides 11-month, week-by-week forecasts for specific areas at www.weathertrends360.com. Many farmers use this service when developing planting and harvesting strategies.
8. See *scientific consensus on climate change* (https://en.wikipedia.org/wiki/Scientific_consensus_on_climate_change).
9. Ojima et al. 2015; Jones 2019; Gibson and Newman 2019. Studies in adjacent provinces and states have made similar projections: see Barrow (2010), Byrne et al. (2010), and Henderson and Thorpe (2010) for Saskatchewan and Manitoba; Shuman (2012) for Wyoming. The U.S. Environmental Protection Agency (2016) projects the implications of climate change for the Dakotas.
10. Williams and Jackson 2007; Williams et al. 2007. See Jackson (2021) for a discussion of three options that managers have for adapting to climate change, namely, resisting changes where feasible, accepting the new ecosystems that develop, or intervening to direct changes toward desirable or acceptable new ecosystems.

Chapter 4: Grasslands

1. Madson 1985.
2. Severson and Sieg (2006) provide a fascinating review of the journals written by early explorers in eastern North Dakota. They discuss the causes of bison movements from one place to another, noting that early explorers usually saw wolves within a day of having seen bison. Elk were mentioned nearly as often as bison.
3. Madson 1982.
4. Van Tramp 1868.
5. Goodwin 2013, p. 110.
6. Wilkins and Wilkins 1977.
7. Küchler 1964; Singh et al. 1983; Barker and Whitman 1988; Lauenroth et al. 1994, 1999. Shortgrass prairies are widespread south of the Dakotas.
8. Hanson and Whitman 1938; Cosby 1965; Dix and Smeins 1967; Ralston 1968; Redman 1972; Godfread and Barker 1975; Whitman and Wali 1975.
9. Flesland and Whitman (1963). Salt-desert shrublands occur near the badlands; see table 12.1.

10. Cool-season grasses have C_3 metabolism; warm-season species have C_4 metabolism and are more drought tolerant.
11. Dandom 2014 (pp. 41–42); for eastern South Dakota, see *Clear Lake Courier* (June 25, 1959).
12. Higgins (1984, 1986) reviews fire records dating back to the early 1800s; frequency of lightning fires in mixed-grass prairie in eastern North Dakota was 6.0 fires per year per 10,000 square kilometers (22.4 in south-central North Dakota, 24.7 in western North Dakota, and 91.7 in the ponderosa pine savannas of northwest South Dakota and southeast Montana). Brown et al. (2005) found that fires during the past 4,500 years were more frequent during wet periods that produced more fuel (see box 3.1). See also Wright and Bailey 1982, Collins and Wallace 1990, Courtwright 2011, Pyne 2017, and Yurkonis et al. 2019.
13. Severson and Sieg (2006, p. 204) concluded that the effect of bison on woodlands was limited in area, but that they could graze grasslands heavily (p. 212). They also concluded (p. 191), on the basis of the historical literature they reviewed, that bison were extirpated from eastern North Dakota by 1883.
14. Weaver 1943, 1954.
15. Bailey 2013. Urbanhower (1992) compared the effects of bison wallows and burrowing animals in northcentral South Dakota.
16. Knapp et al. 1999; Shelby 2003; Bailey 2013. The bones of bison sometimes were shipped to eastern states for the fertilizer industry. Severson and Sieg (2006, p. 187) reported that "bone fields" were mentioned frequently in the journals of early explorers.
17. Grizzlies and gray wolves still live in the Greater Yellowstone Ecosystem and northward into the Canadian Rocky Mountains; gray wolves also are found in northern North Dakota, northern Minnesota, and adjacent provinces.
18. Frémont 1887.
19. Bray and Bray 1993.
20. Knapp et al. 1999; Bailey 2013; O'Brien 2017.
21. Biondini et al. 1999; Helzer 2010; Fuhlendorf et al. 2012.
22. Kotliar et al. 2006.
23. Davidson et al. 2012.
24. Larson 2002.
25. Other weedy plants found on some prairie dog towns in Wind Cave National Park include cheatgrass, various thistles, and common mullein; see Agnew et al. (1986) and Drazkowski et al. (2011).
26. Fahnestock and Detling 2002; Miller et al. 2007.
27. Buskirk 2016.
28. Becher et al. 2013.
29. De Trobriand 1868 (quoted in Wilkins and Wilkins 1977); Allred 1941; Lockwood 2004; Dandom 2014.
30. Zimmerman et al. 2004.
31. Pete Bauman, South Dakota State University Extension Service, personal communication
32. Samson et al. 2004; Brennan and Kuvlesky 2005; Lavelle 2015.
33. Helzer 2010. DeKeyser et al. (2015) describe the long history and effects of Kentucky bluegrass; Grant et al. (2009, 2020) and Dixon et al. (2017, 2019) summarize invasive plant problems in Dakota wildlife habitat.
34. Brown et al. 2005; Baer et al. 2019; Gibson and Newman 2019a, 2019b; Yurkonis and Harris 2019; Grant et al. 2020. Biondini and Manske (1996) evaluated the effect of drought and grazing, Nordt et al. (2008) used stable carbon isotopes in buried soil profiles to describe changes in relative C_4 plant growth in North American prairies as climate changed during the past 10,000 years.
35. Hufkens et al. 2016; Fraser 2019. Morgan et al. (2011) found that elevated carbon dioxide can moderate the desiccating effects of warming.
36. Knapp et al. 2008; Jones 2019.
37. Tolstead 1941; Weaver 1943.
38. Catford and Jones 2019; Yurkonis and Harris 2019.
39. Bardgett and Semchenko 2019; Henry 2019. George et al. (1992) studied the effect of drought on grassland birds in western North Dakota.
40. Tilman et al. 2006; Isbell et al. 2015.
41. Thebault et al. 2014.
42. Knapp et al. 2008; Fraser 2019; Bradford et al. 2019. Henry (2019) discusses how nutrient availability is affected by changes in the pattern of precipitation events caused by climate change; Briske et al. (2015), Lavorel (2019), and Bradford et al. (2019) propose practices that mitigate the adverse effects of climate change on rangeland health and forage production.

Chapter 5: Agriculture and Agroecology

1. Handy-Marchello and Swenson 2018; Wilson 1987; Fenn 2014.
2. Lehmer 1971; Wood 1998; Johnson et al. 2007. Early settlements of the Mandan, Hidatsa, and Arikara were concentrated along the Missouri River between the Niobrara and Knife Rivers; the three tribes are now known as the MHA Nation or Three Affiliated Tribes.
3. Gilmore 1991; Fenn 2014.
4. Corn, the crop that would become so widespread across the Dakotas, was domesticated by indigenous people in southern Mexico beginning about 10,000 years ago, from a wild grass called teosinte (Beadle 1980). A plant recognizable as modern corn appeared about 5,000 years later. Corn seed was widely traded throughout the Americas and was introduced to Spain shortly after 1492.
5. Handy-Marchello and Swenson 2018.
6. Moulton 1988.
7. Lehmer 1971; Fenn 2014.
8. In 2013 income in North Dakota was about \$3.6 billion from agriculture and about the same from oil and gas. In 2014, income was about \$3 billion from agriculture and \$3.7 billion from oil and gas (National Agricultural Statistics Service 2017).
9. Hunt 1951.
10. Nelson 1986; Tweton and Albers 1996.
11. Nelson 1986; Dandom 2014.
12. Severson and Sieg 2006.
13. Manning 1995.

14. Nelson 1986.
15. Nelson 1986; Nelson 1996; Schell and Miller 2004; Handy-Marchello 2005; Brokke 2015. Garden vegetables on homesteads included peas, beans, carrots, beets, cabbages, onions, parsnips, radishes, rutabagas, asparagus, and rhubarb.
16. Nelson 1986, 1996.
17. Nelson 1996.
18. Nelson 1996.
19. Burns 2012; Nelson 1996; Robinson 1966; Drache 1970. Wilkins and Wilkins (1977) described the three dreadful *Ds*—drought, depression, and dust—in North Dakota, writing. "Coincidental with the drying up of the countryside came the grasshoppers in numbers that frustrated all efforts at control. At Mott in 1933, the sky was so filled with them that streetlights had to be turned on in the middle of the day" (pp. 101–2).
20. Burns 2012; Nelson 1996.
21. Montgomery 2017; Anderson 2019.
22. National Agricultural Statistics Service 2017.
23. Licht 1997.
24. National Agricultural Statistics Service 2017.
25. Wright and Wimberly 2013; Reitsma et al. 2015; Johnston 2014.
26. Doherty et al. 2013. Reitsma et al. (2015, 2016) suggested other factors that may have contributed to cropland expansion: changes in land ownership, shifts in crop rotation, technology improvements and industry investments, government policies such as subsidized crop insurance, climate change, and an aging workforce.
27. Tiner 2009; Johnston 2013; Doherty et al. 2013.
28. Ojima et al. 2015; Hartman et al. 2011 (cited in Ojima et al. 2015); Montgomery 2017; Anderson 2019.
29. Galatowitsch and van der Valk 1998.
30. Montgomery 2017. The enormous area in the Great Plains treated with glyphosate, for example, has led to the rapid evolution of glyphosate-resistant weeds that require use of different herbicide treatments that will not kill the crop plants. Common resistant weeds in Dakota fields are barnyard grass, Palmer amaranth, water hemp, and horseweed.
31. Pollan (2006) describes how industrial agriculture has adversely affected human eating habits.
32. Gliessman 2007. Schulte et al. (2017) demonstrated the value of prairie strips in corn-soybean croplands.
33. Odum 1969.
34. Montgomery 2012.
35. Geertsema et al. 2016.
36. Quoted in Montgomery 2017, p. 93.
37. Claassen et al. 2018; Montgomery 2017; Gliessman 2007.
38. Numerous organizations in the Dakotas have active programs to conserve remaining grasslands, wetlands, and their biodiversity. The larger nongovernmental organizations are The Nature Conservancy, Ducks Unlimited, Pheasants Forever, Elk Foundation, South Dakota and North Dakota Wildlife Federations, and the Grassland Coalition. Government conservation agencies include U.S. Fish and Wildlife Service, Natural Resources Conservation Service, North Dakota Department of Fish and Game, and South Dakota Department of Game, Fish, and Parks. Without them the losses of natural ecosystems would have been much higher.
39. Jackson and Jackson 2002; Imhoff and Baumgardner 2006; Montgomery 2017; Anderson 2019.
40. Tschumi et al. 2015; Otto et al. 2016; Koh et al. 2016; Isaacs 2009.
41. Thogmartin et al. (2017) propose conservation plans for the monarch butterfly.
42. Johnson and Volke 2015.
43. Earthen dams were built on Foster Creek in the 1980s and 1990s to fill gullies with sediment and rebuild the floodplain and adjacent aquifers.
44. Savory 2016.
45. The National Audubon Society now assists ranchers who adopt livestock grazing practices favorable for birds, pollinators, and other wildlife, as described in the summer 2019 issue of *Audubon* magazine.
46. Beck 2017.
47. Zilverberg et al. 2015.
48. L. Jackson 2006.
49. Brown 2018.
50. To assist with marketing, the National Audubon Society now certifies meat raised on bird-friendly land (audubon.org/meat).
51. Anderson 2019.
52. Kernza, a perennial wheat, was produced after several decades of selective breeding by Wes Jackson's Land Institute and the University of Minnesota using its parent plant, *Thinopyrum intermedium*, a wheatgrass native to Turkey. Kernza has the taste and quality of a cereal grain. It remains productive for five years before needing to be replanted, thereby reducing production costs, including soil erosion. It also needs less fertilizer than other grain crops. Supplementing or replacing annual wheat with Kernza will depend on its palatability and nutritional value, as well as its yield. The Birchwood Café in Minneapolis is promoting Kernza. The grain's future brightened when the Kernza éclair sold out every day of the Minnesota State Fair. Agronomists expect Kernza to grow well in the soils and climate of the Dakotas. Trials in various states are under way.
53. Gliessman 2007; Geertsema et al. 2016. Schulte et al. (2017) write about corn-soybean croplands.
54. Montgomery 2017.
55. L. Jackson (2006) describes economic factors that have forced some farmers to abandon sustainable, low-input practices; DeVuyst et al. (2006) concluded that government payments impede adoption of more environmentally friendly farming in the Northern Great Plains.
56. Karl 2009; Wuebbles et al. 2017.
57. See online reports of the Intergovernmental Panel on Climate Change (2018).
58. See U.S. Department of Agriculture, Census of Agriculture 2017.
59. About 4 million acres of crop land in South Dakota were too wet to plant in 2019.
60. U.S. Environmental Protection Agency 2016.

61. Pearsall 2013; Montgomery 2017. Fred Kirschenmann is a past director of the Leopold Center for Sustainable Agriculture at Iowa State University (Jackson 2006).

Chapter 6: Trees on the Farm

1. Stoeckeler and Williams 1949; Bratton et al. 1995; Meneguzzo et al. 2018. Early farmers brought tree seed with them, and in the late 1800s state and federal programs provided seeds of species that might be good for shelterbelts. Before planting, some counties had essentially no trees except along creeks and rivers. Piva et al. (2013) describe various kinds of "treed" land in South Dakota.
2. The Clarke-McNary Act of 1924 funded the production and distribution of saplings for shelterbelts; few trees were planted in the Dakotas until after Congress passed the Timber Culture Act of 1873, which predated the arrival of most immigrants of European origin. Droze (1977), Baer (1989), Cable (1999), Orth (2007), Gardner (2009), and Karle and Karle (2017) describe the Prairie State Forestry Project.
3. Gardner 2009; Orth 2007.
4. Zon 1935.
5. Bratton et al. (1995) estimated 18,813 miles of windbreaks for South Dakota. See also Meneguzzo et al. 2018.
6. Stoeckeler and Williams 1949; Baer 1989; Cable 1999; Brandle et al. 2004; Gardner 2009; Patel-Weynand et al. 2017. Wind speeds can be reduced from 25 to 5 miles per hour, but this benefit declines with distance from windbreak.
7. Read 1958; Orth 2007.
8. Stoeckeler and Williams 1949; Read 1958; Emmerich and Vohs 1982; Cable 1999.
9. Emmerich and Vohs 1982; Johnson and Beck 1988; Dix et al. 1995; Swanson et al. 2003; Brandle et al. 2004; Liu and Swanson 2014.
10. Johnson and Beck 1988; Dix et al. 1995. Pejcar et al. (2018) discuss the services and disservices of birds in agroecosystems.
11. Tack et al. 2017; Samson et al. 2004.
12. Johnson and Temple 1990; Pierce et al. 2001; Ellison et al. 2013; Thompson et al. 2016; Greer et al 2016. Pietz et al. (2009) and Benson et al. (2013) cautioned that the removal of trees and the effects of cowbirds vary geographically. Tack et al. (2017) found that some grassland birds became more abundant after shelterbelt removal.
13. USDA Northern Great Plains Research Laboratory in Mandan has produced quantitative tools to facilitate precision agriculture; see chapter 5.
14. Burke et al. (2019) found that shelterbelt land area increased from 1962 to 2014 in Grand Forks County, North Dakota, but declined from 2014 to 2016.
15. Hocking et al. 2010; Meneguzzo et al. 2018. For an overview on the value of maintaining shelterbelts, see https://thefern.org/2017/11/uprooting-fdrs-great-wall-trees/.
16. Baer 1989; Grala and Colletti 2003; Brandle et al. 2004; Gardner 2009; Amichev et al. 2015; Ballesteros-Possu et al. 2017; Patel-Weynand et al. 2017.
17. Grala and Colletti 2003.
18. Tyndall and Colletti 2007; Amichev et al. 2015.
19. Amichev et al. 2015.
20. Kort and Turnock 1999; Amichev et al. 2015; Ballesteros-Possu et al. 2017. The USDA Natural Resouces Conservation Service has developed the Whole Farm and Ranch Carbon and Greenhouse Gas Accounting System (http://comet-farm.com/Home).
21. Montgomery 2017; Patel-Weynand et al. 2017; see also chapters 5 and 13.

Chapter 7: Upland Deciduous Forests, Woodlands, and Parklands

1. Rumble et al. (1998) list prairie woodland birds and include landscape photographs taken in 1902 and 1909 in South Dakota along the Missouri River, at Slim Buttes, and elsewhere.
2. Bluemle (2016) refers to Turtle Mountain as a "glaciated butte."
3. Potter and Moir 1961; Disrud 1968; Henderson et al. 2002.
4. Potter and Moir 1961. The nearest naturally occurring spruce is in the Spruce Woods Forest Preserve of Manitoba, located in the Assiniboine River Valley. Paleoecologists have determined that spruce disappeared from Turtle Mountain in the late 1800s, when the area experienced a 14-year drought and fires occurred more often (1877–1891). This period also coincides with the arrival of the railroad and fire-igniting steam engines.
5. Henderson et al. 2010.
6. Other "island forests" in the Northern Great Plains are the Black Hills, the Sweet Grass Hills of Montana, the Cypress Hills of Alberta and Saskatchewan, and the Spruce Woods of Manitoba.
7. Bluemle (2016) describes the geologic history of the Killdeer Mountains, a high ridge between two buttes.
8. Crompton et al. 1977. Common juniper (*Juniperus communis*), horizontal juniper (*J. horizontalis*), bearberry (*Arctostaphlos uva-ursi*), and snowbrush (*Ceanothus velutinus*) are found at higher elevations on the Killdeer Mountains.
9. Severson and Sieg 2006.
10. Savage 2004. Fires don't burn well downslope, and the often-dense shrub and tree growth reduce the amount of flammable grass cover.
11. Brokke 2015.
12. Tolstead 1941b; Uresk et al. 2015.
13. Thwaitess 1905.
14. Shapley et al. 2005. Haugen et al. (2005) surveyed North Dakota forests; Moore and Flake (1994) described the forests of eastern and central South Dakota; Piva et al. (2013) describe various kinds of "treed" land in South Dakota. Dense forests are found on islands in Waubay Lake and Devils Lake.
15. Tieszen and Pfau 1995; Spencer et al. 2009, 2015.
16. The wood of eastern redcedar is good for construction because of its resistance to decay. It also is a better fuel than the soft wood of cottonwood. Most redcedar trees may have been harvested in some places, such as near Mandan villages along the Missouri River. Settlers of European descent probably harvested this tree for the same reasons, leading to the local extirpation of the species.

17. Bird 1930; Archibold 1999; Savage 2004.
18. Archibold 1999.
19. Plant opal (phytolith) comprises microscopic, stonelike particles that are produced by grasses and persist in the soil where grasses once grew. Plant opal is not found in forest soils unless the trees have recently invaded grasslands (Fredlund and Tieszen 1997).
20. Rapid spouting from aspen roots begins after the aboveground buds of the trees are killed or removed by browsing, a phenomenon known as apical dominance and regulated by plant hormones. The clonal nature of aspen groves, called "bluffs" in Canada, is conspicuous in autumn, when all the "trees" in a clone turn leaf color simultaneously while a neighboring clone is green.
21. Anderegg et al. (2012, 2013) suggest that if aspen roots are deprived of carbon reserves by drought-induced mortality, directly or indirectly, then water uptake is less and plant water stress becomes more severe.
22. Boetcher and Johnson 2005.
23. Archibold 1999, p. 414; Grant and Berkey 1999.
24. Bird (1930) illustrates food webs and succession in aspen parklands; Svedarsky and Buckley (1975) report on the effects of fire; Bork et al. (2013) studied the effects of browsing by elk, bison, deer, and cattle.
25. See Bergdahl and Hill (2016) and periodic forest health reports of the North Dakota Forest Service and South Dakota Division of Resource Conservation and Forestry for management recommendations and updates on problematic insects and pathogens.
26. Tolstead 1941b; Girard et al. 1987, 1989. Henderson and Thorpe (2010) predicted that future droughts will cause shrinking aspen groves and less shrub cover.
27. Thurman et al. (2020) discuss ways of assessing the adaptive capacity of species to climate change.

Chapter 8: Ponderosa Pine, Pine Ridge, and the Black Hills

1. Hadley (1969) concluded that grassland soils desiccate more rapidly than woodland soils during summer droughts, and that pine is at a competitive disadvantage with herbaceous plants that have more surface roots. Potter and Green (1964a, 1964b) and Hadley (1969) described pine woodlands in southwestern North Dakota, where they occur largely on ridges, buttes, and knolls. Tolstead (1941b, 1942, 1947) and Weaver (1965) described woodlands in northern Nebraska and south-central South Dakota, including Pine Ridge.
2. Gartner and Thompson 1972; management plans prepared for the Black Hills and Nebraska National Forests.
3. Tolstead 1941b, 1947; Johnsgard 2001.
4. Fiedler and Arno (2015) review the ecology of ponderosa pine; Piva et al. (2013) document that 78 percent of the timberland in South Dakota is dominated by ponderosa pine.
5. Froiland 1990; Hoover 2009.
6. Kornfeld and Osborn 2003; Estes (2021) writes about the Oceti Sakowin, Dakota, Nakota, and Lakota who think of the Black Hills as their home.
7. Freeman (2015) reviews the history of forestry in the Black Hills. Shepperd and Battaglia (2002) describe the forest types; see also Larson and Johnson (1999); Marriott and Faber-Langendoen (2000a), and Wienk et al. (2004).
8. Custer 1875, p. 506; Donaldson 1875, p. 564; Ludlow 1875; Raventon 1994; Grafe and Horsted 2005.
9. William E. Curtis from Chicago, quoted by Freeman 2015, p. 16.
10. Cougars (mountain lions) were essentially extirpated from the Hills in the early 1900s, but they are now common.
11. Buttrick 1914; McIntosh 1931; Raventon 1994; Larson and Johnson 1999.
12. Lodgepole pine and limber pine are common in the Rocky Mountains to the west but are found in only one or two localities in the Black Hills (Thilenius 1970); Engelmann spruce, subalpine fir, and Douglas fir are not found in the Dakotas, except where planted.
13. Buttrick 1914; Wright 1970.
14. Marriott and Faber-Langendoen (2000a) describe the riparian and wetland plant communities of the Black Hills.
15. McIntosh 1931.
16. Severson and Thilenius 1976.
17. MacCracken et al. 1983; Marriott and Faber-Langendoen 2000b; see chapter 4.
18. Marriott and Faber-Langendoen 2000b; Marriott 2012.
19. MacCracken et al. 1983; Hoffman and Alexander 1987. Mountain mahogany is found in the Black Hills east of Newcastle and is rare in the Bear Lodge Mountains.
20. The Lakota name for the Black Hills is *Paha Sapa*, which translates to "hills that are black," an impression created by the appearance of ponderosa pine forests from a distance. Lageson and Spearing (1988) and Gries (1996) summarize the geologic history of the Hills and include road logs that identify geologic features; Froiland (1990) and Raventon (1994) review the natural history of the Black Hills. The geologic origin of Devils Tower is unclear, but it is one of several igneous intrusions in the area.
21. Larson and Johnson (1999) describe six geomorphic regions rather than five, adding the gray shale foothills common around the Bear Lodge Mountains.
22. The Spearfish Formation is known as the Chugwater Formation in Wyoming; mammoth bone excavations from this formation can be observed at the Mammoth Site in Hot Springs, South Dakota.
23. Norris et al. (2016) examined plant remnants in present-day packrat middens in the Black Hills with layers that are 11,000 years old or more. Three 9,000-year-old middens in the nearby Bighorn Mountains contained ponderosa pine macrofossils dated at 2,630 years old or younger. The more moderate Black Hills climate may have enabled ponderosa pine to become established earlier than in the Bighorns.
24. Estes (2021) writes about the Oceti Sakowin, Dakota, Nakota, and Lakota who think of the Black Hills as their home.
25. Raventon 2003; Freeman 2015. Another significant effect was the introduction of human diseases, as described in chapters 2 and 11.

26. Freeman 2015.
27. Freeman 2015. Weaver (1965) reported that much of the woodlands along Pine Ridge were harvested from 1880 to 1900; Potter and Green (1964a) found that ponderosa pine stands in North Dakota were young, probably because most of the trees were harvested in the 1880s, and that tree growth was slow (annual ring widths mostly less than 1.25 mm).
28. Freeman 2015.
29. Sargent 1897; Graves 1899; Freeman 2015.
30. Freeman 2015.
31. Boldt and Van Deusen 1974; Shepperd and Battaglia 2002.
32. Shepperd and Battaglia 2002; Spiering and Knight 2005. Twenty-three birds and 10 mammals in the Black Hills depend on dead trees for food, nest sites, roosts, perches, and dens.
33. Wienk et al. (2004) found that a combination of thinning and prescribed burning stimulates understory plant growth.
34. Uresk and Painter (1985) found that shrubs and trees in the northern Black Hills provide 37 percent of cattle diets during September; in summer the diet was 54 percent grasses, 17 percent forbs, and 28 percent shrubs and trees. Sedgwick and Knopf (1991) reported that cattle consume recently fallen cottonwood leaves in September and October along the South Platte River in Colorado.
35. Freeman 2015.
36. See the 2018 Final Environmental Impact Statement for the Black Hills Resilient Landscapes Project on the website of the Black Hills National Forest; Ziegler et al. (2017) recommended strategies for accelerating regeneration following disturbances.
37. Lovaas 1976; Shinneman and Baker 1997; Baker 2009. Fisher et al. (1987) concluded that fire frequency in the northern Black Hills increased when the Lakota Sioux from the east took possession of the Hills from the Cheyenne, Kiowa, and Crow in about 1770.
38. Fisher et al. (1987) concluded that the woodlands, savanna, and prairie had been more or less stable for "hundreds if not thousands of years" in the broken terrain near Devils Tower. See also Wienk et al. (2004), Brown (2006), Brown and Cook (2006), and Brown et al. (2008).
39. Shinneman and Baker 1997.
40. Newton and Jenney 1880, p. 322.
41. Quotation from Dodge (1876, p. 62); see photos in Progulske (1974) and Grafe and Horsted (2002).
42. Baker et al. (2007) and Baker (2018) concluded that a variable severity fire model is more realistic and that forest thinning with subsequent prescribed surface fires is not appropriate for some forests in the Black Hills. For the details of this debate, see Baker (2009), Williams and Baker (2012), and Fulé et al. (2014).
43. Shinneman and Baker 1997; Brown and Cook 2006; Baker et al. 2007; Brown and Schoettle 2008; Brown et al. 2008.
44. Brown (2006) found that fires were more frequent during La Niña years, cool phases of the Pacific Decadal Oscillation, and warm phases of the Atlantic Multidecadal Oscillation.
45. Gartner and Thompson 1972; Shepperd and Battaglia 2002.
46. Brown 2006; Brown et al. 2008.
47. Orr 1975.
48. Early journals are reviewed in Shepperd and Battaglia (2002).
49. Wienk et al. 2004.
50. Lovaas 1976. Wind Cave National Park used prescribed burns in the early 1970s.
51. Battaglia et al. 2008.
52. Lentile et al. 2006; Keyser et al. 2008; Stark et al. 2011; Roberts et al. (2019) studied the effects of fires of different intensity on pine ridges in northern Nebraska.
53. Symstad et al. 2014. Managers now estimate the potential for the spread of invasive plants before proceeding with a prescribed burn, which involves surveying the understory plants.
54. Shepperd and Battaglia (2002) describe the insects and fungal diseases common in the area, namely, mountain pine beetle, pine engraver, red turpentine beetle, Armillaria root rot, red rot, western gall rust, needle cast, and diplodia tip blight; Graham et al. (2016, 2019) review the ecology and management of mountain pine beetles; Freeman (2015) reports how settlers observed many dead trees in the late 1800s.
55. Graham et al. 2016.
56. Graham et al. 2016.
57. Graham et al. 2016; Negrón et al. 2017.
58. Kaufmann and Watkins (1990).
59. Some success in controlling beetles has been achieved on small properties where the pines are highly valued.
60. Raffa et al. 2008; Bentz et al. 2010.
61. Bentz et al. 2010.
62. The blue-stain fungus does not affect wood strength and is named for the bluish color it gives to the sapwood.
63. Bentz et al. 2010; Edburg et al. 2012; Graham et al. 2016.
64. Regniere and Bentz 2007; Chapman et al. 2012. Preisler et al. (2012) concluded that warmer winter temperatures and drought were the strongest predictors of a beetle outbreak.
65. Kayes and Tinker 2012; Mercado et al. 2014. Collins et al. (2011) found that, even in the most severely affected stands of lodgepole pine, the smaller trees that survived were generally numerous enough to meet or exceed U.S. Forest Service requirements for stand regeneration. Less competition from older trees enables the establishment of new tree seedlings.
66. Romme et al. (1986) found that annual wood production per unit area returned to previous levels or even higher 10–15 years after a bark beetle outbreak in lodgepole pine forests. Also, Brown et al. (2010) concluded that some forests in British Columbia were still carbon sinks even during the first several years after severe mortality caused by bark beetles; and Pfeifer et al. (2011) projected that the amount of ecosystem carbon in Idaho forests would recover in 25 years or less after a bark beetle outbreak.
67. Morehouse et al. 2008; Clow et al. 2011; Griffin and Turner 2012; Edburg et al. 2012.
68. Freeman 2015.
69. Graham et al. 1994; Stevens-Rumann et al. 2018; Coop et al. 2020. There may be more downed timber (coarse woody debris) now than ever before.

70. U.S. Forest Service 2018, Final Environmental Impact Statement, Black Hills Resilient Landscapes Project.
71. Battaglia et al. (2008) report that prescribed fires have helped reduce some exotic plants.
72. For climate projections, see websites of the National Center for Environmental Information in NOAA (https://www.ncdc.noaa.gov/cag/) and the Intergovernmental Panel on Climate Change (IPCC) of the United Nations (https://www.ipcc.ch).
73. Rehfeldt et al. 2006
74. King et al. 2013; Petrie et al. 2017; Stevens-Rumann et al. 2018; Coop et al. 2020. Kaye et al. (2010) observed less ponderosa pine seedling establishment in the west-central Great Plains during drought years, which are likely to become more frequent; Anderegg et al. (2013) studied the effect of drought on aspen.
75. U.S. Forest Service 2018, Final Environmental Impact Statement, Black Hills Resilient Landscapes Project, p. 78. A drought occurred in the Black Hills from 2000 to 2007.
76. Bachelet et al. 2000; Shepperd and Battaglia 2002; King et al. 2013; Graham et al. 2016; Nagel et al. 2017.

Chapter 9: Rivers and Riparian Ecosystems

1. Chapter 10 focuses on the Missouri River; Stoner et al. (1993) and Koel and Peterka (2003) write about the Red River of the North; Berry and Duffy (1993) describe the James River.
2. Schumm 2005.
3. North Dakota Department of Health 2018; South Dakota Department of Environment and Natural Resources 2018.
4. Rosenberg et al. 2005.
5. Bluemle 2016.
6. Ley (2012) studied the vegetation and shifting channel of the Big Sioux River.
7. Schumm 1977.
8. Leopold 1994.
9. https://en.wikipedia.org/wiki/James_River_(Dakotas).
10. Wanek 1967. The longest dry period was in 1937.
11. Major recorded floods occurred on the Red River in 1826, 1897, 1950, 1997, 2009, and 2011.
12. Shelby 2003.
13. Ley 2012.
14. Naiman et al. 2005.
15. Rumble et al. 1998.
16. Naiman et al. (2005) explain the ecological basis for the high biological diversity and productivity of riparian zones worldwide, including the importance of the natural flood regime. Marriott and Faber-Langendoen (2000a) describe the riparian plant communities in the vicinity of the Black Hills; Uresk (2015) emphasized the importance of riparian succession for wildlife.
17. Braatne et al. 1996; Johnson 2000; Severson and Sieg 2006.
18. Everitt 1968.
19. Edmondson et al. 2014.
20. Reily and Johnson 1982; Schook 2017.
21. Number of fish species (native, endangered) in different rivers: Missouri main stem 85 (59, 26); White 40 (30, 10); Big Sioux 68 (58, 10); James 57 (45, 12); Red 94 (80, 14). See Koel and Peterka (2003), Shearer and Berry (2003), Galat et al. (2005), Rosenberg et al. (2005), Kaeming et al (2007), and Berry et al. (2007). Data from these sources may differ due to different sample protocols and dates; data for the Red River include the Assiniboine River tributary.
22. Berry et al. 2007.
23. Kaeming et al. 2007.
24. Johnston 2012; Higgins et al 2000.
25. State Historical Society of North Dakota.
26. Harris and Aldous 1946.
27. Airborne mercury from coal burning that is deposited in beaver ponds is released as methylmercury.
28. Striped Face-Collins and Johnston 2007.
29. Johnson and Volke 2015.
30. Johnston 2012.
31. Gough 1988.
32. Severson and Sieg 2006.
33. Stohlgren et al. (1998) found that riparian zones are "havens" for invasive plants, which in the Dakotas include Kentucky bluegrass, smooth brome, reed canarygrass, Russian olive, buckthorn, Canada thistle, bindweed, and garlic mustard. Invasive aquatic animals include bighead carp, common carp, grass carp, and zebra and quagga mussels. Exotic tree diseases also are of concern: Dutch elm disease has killed most American elm trees and the emerald ash borer could kill most or all of the green ash during the next few decades. Collette and Pither (2015) write about Russian olive.
34. http://elpc.org/tag/des-moines-water-works/.
35. A 2018 law in Minnesota requires landowners near specified streams and rivers to plant a natural buffer strip from 16.5–50 feet wide to reduce input of sediment and farm chemicals (especially nitrate and phosphorus) into the adjacent water body.
36. Johnson and Volke 2015.
37. McMillion (2006) describes the beneficial effects of beavers.
38. American Rivers Dam Removal Database; O'Conner et al. 2015.
39. Brown 2015.
40. Foley et al. 2017.
41. Perry et al. 2012.
42. Ojima et al. 2015.

Chapter 10: The Missouri River

1. Hoffman and Graham (1998) review the archeological data used to determine when the first people arrived in the Great Plains. The Paleo-Indian period of the Middle Missouri Tradition covers the time up to about 6,000 BC; the Foraging Period from 6,000 BC to 500 BC; The Plains Woodland Period began about the time of the Christian Era and ended a millennium later; the Plains Village period spans the period from about AD 900 to 1780; the Equestrian period began about AD 1720. This classification system is based on Lehmer (1971). See also Toom

(1992), Johnson (2007), and Handy-Marchello and Swenson (2018).

2. Ewers 1968.
3. Ewers 1968. During one of two trips the Vérendrye party buried a lead plate with the embossed arms and inscription of the king of France, to establish their presence along the Missouri. The plate was discovered in 1913 by three teenagers climbing the prairie bluffs above Fort Pierre, South Dakota, and is now displayed in South Dakota's Cultural Heritage Center in Pierre.
4. Lehmer (1971) describes how house shapes and methods of construction changed over time during cultural periods.
5. Fawcett 1988.
6. Lehmer 1971. Carter Johnson and Warren Keammerer, graduate students at North Dakota State University, fortuitously crossed paths with Donald Lehmer in 1970 near Cross Ranch. They showed him floodplain plants that he had known only from seeds and plant fragments found in garbage pits.
7. Fawcett (1988) and Griffin (1977) are comprehensive studies of the use of timber by the Missouri River tribes and the extent to which overexploitation may have triggered movement of villages.
8. Lass 2008.
9. Lehmer 1971.
10. Moulton (1987) is the most recent, easily accessible source for Lewis and Clark's journal entries.
11. Isenberg 2000; Lott 2002.
12. Lewis's description of his "perogue" (spelled several ways) suggests that it was a large open boat that could carry more cargo than a dugout canoe; see http://www.lewis-clark.org.
13. Ambrose 1996; Johnsgard 2003.
14. Moulton (1987) is the source of all quotations in this chapter that are attributed to Lewis and Clark.
15. Gilmore 1991.
16. Moulton 1987, vol. 3, pp. 179–183. Other quotations also are from this volume or volume 4.
17. Ambrose 1996.
18. Burroughs 1961.
19. Lass 2008.
20. Chittenden 1962.
21. Lass 2008.
22. Fenn 2014.
23. Lass 2008.
24. Ambrose 1996.
25. Catlin 1973.
26. Ewers 1968; Witte and Gallagher 2012.
27. Herrick 1968; Patterson 2016.
28. Lott 2002.
29. Gilmore 1991.
30. Johnson 1992; Dixon et al. 2012; Wilson 1970; Johnson et al. 1976; Friedman et al. 2018.
31. Naiman et al. (2005) explain the high biological diversity and productivity of riparian zones worldwide, including the importance of the natural flood regime.
32. Naiman et al. 2005.
33. Junk et al. (1989) described the concept of pulse flow.
34. Hibbard 1972; Higgins et al. 2000; Gentry et al. 2006.
35. Johnson 1992; Dixon et al. 2012.
36. National Research Council (2002), the most comprehensive review of the Missouri River past, present, and future with emphasis on the effects of engineering projects such as damming, cessation of floods, snagging, and bank stabilization.
37. Lawson (2009) and Carrels (1999) review the damage done to Indians and other settlements; a seventh dam, the Canyon Ferry Dam, was constructed by the U.S. Bureau of Reclamation.
38. National Research Council 2002; Kaeming et al. 2007.
39. Osterkamp et al. 1998. Johnson et al. (2015) review the ecological problems in Missouri River remnant forests caused by dam and reservoir operations and the absence of river management to correct or lessen their effects.
40. National Research Council 2011; Skalak et al. 2013.
41. Greene and Knox 2014.
42. Johnson et al. 1976.
43. Johnson et al. 2012.
44. Berry 2007.
45. Hesse 1989
46. Missouri National Recreational River Act of 1978.
47. U.S. Army Corps of Engineers 2010.
48. S. Chipps, personal communication
49. National Research Council 2002, 2011.
50. Volke et al. (2015) and Volke et al. (2019) studied the White River delta; Beall (2019) studied the Niobrara delta; Kaeming et al. (2007) examined the effect of deltas on fish.
51. Dixon et al. (2015) reviewed the causes and characteristics of the flood.
52. Davidson (2014) evaluates whether the waters of the Missouri River are a national or a local resource, concluding: "When faced with some future critical shortage (of water) in another region, the national interest will have a fair call on the River." The Corps of Engineers' Master Manual prescribes the allocation of water and storage among the reservoirs.

Chapter 11: Prairie Wetlands and Lakes

1. Pierce 2012.
2. Kantrud et al. 1989; Weller 1981, 1999.
3. U.S. Fish and Wildlife Service 2018.
4. Smith et al. 1964; Keddy et al. 2009. Subspecies of Canada geese nest in Canada.
5. Pool and Austin 2006; Reiger et al. 2006.
6. Deutschman and Peterka 1988; Mushet et al. 2015.
7. Hanten and Hanten 2007. Fathead minnow abundance in Dakota lakes can reach over 200 pounds per acre of water.
8. Zilverberg et al. (2014) found that income per acre from wetland hay and prairie cordgrass seed can exceed that from nearby fields of corn and soybeans.
9. The Dingell-Johnson Sport Fish Restoration Program and the Pittman-Robertson Wildlife Restoration Program contributed about $32 million to South and North Dakota combined (U.S.

Department of Interior 2014). The sale of federal duck stamps totaled $40 million in 2015–2016 (U.S. Department of Interior Migratory Bird Conservation Fund 2017), 60 percent of which was directed to PPR states for wetland protection and restoration.

10. Zedler 2003.
11. Voldseth et al. 2007; Mitsch and Gosselink 2000; Kharel et al. 2016.
12. Zedler 2003; Zedler and Kercher 2005; Mann 2017.
13. Zilverberg et al. 2018.
14. Maslin 2014.
15. Berheim et al. 2019; Main et al. 2014; Royte 2017; National Research Council 1995.
16. Cowardin et al. 1979.
17. Hydric soils are unique because of prolonged submergence.
18. Stewart and Kantrud 1971. Van der Valk and Mushet (2016) caution that the highly variable Great Plains climate challenges inflexible classification systems because wetlands can shift classes under extreme weather conditions.
19. Kantrud et al. 1989; Gleason et al. 2004; Wagner 1997.
20. Winter and Carr 1980; Johnson et al. 2004.
21. Kantrud and Stewart 1977.
22. Johnson et al. 2004; Weller 1981.
23. Bedford and Godwin 2003. Fens have a springy nonacidic peat substrate and a reliable source of water with high concentrations of calcium and magnesium bicarbonates. Fens often appear as small terraces or domes.
24. Hill and Dixon 2011. Fens in Iowa now make up only 0.01 percent of the land area but harbor 12 percent of the state's total number of rare plants. No comparable inventory has been done for the Dakotas, but rare vascular plants in the Dakotas associated with fens include lesser fringed gentian (*Gentianopsis procera*), Loesel's twayblade (*Liparis loeselii*) Kalm's lobelia (*Lobelia kalmia*), waxy bog-star (*Parnassia glauca*), slender beakrush (*Rynchospora capillacea*) and Riddell's goldenrod (*Oligoneuron riddellii*), seaside arrowgrass (*Triglochin maritima*), marsh arrowgrass (*Triglochin palustris*), and lesser bladderwort (*Utricularia minor*). Other fen plants are listed on the website of the Minnesota Department of Natural Resources.
25. Nekola 1994. Leoscke and Pearson (1988) describe Iowa's fens.
26. Bedford and Godwin 2003.
27. van der Valk and Davis 1978; Johnson et al. 2010; Cressey et al. 2016.
28. Weller and Fredrickson 1974.
29. Errington 1957, 1963.
30. Henry and Thompson 1965; van der Valk 1989; Higgins 1984, 1986; Higgins et al. 1989.
31. Weller 1999.
32. Dixon et al. 2019.
33. The Duck Stamp Act was passed in 1934, creating a way for hunters to help conserve waterfowl habitat.
34. Prince 1997.
35. Dahl 1990, 2014; Johnston 2013. The Conservation Compliance Provision of the 1985 Food Security Act became known as the "Swampbuster Provision."
36. Finocchairo 2016.
37. Werner et al. 2016.
38. Galatowitsch 2012; Werner et al. 2016; Tangen and Finocchario 2017.
39. Galatowitsch and van der Valk 1998; Galatowitsch 2012; James and Herbert 2018; Ducks Unlimited works with landowners to restore wetlands and prairies.
40. Galatowitsch and van der Valk 1998; Tulbure et al. 2008.
41. Federal Register 1995; see also chapter 13.
42. Young 1924.
43. Cowardin et al. 1979.
44. Mushet et al. 2015.
45. Mushet et al. 2015.
46. Willis et al. 2009.
47. Swanson et al. 1988; Mushet et al. 2015. Seven salts were identified: calcium bicarbonate, magnesium bicarbonate, sodium bicarbonate, magnesium sulfate, sodium sulfate, sodium chloride, and magnesium chloride.
48. Swanson et al. 1988; Mushet. et al. 2015. Medicine Lake in South Dakota (Prairie Coteau) and Chase Lake in North Dakota (Missouri Coteau) are examples of well-studied brackish lakes.
49. Swanson et al. 1988.
50. Swanson et al. 1988.
51. Niehus et al. 1999.
52. Niehus et al. 1999.
53. Shapley et al. 2005.
54. Shapley et al. 2005. Trees of an age exceeding the length of meteorological records (100–125 years) are rare in the region because of heavy cutting of the scattered forests by Euro-American settlers and soldiers building and heating forts in the late 1800s.
55. Two tree cores were collected in 1992 by David Meko of the Laboratory of Tree Ring Research (Tucson, AZ).
56. Zheng et al. 2014. Prior to the most recent flooding, Devils Lake flowed into Stump Lake sometime between 1820 and 1840.
57. Todhunter and Rundquist 2004; Todhunter and Knish 2014.
58. Haskell et al. 1996; Bluemle et al. 2000.
59. Larson 2012; Wiche et al. 2000.
60. For information on Devils Lake economy, see http://tourism.devilslakend.com/impact-of-fishing-hunting-reverberates-throughout-Devils-Lake-economy.
61. Johnson et al. 2005; Liu et al. 2016; van der Valk and Mushet 2016; Withey and van Kooten 2011; Shjeflo 1968. Average evapotransipiration rate in midsummer is about 1.5 inches per week.
62. Millett and Johnson 2009; Mushet et al. 2015.
63. Ballard et al. 2014.
64. Johnson et al. 2005, 2010.
65. U.S. Fish and Wildlife Service data.
66. Steen et al. 2016; Niemuth et al. 2014.
67. Johnson et al. (2005, 2010) and Johnson and Poiani (2016) concluded that wetlands on the Missouri Coteau have a drier climate than farther east and may become too dry to produce waterfowl as in the past.
68. Doherty et al. 2013.

69. Johnson et al. 2010; McIntyre et al. 2019.
70. McIntyre et al. 2019.
71. McLean et al. 2016.

Chapter 12: Buttes, Badlands, and Sandhills

1. Buttes and mesas are formed in the same way, but the average distance across the top of a mesa is greater. The term *mesa* is not commonly used in the Dakotas. Well-known buttes include Slim Buttes, Eagle Nest Butte, White Butte, Chimney Butte, Bullion Butte, North and South Killdeer Buttes, Thunder Butte, Sentinel Butte, and Black Butte. Bear Butte northeast of Sturgis, South Dakota, is a volcanic neck, not a butte, and was formed in about the same way as Devils Tower.
2. Caprock is commonly limestone, sandstone, sandy shales, clinker (scoria), and silcrete. Bluemle (2016) describes the origin of buttes in North Dakota, noting that Turtle Mountain was formed when glacial deposits covered a group of buttes in the area. Some of the topography of the Missouri Coteau and Prairie Coteau was created in the same way.
3. Potter and Green (1964b) found about 200 acres of savannalike limber pine on breaks above the Little Missouri River at about 2,850 feet elevation.
4. Quinnild and Cosby 1958; Larson and Whitman 1942; Whitman and Hanson 1939.
5. Bluemle (2016) and Kaye and Schoch (1993) describe the geologic history of North Dakota badlands. In the White River Badlands of South Dakota, the oldest exposed strata are Cretaceous Pierre Shale, but most surface rocks are in the much younger Sharps Formation and underlying White River Group, as summarized by Froiland (1990).
6. Bluemle (2016) notes that the southernmost badlands in southwest North Dakota consist mainly of the Cretaceous Hell Creek Formation, where dinosaur fossils have been found. A few other tracts of badlands in North Dakota cover less area and have the same fossils as Badlands National Park, such as near the town of South Heart (John Bluemle, personal communication).
7. Amberg et al. 2012, 2014; Von Loh et al. 2000; Olson 1988; Godfread 1994; Hansen et al. 1984; Girard et al. 1987, 1989; Brown 1971.
8. Brown 1971. Butler et al. (1986) described woodlands in concave depressions that receive water from above, especially on north slopes.
9. The average annual precipitation in the South Dakota badlands is 3.5 inches higher than in the North Dakota badlands (18.5 and 15 inches), but apparently this does not compensate for increased evapotranspiration caused by the higher temperature in South Dakota.
10. Dibner et al. 2017.
11. Plumb and Sucec 2006.
12. Badlands National Park covers 242,756 acres (379 square miles). Of that, 64,144 acres (100 square miles) was designated by Congress as the Badlands Wilderness. Theodore Roosevelt National Park has 70,445 acres (110 square miles) divided into three units: South Unit with 46,157 acres (72 square miles), North Unit with 24,070 acres (38 square miles), and the Elkhorn Ranch Unit with 218 acres; Theodore Roosevelt Wilderness covers 29,920 acres (47 square miles) in the south unit.
13. Dibner et al. 2017 (section 4.15); see chapter 4; Blake McCann, personal communication.
14. Amberg et al. 2014; Von Loh et al. 2000; Ralston 1960.
15. Dibner, et al. 2017; Amberg et al. 2012, 2014; Von Loh et al. 2000; Butler et al. 1986; Butler and Cogan 2004.
16. Dibner et al. (2017) reported declines in prairie dog population size (in Badlands National Park) from 2007 to 2011, but relatively stable or increasing populations from 2013 to 2015.
17. Amberg et al. 2014.
18. Amberg et al. 2012, 2014.
19. Bluemle 2016; Muhs et al. 1997; Gries 1996; Bleed and Flowerday 1990; Johnsgard 2001; Anderson 2011. Bradbury (1980) writes about the paleoecology of the Nebraska Sandhills.
20. Mangan et al. 2004; Muhs et al. 1997.
21. For studies on sand dune plants in the Dakotas, see Tolstead (1941a, 1941b, 1942), Burgess (1965), Wanek and Burgess (1965), Hulett et al. (1966), Manske (1980), and Uresk et al. (2010).
22. Bleed and Flowerday 1990; Johnsgard 2001.
23. Denbigh Experimental Forest is a demonstration of how trees grow well on sandy soils in North Dakota; similar demonstrations were successful in Nebraska, namely, the Nebraska National Forest.
24. Nelson (1964) and Hanson (1976) describe sandhill forests; Manske and Barker (1988) describe the habitat preferences of the greater prairie chicken.
25. Muhs and Maat 1993; Muhs and Wolfe 1999; Forman et al. 2001.

Chapter 13: Working toward Sustainability

1. Mann 2005.
2. Bison harvesting increased sharply when horses became available; Isenberg 2000.
3. Ojima et al. 2015. Increasing yields are the result of more irrigation, pest management, fertilizer, better tillage practices, and improved crop varieties. Most cultivated fields have lost soil organic matter, compared to prairie soils (Haas et al. 1957; Burke et al. 1991, 1997; Hartman et al. 2011). Stephens et al. (2008) studied conversion of grassland to cropland on the Missouri Coteau from 1989 to 2003. Knopf and Samson (1997) write about vertebrate conservation.
4. Smith et al. 2010; Kurtz 2013; Palmer et al. 2016. Geertsema et al. (2016) describe the ecological intensification of agriculture, which enables ecosystem services and resource-efficient production; Schulte et al. (2017) describe the value of prairie strips in corn-soybean croplands.
5. Management of the national grasslands in the Dakotas is now coordinated by the USDA Dakota Prairie Grasslands Supervisor's Office in Bismarck.
6. The concept of ecosystem management is reviewed by Grumbine (1994), Dale et al. (2000), and others.
7. Amberg et al. 2014

8. U.S. Secretary of Agriculture Henry A. Wallace, in 1937, attributed the dirty thirties on the Great Plains to "a system of agriculture not adapted to the region. . . . On the western Plains, it was both a stimulus to over-cultivation and a condemnation of the cultivators to poverty." Farmers were encouraged to resettle on better lands. Generally, successful settlers had been able to claim some riparian land and had adopted new practices after previous droughts.
9. Manning 2009; Aune and Plumb 2019.
10. For more about the America Prairie Preserve, see https://www.americanprairie.org.
11. Licht 1997, p. 179; Cochrane and Runge 1992.
12. Banerjee 2013; Rodewald et al. 2020. Epanchin-Niell and Boyd (2020) describe how conservation (mitigation) banks and other incentives can be used to help private landowners conserve imperiled species. Ducks Unlimited uses this approach to assist landowners desiring to protect and restore wetlands. The 2008 Farm Bill called for measuring the environmental benefits of conservation.
13. Klebnikov 2018.
14. Bailey 2013. The Wolakota Buffalo Range on the Rosebud Reservation is described at https://rosebudbuffalo.org.
15. Licht 1997; Knapp et al. 1999; Cochrane and Runge 1992; Bailey 2013; Alston et al. 2019. African wildlife reserves commonly include lions (D. Licht, personal communication); Licht et al. (2010) discuss the potential for using small populations of wolves.
16. See chapter 5; for the USDA farm and ranch carbon and greenhouse gas accounting system, see http://comet-farm.com/Home; Geertsema et al. (2016) describe ways of restoring ecosystem services to industrial farms.
17. Henderson and Thorpe 2010.
18. Tilman et al. (2011) discuss how carefully designed agricultural intensification could reduce the need to cultivate remaining prairies, thereby diminishing greenhouse gas emissions. Numerous corporations now promote carbon sequestration and the sale of carbon credits. Apfelbaum (2007) compares prairies and forests for global carbon management.
19. Sauchyn et al. 2010; Henderson and Thorpe 2010.
20. Morgan et al. (2011) found that water-use efficiency may increase with higher CO_2 concentrations, potentially moderating the desiccating effects of warming.
21. Galatowitsch 2009; Henderson and Thorpe 2010; Stein and Shaw 2013; Ojima et al. 2015.
22. Climate refugia are described in a 2020 special issue of *Frontiers in Ecology and the Environment* (vol. 18, no. 5).
23. Jackson (2021) discusses three options for adapting to climate change, namely, resisting changes where feasible, accepting the new ecosystems that develop, or intervening to direct changes toward desirable or acceptable new ecosystems. Thurman et al. (2020) discuss ways of assessing the adaptive capacity of different kinds of species to climate change.
24. Geertsema et al. 2016; Montgomery 2017; Brown 2018.
25. Berry 2006.

References

Agnew, W., D. W. Uresk, R. M. Hansen, et al. 1986. Flora and fauna associated with prairie dog colonies and adjacent ungrazed mixed-grass prairie in western South Dakota. J. Range Manage. 39:135–39.

Ahler, S. A., T. D. Thiessen, and M. K. Trimble. 1991. People of the willows. The prehistory and early history of the Hidatsa Indians. Univ. North Dakota Press, Grand Forks.

Allred, B. W. 1941. Grasshoppers and their effect on sagebrush on the Little Powder River in Wyoming and Montana. Ecology 22:387–92.

Alroy, J. 2001. A multispecies overkill simulation of the end-Pleistocene megafauna mass extinction. Science 252:1893–96.

Alston, J. M., B. M. Maitland, B. T. Brito, et al. 2019. Reciprocity in restoration ecology: when might large carnivore reintroduction restore ecosystems? Biol. Conserv. 234:82–89.

Amberg, S., K. Kilkus, M. Komp, et al. 2014. Theodore Roosevelt: National Park: Natural resource condition assessment. Natural Resource Report NPS/THRO/NRR—2014/776. National Park Service, Fort Collins, Colorado.

Amberg, S., K. Kilkus, S. Gardner, et al. 2012. Badlands National Park Climate Change Vulnerability Assessment: Natural Resource Report NPS/BADL/NRR—2012/505. National Park Service, Ft. Collins, CO.

Ambrose, S. E. 1996. Undaunted Courage. Touchstone Publ., New York.

Amichev, B. Y., M. J. Bentham, D. Cerkowniak, et al. 2015. Mapping and quantification of planted tree and shrub shelterbelts in Saskatchewan, Canada. Agrofor. Syst. 89:49–65.

Anderegg, W. R. L., J. A. Berry, D. D. Smith, et al. 2012. The roles of hydraulic and carbon stress in a widespread climate-induced forest die-off. P. Natl. Acad. Sci. 109:233–37.

Anderegg, W. R. L., L. Plavcova, L. D. L. Anderegg, et al. 2013. Drought's legacy: Multiyear hydraulic deterioration underlies widespread aspen forest die-off and portends increased future risk. Global Change Biol. 19:1188–96.

Anderson, F. J. 2011. Dunes on the wind-swept prairie: North Dakota's eolian sands. Geo News, July. North Dakota Geological Survey, Bismarck. https://www.dmr.nd.gov/ndgs/documents/newsletter/2011Summer/DunesontheWind-SweptPrairie.pdf.

Anderson, S. 2019. One Size Fits None. Univ. Nebraska Press, Lincoln.

Apfelbaum, S. I. 2007. Prairies, savannas, and forests and global carbon management: The challenges. In J. M. Kimble, C. W. Rice, D. Reed, et al., eds., Soil Carbon Management: Economic, Environmental, and Societal Benefits, 251–62. CRC Press, Boca Raton, Florida.

Archibold, O. W. 1999. The aspen parkland of Canada. In R. C. Anderson, J. S. Fralish, and J. M. Baskin, eds. Savannas, barrens, and rock outcrop plant communities of North America, 406–19. Cambridge Univ. Press, Cambridge.

Aune, K., and G. Plumb. 2019. Theodore Roosevelt and Bison Restoration on the Great Plains. The History Press, Charleston, South Carolina.

Axelrod, D. I. 1985. Rise of the grassland biome: Central North America. Botanical Rev. 51:163–201.

Bachelet, D., J. M. Lenihan, C. Daly, et al. 2000. Interactions between fire, grazing and climate change at Wind Cave National Park, SD. Ecol. Modelling 134:229–44.

Baer, N. W. 1989. Shelterbelts and windbreaks in the Great Plains. J. Forestry 87:32–36.

Baer, S. G., D. J. Gibson, and L. C. Johnson. 2019. Restoring grassland in the context of climate change. In D. J. Gibson and J. A. Newman, eds. Grasslands and Climate Change, 310–22. Cambridge Univ. Press, Cambridge.

Bailey, J. A. 2013. American Plains Bison: Rewilding an Icon. Sweetgrass Books, Helena, Montana.

Bailey, V. 1926. A Biological Survey of North Dakota. North American Fauna No. 49. Bureau of Biological Survey, U.S. Dept. Agriculture.

Baker, R. G., E. A. Bettis III, D. P. Schwert, et al. 1996. Holocene paleoenvironments of northeast Iowa. Ecol. Monogr. 66:203–34.

Baker, W. L. 2009. Fire Ecology in Rocky Mountain Landscapes. Island Press, Washington, DC.

Baker, W. L. 2018. Transitioning western US dry forests to limited committed warming with bet hedging and natural disturbances. Ecosphere 9, e02288.

Baker, W. L., T. T. Veblen, and R. L. Sherriff. 2007. Fire, fuels, and restoration of ponderosa pine-Douglas-fir forests in the Rocky Mountains, USA. J. Biogeogr. 34:251–69.

Ballard, T., R. Seager, J. E. Smerdon et al. 2014. Hydroclimate variability and change in the Prairie Pothole Region, the "Duck Factory" of North America. Earth Interact. 18, Paper No. 14.

Ballesteros-Possu, W., J. R. Brandle, and M. Schoeneberger. 2017. Potential of windbreak trees to reduce carbon emissions by agricultural operations in the U.S. Forests 8:137–38.

Banerjee, S., S. Seechi, J. Fargione, et al. 2013. How to sell ecosystem services: A guide for designing new markets. Front. Ecol. Env. 11:297–304.

Bardgett, R. D., and M. Semchenko. 2019. Grassland belowground feedbacks and climate change. In D. J. Gibson and J. A. Newman, eds. Grasslands and Climate Change, 203–17. Cambridge Univ. Press, Cambridge.

Barker, W. T., and W. C. Whitman. 1988. Vegetation of the Northern Great Plains. Rangelands 10:266–72.

Barkley, T. M., ed. 1986. Flora of the Great Plains. Univ. Press Kansas, Lawrence.

Barras, C. 2019. Does fossil site record dino-killing impact? Science 364:10–11.

Barrow, E. 2010. Climate change scenarios for the prairie provinces. In D. Sauchyn, H. Diaz, and S. Kulshreshtha, eds. The New Normal: The Canadian Prairies in a Changing Climate, 41–58. Canad. Plains Res. Center Press, Univ. Regina, Regina, Saskatchewan.

Battaglia, M. A., F. W. Smith, and W. D. Shepperd. 2008. Can prescribed fire be used to maintain fuel treatment effectiveness over time in Black Hills ponderosa pine forests? For. Ecol. Manage. 256: 2029–38.

Beadle, G. W. 1980. The ancestry of corn. Sci. Am. 242:112–19.

Beall, C. 2019. Recent expansion of native woody vegetation within the Lewis and Clark Reservoir delta. Thesis, University of South Dakota-Vermillion.

Becher, M. A., J. L. Osborne, P. Thorbek, et al. 2013. Towards a systems approach for understanding honeybee decline. J. Applied Ecol. 50:868–80.

Beck, D. 2017. History and structure of Dakota Lakes Research Farm: Executive summary (unpublished report).

Bedford, B. I., and K. S. Godwin. 2003. Fens of the United States: Distribution, characteristics, and scientific connection versus legal isolation. Wetlands 23:608–29.

Benson, T. J., S. J. Chiavacci, and M. P. Ward. 2013. Patch size and edge proximity are useful predictors of brood parasitism but not nest survival of grassland birds. Ecol. Appl. 23:879–87.

Bentz, B. J., J. Regniere, C. J. Fettig, et al. 2010. Climate change and bark beetles of the western United States and Canada: Direct and indirect effects. BioScience 60:602–13.

Bergdahl, A. D., and A. Hill, tech. coords. 2016. Diseases of trees in the Great Plains. USDA Forest Service RMRS-GTR-335.

Berheim, E. H., J. A. Jenks, J. Lundgrun et al. 2019. Effects of neonicotinoid insecticides on physiology and reproductive characteristics of captive female and fawn white-tailed deer. Sci. Rep. 9, 4534. https://doi.org/10.1038/s41598-019-40994-9

Berry, C. R., Jr., K. F. Higgins, D. W. Willis, et al., eds. 2007. History of Fisheries and Fishing in South Dakota. South Dakota Dept. Game, Fish, and Parks., Pierre.

Berry, C. R., Jr., W. G. Duffy, R. Walsh, et al. 1993. The James River of the Dakotas. U.S. Fish and Wildlife Service, Biological Rep. 19, 70–86.

Berry, W. 2006. Conservationist and agrarian. In D. Imhoff and A. J. Baumgartner, eds. Farming and the Fate of Wild Nature, pp. 3–13. Watershed Media, Healdsburg, California.

Biondini, M. E., A. A. Steuter, and R. Hamilton. 1999. Bison use of fire-managed remnant prairies. J. Range Manage. 52:454–61.

Biondini, M. E., and L. Manske. 1996. Grazing frequency and ecosystem processes in a northern mixed prairie. Ecol. Appl. 6:239–56.

Bird, R. D. 1930. Biotic communities of the aspen parkland of central Canada. Ecology 11:356–442.

Blanchard, B. V., and L. Briefer. 2015. The lost narrative: Ecosystem services and the missing Wasatch watershed conservation story. Ecosyst. Serv. 16:105–11.

Bleed, A. S., and C. Flowerday, eds. 1990. An Atlas of the Sandhills. Institute of Agriculture and Natural Resources, Univ. Nebraska, Lincoln.

Bluemle, J. P. 2000. The Face of North Dakota (3rd ed.). Educational Series 26, North Dakota Geological Survey, Bismarck.

Bluemle, J. P. 2016. North Dakota's Geological Legacy: Our Land and How It Formed. North Dakota State Univ. Press, Fargo.

Boettcher, S. E., and W. C. Johnson. 2005. Cattle and wooded draws: A second look. Rangelands 27:40–42.

Boldt, C. E., and J. L. VanDeusen. 1974. Silviculture of ponderosa pine in the Black Hills: The status of our knowledge. USDA Forest Serv. Res. Paper RM-124.

Bork, E. W., C. N. Carlyle, J. F. Cahill, et al. 2013. Disentangling herbivore impacts on *Populus tremuloides*: A comparison of native ungulates and cattle in Canada's aspen parkland. Oecologia 173:895–904.

Botkin, D. B. 1995. Our Natural History: The Lessons of Lewis and Clarke. G. P. Putnam's Sons, New York.

Braatne, J. H., S. B. Rood, and P. E. Heilman. 1996. Life history, ecology and conservation of riparian cottonwoods in North America. In R. F. Stettler et al., eds. Biology of *Populus* and its implications for management and conservation, pp. 57–85. NRC Research Press, Ottawa.

Bradbury, J. P. 1980. Late quaternary vegetation history of the central Great Plains and its relationship to eolian processes in the Nebraska sand hills. U.S. Geol. Surv. Prof. Pap.1120-C.

Bradford, J., M. Duniway, and S. Munson. 2019. Assessing rangeland health under climate variability and change. In D. J.

Gibson and J. A. Newman, eds. Grasslands and Climate Change, 293–309. Cambridge Univ. Press, Cambridge.

Brandle, J. R., L. Hodges, and X. H. Zhou. 2004. Windbreaks in North American agricultural systems. Agrofor. Syst. 61:65–78.

Bratton, G. F., P. R. Schaefer, and J. R. Brandle. 1995. Conservation forestry on the Great Plains. In S. R. Johnson and A. Bouzaher, eds., Conservation of Great Plains Ecosystems, 211–27. Kluwer Academic Publishers, New York.

Bray, E. C., and M. C. Bray, trans. 1993. Joseph N. Nicollet on the Plains and Prairies. Minnesota Historical Society, St. Paul.

Brennan, L. A., and W. P. Kuvlesky. 2005. North American grassland birds: An unfolding conservation crisis? J. Wildl. Manage. 69:1–13.

Briske, D. D., L. A. Joyce, H. W. Polley, et al. 2015. Climate-change adaptation on rangelands: Linking regional exposure with diverse adaptive capacity. Front. Ecol. Environ. 13:249–56.

Brokke, K. R. G. 2015. Transformation of the Red River Valley of the North: An environmental history. Diss., North Dakota State Univ., Fargo.

Brown, G. 2018. Dirt to Soil. Chelsea Green Publishing, White River Junction, Vermont.

Brown, K. J., J. S. Clark, E. C. Grimm, et al. 2005. Fire cycles in North American interior grasslands and their relation to prairie drought. Proc. Natl. Acad. Sci. 102:8865–70.

Brown, M., T. A. Black, Z. Nesic, et al. 2010. Impact of mountain pine beetle on the net ecosystem production of lodgepole pine stands in British Columbia. Agric. For. Meteorol. 150:254–64.

Brown, P. M, and B. Cook. 2006. Early settlement forest structure in Black Hills ponderosa pine forests. For. Ecol. Manage. 223: 284–90.

Brown, P. M. 2006. Climate effects on fire regimes and tree recruitment in Black Hills ponderosa pine forests. Ecology. 87: 2500–10.

Brown, P. M., and A. W. Schoettle. 2008. Fire and stand history in two limber pine (*Pinus flexilis*) and Rocky Mountain bristlecone pine (*Pinus aristata*) stands in Colorado. Int. J. Wildland Fire 17:339–47.

Brown, P. M., C. L. Wienk, and A. J. Symstad. 2008. Fire and forest history at Mount Rushmore. Ecol. Appl. 18:1984–99.

Brown, R. W. 1971. Distribution of plant communities in southeastern Montana badlands. Am. Midl. Nat. 85:458–77.

Brussard, L. 2012. Ecosystem services provided by the soil biota. In D. H. Wall, ed., Soil Ecology and Ecosystem Services, 45–58. Oxford Univ. Press, Oxford.

Bryce, S. A., J. M. Omernik, D. E. Pater, et al. 1998. Ecoregions of North Dakota and South Dakota. U.S. Geol. Surv., Northern Prairie Wildlife Research Center, Jamestown, North Dakota.

Burgess, R. L. 1964. Ninety years of vegetational change in a township in southeastern North Dakota. Proc. North Dakota Acad. Sci. 18:84–94.

Burgess, R. L. 1965. A study of plant succession in the sandhills of southeastern North Dakota. Proceedings, North Dakota Acad. Science 19:62–80.

Burke, I. C., T. G. F. Kittel, W. K. Lauenroth, et al. 1991. Regional analysis of the central Great Plains. BioScience 41:685–92.

Burke, I. C., W. K. Lauenroth, and D. G. Milchunas. 1997. Biogeochemistry of managed grasslands in central North America. In E. A. Paul, et al., eds. Soil Organic Matter in Temperate Agroecosystems, 85–102. CRC Press, Boca Raton, Florida.

Burke, M. W. V., B. C. Rundquist, and H. Zheng. 2019. Detection of shelterbelt density change using historic APFO and NAIP aerial imagery. Remote Sens. 11:218; https://doi.org/10.3390/rs11030218.

Burroughs, R. D. 1961. The Natural History of the Lewis and Clark Expedition. Michigan State Univ. Press.

Buskirk, S. W. 2016. Wild Mammals of Wyoming and Yellowstone National Park. Univ. California Press, Oakland.

Burns, K. 2012. The Dust Bowl. Public Broadcasting Service (video).

Butler, J. L., and D. R. Cogan. 2004. Leafy spurge effects on patterns of plant species richness. Rangeland Ecol. Manage. 57:305–11.

Butler, J., H. Goetz, and J. L. Richardson. 1986. Vegetation and soil-landscape relationships in the North Dakota badlands. Am. Midl. Nat. 116:378–86.

Buttrick, P. L. 1914. The probable origin of the forests of the Black Hills of South Dakota. For. Quart. 12:223–27.

Byrne, J., S. Kienzle, and D. Sauchyn. 2010. Prairies, water, and climate change. In D. Sauchyn, H. Diaz, and S. Kulshreshtha, eds. The New Normal: The Canadian Prairies in a Changing Climate, 61–79. Canad. Plains Res. Center Press, Univ. Regina, Saskatchewan.

Cable, T. T. 1999. Nonagricultural benefits of windbreaks in Kansas. Great Plains Res. 9:41–53.

Carrels, P. 1999. Uphill Against Water: The Great Dakota Water War. Univ. Nebraska Press, Lincoln.

Catford, J. A., and L. P. Jones. 2019. Grassland invasion in a changing climate. In D. J. Gibson and J. A. Newman, eds. Grasslands and Climate Change, 149–71. Cambridge Univ. Press, Cambridge.

Catlin, G. 1973 (1841). Letters and Notes on the Manners, Customs, and Conditions of the North American Indians. 2 volumes. Dover Publications, Inc. New York.

Chapman, K. A., A. Fischer, and M. K. Ziegenhagen. 1998. Valley of Grass: Tallgrass Prairie and Parkland of the Red River Region. North Star Press of St. Cloud Inc., St. Cloud, Minnesota.

Chapman, T. B., T. T. Veblen, and T. Schoennagel. 2012. Spatiotemporal patterns of mountain pine beetle activity in the southern Rocky Mountains. Ecol. 93:2175–85.

Chittenden, M. H. 1962. History of Early Steamboat Navigation on the Missouri River: Life and Adventures of Joseph La Barge. Ross and Haines, Minneapolis.

Clambey, G. 1980. Vegetation studies in the prairie-forest transition region. Proc. North Dakota Acad. Sci., vol. 34.

Clark, J. S., E. C. Grimm, J. J. Donovan, et al. 2002. Drought cycles and landscape responses to past aridity on prairies of the northern Great Plains, USA. Ecol. 83:595–601.

Clark, J. S., E. C. Grimm, J. Lynch, and P. G. Mueller. 2001. Effects of Holocene climate change on the C4 grassland/woodland boundary in the northern plains, USA. Ecol. 82:620–36.

Clow, D. W., C. Rhoades, J. Briggs, et al. 2011. Responses of soil and water chemistry to mountain pine beetle induced tree mortality in Grand County, Colorado, USA. Appl. Geochem. 26:5174–78.

Cochrane, W. W., and C. F. Runge. 1992. Reforming farm policy. Iowa State Univ. Press, Ames.

Collette, L. K. D., and J. Pither. 2015. Russian olive (Eleagnus angustifolia) biology and ecology and its potential to invade northern North American riparian ecosystems. Invasive Plant Sci. Manage. 8:1–14.

Collins, B. J., C. C. Rhoades, R. M. Hubbard, et al. 2011. Tree regeneration and future stand development after bark beetle infestation and harvesting in Colorado lodgepole pine stands. For. Ecol. Manage. 261:2168–75

Collins, S. L., and L. L. Wallace, eds. 1990. Fire in North American tallgrass prairies. Univ. Oklahoma Press, Norman.

Cook, E. R., C. A. Woodhouse, C. M. Eakin, et al. 2004. Long-term aridity changes in the western United States. Science 306:1015–18.

Coop, J. D., S. A. Parks, C. S. Stevens-Rumann, et al. 2020. Wildfire-driven forest conversion in western North America landscapes. BioScience 70:659–73.

Cosby, H. E. 1965. Fescue grassland in North Dakota. J. Range Manage. 18:284–85.

Costanza, R., R. D'Arge, R. de Groot, et al. 1997. The value of the world's ecosystem services and natural capital. Nature 38:253–60.

Costanza, R., R. de Groot, P. Sutton, et al. 2014. Changes in the global value of ecosystem services. Global Environ. Change 26:152–58.

Courtwright, J. 2011. Prairie Fire: A Great Plains History. Univ. Press Kansas, Lawrence.

Cowardin, L. M., V. Carter, F. C. Golet, et al. 1979. Classification of Wetlands and Deepwater Habitats of the United States. U.S. Dept. of the Interior, Fish and Wildlife Service, Washington, DC.

Cressey, R. I., J. E. Austin, and J. D. Stafford. 2016. Three responses of wetland conditions to climate extremes in the Prairie Pothole Region. Wetlands 36 (suppl. 2):S357–S370.

Crompton, J., K. Killingbeck, and R. Bares. 1977. Tree and shrub stratum composition on the Killdeer Mountains. Proc. North Dakota Acad. Sci. 31:3 (abstract).

Custer, G. A. 1875. Expedition to the Black Hills. 43d Congress, 2d sess. Senate Exec. Doc. 32.

Cutright, P. E. 1969. Lewis and Clark: Pioneering Naturalists. Univ. Illinois Press, Urbana.

Dahl, T. E. 1990. Wetland Losses in the United States 1780s to 1980s. U.S. Dept. of the Interior, Fish and Wildlife Service, Washington, DC.

Dahl, T. E. 2014. Status and Trends of Prairie Wetlands in the United States 1997–2009. U.S. Department of the Interior, Fish and Wildlife Service, Ecological Services, Washington, DC.

Dale, V. H., S. Brown, R. A. Haeuber, et al. 2000. Ecological principles and guidelines for managing the use of land. Ecol. Applic. 10:639–70.

Danbom, D. B. 2014. Sod Busting: How families made farms on the nineteenth-century plains. Johns Hopkins Univ. Press, Baltimore.

Davidson, A. D., J. K. Detling, and J. H. Brown. 2012. Ecological roles and conservation challenges of social, burrowing, herbivorous mammals in the world's grasslands. Front. Ecol. Environ. 10:477–86

Davidson, J. H. 2014. Missouri River reservoirs in a century of climate change: National or local resource? J. Environ. Law 20(2)/2.

DeKeyser, E. S., L. A. Dennhardt, and J. Hendrickson. 2015. Kentucky bluegrass invasion in the Northern Great Plains: A story of rapid dominance in an endangered ecosystem. Invasive Plant Sci. Manage. 8:255–61.

DePalma, R. A., J. Smit, D. A. Burnham, et al. 2019. A seismically induced onshore surge deposit at the KPg boundary, North Dakota. P. Natl. Acad. Sci. 116:8190–99.

Deutschman, M. R., and J. J. Peterka. 1988. Secondary production of tiger salamanders (*Ambystoma tigrinum*) in three North Dakota prairie lakes. Can. J. Fish. Aquat. Sci. 45:691–97.

DeVuyst, E. A., T. Foissey, and G. O. Kegode. 2006. An economic comparison of alternative and traditional cropping systems in the northern Great Plains, USA. Renew. Agric. Food Syst. 21:68–73

Dibner, R. R., N. Korfanta, G. Beauvais, et al. 2017. Badlands National Park: Natural Resource Condition Assessment. Natural Resource Report NPS/NGPN/NRR.

Disrud, D. T. 1968. Wetland vegetation of the Turtle Mountains of North Dakota. Diss., North Dakota State Univ., Fargo.

Dix, M. E., R. J. Johnson, M. O. Harrell, et al. 1995. Influences of trees on abundance of natural enemies of insect pests: A review. Agrofor. Syst. 29:303–11.

Dix, R. L., and F. E. Smeins. 1967. The prairie, meadow, and marsh vegetation of Nelson County, North Dakota. Can. J. Bot. 45:21–58. (see also Godfread and Barker 1975)

Dixon, C., P. Comeau, K. Askerooth, et al. 2017. Prairie reconstruction guidebook for North Dakota. R1840, North Dakota State Univ. Extension Service, Fargo.

Dixon, C., S. Vacek, and T. Grant. 2019. Evolving management paradigms on U.S. Fish and Wildlife Service lands in the Prairie Pothole Region. Rangelands 41:36–43.

Dixon, M. D., C. J. Boever, V. L. Danzeisen, et al. 2015. Effects of a "natural" flood event on the riparian ecosystem of a regulated large-river system: the 2011 flood on the Missouri River, USA. Ecohydrol. Hydrobiol. 8:18–24.

Dixon, M. D., W. C. Johnson, M. L. Scott, et al. 2012. Dynamics of plains cottonwood (*Populus deltoides*) forests and historical landscape change along unchannelized segments of the Missouri River, USA. Environ. Manage. 49:990–1008.

Dodge, R. I. 1876. The Black Hills. James Miller Publ., New York. (Reprinted 1965 by Ross and Haines, Minneapolis, Minn.), p. 62

Doherty, K. E., A. J. Ryba, C. L. Stemler, et al. 2013. Conservation planning in an era of change: state of the US Prairie Pothole Region. Wildl. Soc. Bull. 37:546–63.

Donaldson, A. B. 1875. The Black Hills Expedition. In South Dakota State Hist. Soc. 1914, vol. 7, pp. 554–80. South Dakota Historical Collections, Pierre.

Drache, H. M. 1970. The Challenge of the Prairie: Life and Times of Red River Pioneers. North Dakota Inst. Regional Studies, Fargo.

Drazkowski, B., M. R. Komp, K. Stark, et al. 2011. Wind Cave National Park: Natural Resource Condition Assessment. Natural Resources Report NPS/WICA/NRR—2011/478.

Droze, W. H. 1977. Trees, Prairies, and People: A history of tree planting in the plains states. Texas Woman's University, Denton.

Edburg, S. L., J. A. Hicke, P. D. Brooks, et al. 2012. Cascading impacts of bark beetle-caused tree mortality on coupled biogeophysical and biogeochemical processes. Front. Ecol. Environ. 10:416–24.

Edmondson, J, J. Friedman, D. Meko, et al. 2014. Dendroclimatic potential of plains cottonwood (*Populus deltoides* Subsp. *monilifera*) from the Northern Great Plains, USA. Tree-Ring Research 70:21–30.

Ellison, K. S., C. A. Ribic, D. W. Sample, et al. 2013. Impacts of tree rows on grassland birds and potential nest predators: A removal experiment. PloS ONE 8: e59151.doi:10.1371/journal.pone.0059151.

Emmerich, J. M., and P. A. Vohs. 1982. Comparative use of four woodland habitats by birds. J. Wildl. Manage. 46:43–49.

Epanchin-Niell, R., and J. Boyd. 2020. Private-sector conservation under the US Endangered Species Act: a return-on-investment perspective. Front. Ecol. Environ. 18:409–16.

Errington, F., and D. Gewertz. 2015. Pheasant capitalism: Auditing South Dakota's state bird. Am. Ethnol. 42:399–414.

Errington, P. L. 1957. Of Men and Marshes. Macmillan Co., New York. 150 pp.

Errington, P. L. 1963. Muskrat Populations. Iowa State Univ. Press, Ames.

Estes, N. 2021. The battle for the Black Hills. High Country News 53 (Jan.), 14–17.

Everitt, B. L. 1968. Use of cottonwood in an investigation of the recent history of a flood plain. Amer. J. Sci. 266:417–39.

Ewers, J. C. 1968. Indian Life on the Upper Missouri. Univ. Oklahoma Press, Norman.

Facey, V. L., J. C. La Duke, and A. M. Wycoff. 1986. Vascular plants of Oakville Prairie, North Dakota. Prairie Nat. 18:203–10.

Fahnestock, J. T., and J. K. Detling. 2002. Bison-prairie dog-plant interactions in a North American mixed-grass prairie. Oecologia 132:86–95.

Fawcett, Jr., W. B. 1988. Changing prehistoric settlement along the Middle Missouri River: Timber depletion and historical context. Plains Anthropol. 33:67–94.

Federal Register. 1995. Federal guidance for the establishment, use and operation of mitigation banks. Volume 60, Number 228.

Fenn, E. A. 2014. Encounters at the Heart of the World: A History of the Mandan People. Hill and Wang Publ., New York.

Fiedler, C. E., and S. F. Arno. 2015. Ponderosa: People, Fire, and the West's Most Iconic Tree. Mountain Press Publishing Co., Missoula.

Finocchiaro, R. G. 2016. Agricultural Subsurface Drainage Tile Locations by Permits in North Dakota: U.S. Geological Survey data release, http://dx.doi.org/10.5066/F7QF8QZW.

Fisher, R. F., M. J. Jenkins, and W. F. Fisher. 1987. Fire and the prairie-forest mosaic of Devil's Tower National Monument. Am. Midl. Nat. 117:250–57.

Flake, L., A. Gabbert, T. Kirschenmann, et al. 2012. Ring-necked Pheasants: Thriving in South Dakota. South Dakota Dept. Game, Fish, and Parks, Pierre.

Flesland, J. R., and W. C. Whitman. 1963. A vegetational analysis of the salt-desert shrub type in western North Dakota. Proc. North Dakota Acad. Sci. 18:73–75.

Flint, R. F. 1955. Pleistocene geology of eastern South Dakota. U.S. Geol. Surv. Paper 262.

Foley, M. M., J. R. Bellmore, J. E. O'Connor et al. 2017. Dam removal: Listening in. Reviews of Geophysics/Early View. Commentary doi.org/10.1002/2017WR020457

Forman, S. L., R. Oglesby, and R. S. Webb. 2001. Temporal and spatial patterns of Holocene dune activity on the Great Plains of North America: Megadroughts and climate links. Global and Planet. Change 29:1–29.

Fraser, L. H. 2019. Production changes in response to climate change. In D. J. Gibson and J. A. Newman, eds. Grasslands and Climate Change, 82–97. Cambridge Univ. Press, Cambridge.

Fredlund, G. G., and L. L. Tieszen. 1997. Phytolith and carbon isotope evidence for late Quaternary vegetation and climate change in the southern Black Hills, South Dakota. Quat. Res. 47:206–17.

Freeman, J. F. 2015. Black Hills Forestry: A History. Univ. Press of Colorado, Boulder.

Frémont, J. C. 1887. Memoirs of my life. Belford, Clarke and Co., Chicago.

Friedman, J. M., F. R. Ankney, and J. M. Wolf. 2018. Age and growth of cottonwood trees along the Missouri River, North Dakota. Prairie Naturalist 50:26–35.

Froiland, S. G. 1990. Natural History of the Black Hills and Badlands. Center for Western Studies, Augustana Univ., Sioux Falls, South Dakota.

Fuhlendorf, S. D., D. M. Engle, R. D. Elmore et al. 2012. Conservation of pattern and process: Developing an alternative paradigm of rangeland management. Rangeland Ecol. Manage. 65:579–89.

Fulé, P. A., T. W. Swetnam, P. M. Brown, et al. 2014. Unsupported references of high severity fires in historical dry forests of the western United States: Response to Williams and Baker. Global Ecol. Biogeogr. 23:825–30.

Galat, D. L., C. R. Berry, Jr., E. J. Peters, et al. 2005. The Missouri River Basin. In A. C. Benke and C. E. Cushing. Rivers of North America, 427–464. Elsevier Academic Press, Amsterdam.

Galatowitsch, S. M. 2009. Regional climate change adaptation strategies for biodiversity conservation in a mid-Continental region of North America.

Galatowitsch, S. M. 2012. Ecological Restoration. Sinauer Associates, Inc., Publ. Sunderland, Massachusetts.

Galatowitsch, S. M., and A. G. van der Valk. 1998. Restoring Prairie Wetlands: An Ecological Approach. Iowa State Univ. Press, Ames.

Gardner, R. 2009. Trees as Technology: Planting shelterbelts on the Great Plains. Hist. Technol. 25:325–41.

Gartner, F. R., and W. W Thompson. 1972. Fire in the Black Hills forest-grass ecotone. Proc. Tall Timbers Fire Ecol. Conf. no. 12, pp. 37–68.

Geertsema, W., W. Rossing, D. Landis, et al. 2016. Actionable knowledge for ecological intensification of agriculture. Front. Ecol. Environ. 14:209–16.

Gentry, D. J., D. L. Swanson, and J. D. Carlisle. 2006. Species richness and nesting success of migrant forest birds in natural river corridors and anthropogenic woodlands in southeastern South Dakota. The Condor 108:140–53.

George, T. L., A. C. Fowler, R. L. Knight et al. 1992. Impacts of a severe drought on grassland birds in western North Dakota. Ecol. Appl. 2:275–84.

Gibson, D. J., and J. A. Newman, eds. 2019a. Grasslands and Climate Change. Cambridge Univ. Press, Cambridge.

Gibson, D. J., and J. A. Newman. 2019b. Grasslands in the Anthropocene: Research and Conservation needs. In D. J. Gibson and J. A. Newman, eds. Grasslands and Climate Change, 323–39. Cambridge Univ. Press, Cambridge.

Gilmore, M. R. 1991. Uses of Plants by the Indians of the Missouri River Region. Univ. Nebraska Press, Lincoln.

Girard, M. M., H. Goetz, and A. J. Bjugstad. 1987. Factors influencing woodlands of southwestern North Dakota. Prairie Nat. 19:189–98.

Girard, M. M., H. Goetz, and A. J. Bjugstad. 1989. Native woodland habitat types of southwestern North Dakota. USDA Forest Service Res. Paper RM-281.

Gleason, R., N. Euliss, et al. 2004. Invertebrate egg banks of restored, natural, and drained wetlands in the Prairie Pothole Region of the United States. Wetlands 24:562–72.

Gliessman, S. R. 2007. Agroecology: The Ecology of Sustainable Food Systems. CRC Press, Inc., Boca Raton, Florida.

Godfread, C. 1994. The vegetation of the Little Missouri Badlands of North Dakota. In R. J. Andrascik, ed., Proc.: Leafy spurge strategic planning workshop, 17–24. Dickinson, North Dakota, March 29–30, 1994. U.S. Department of Interior, Theodore Roosevelt Natl. Park, Medora.

Godfread, C., and W. T. Barker. 1975. Vascular flora of Barnes and Stutsman Counties, North Dakota. In M. K. Wali, ed. Prairie: A Multiple View, 333–40. Univ. North Dakota Press, Grand Forks.

Goodwin, D. K. 2013. The Bully Pulpit: Theodore Roosevelt, William Howard Taft, and the Golden Age of Journalism. Simon and Schuster, New York.

Gough, B. M., ed. 1988. The Journal of Alexander Henry the Younger, 1799–1814. Volume I. Red River and the Journey to the Missouri. Publications of the Champlain Society, Toronto.

Grafe, E., and P. Horsted. 2002. Exploring with Custer: The 1874 Black Hills Expedition. Golden Valley Press, Custer, South Dakota.

Graham, R. T., A. E. Harvey, M. F. Jurgensen, et al. 1994. Managing coarse woody debris in forests of the Rocky Mountains. USDA Forest Service Res. Paper INT-RP-477.

Graham, R. T., L. A. Asherin, M. A. Battaglia, et al. 2016. Mountain pine beetles: A century of knowledge, control attempts, and impacts central to the Black Hills. USDA Forest Service Gen. Tech. Rep. RMRS-GTR-353.

Graham, R. T., L. A. Asherin, T. B. Jain, et al. 2019. Differing ponderosa pine forest structures, their growth and yield, and mountain pine beetle impacts: Growing stock levels in the Black Hills. USDA Forest Service Gen. Tech. Rep. RMRS-GTR-393.

Grala, R. K., and J. P. Colletti. 2003. Estimates of additional maize (*Zea mays*) yields required to offset costs of tree-windbreaks in midwestern USA. Agrofor. Syst. 59:11–20.

Grant, T. A., and G. B. Berkey. 1999. Forest area and avian diversity in fragmented aspen woodland of North Dakota. Wildl. Soc. Bull. 27:904–14.

Grant, T. A., B. Flanders-Wanner, T. L. Shaffer, et al. 2009. An emerging crisis across the northern prairie refuges: Prevalence of invasive plants and a plan for adaptive management. Ecol. Restor. 27:58–65.

Grant, T. A., T. L. Shaffer, and B. Flanders. 2020. Patterns of smooth brome, Kentucky bluegrass, and shrub invasion in the Northern Great Plains vary with temperature and precipitation. Natural Areas J. 440:11–22. (see also Rangeland Ecol. Manage. 73:321–28)

Graves, H. S. 1899. The Black Hills Forest Reserve. In U.S. Geol. Surv., 19th Annual Report, 1897–1898, Part V: Forest Reserves, 67–164. Gov. Printing Office, Washington, DC.

Greene, S. I., and J. C. Knox. 2014. Coupling legacy geomorphic surface facies to riparian vegetation: Assessing red cedar invasion along the Missouri River downstream of Gavins Point dam, South Dakota. Geomorph. 204:277–86.

Greer, M. J., K. K. Bakker, and C. D. Dieter. 2016. Grassland bird response to recent loss and degradation of native prairie in central and western South Dakota. Wilson J. Ornith. 128:278–90.

Gries, J. P. 1996. Roadside Geology of South Dakota. Mountain Press Publ. Co., Missoula, Montana.

Griffin, D. E. 1977. Timber procurement and village location in the Middle Missouri subarea. Plains Anthropologist 22:77–185.

Griffin, J. M., and M. G. Turner. 2012. Changes to the N cycle following bark beetle outbreaks in two contrasting conifer forest types. Oecologia 170:551–65.

Grimm, E. C. 2001. Trends and paleoecological problems in the vegetation and climate history of the Northern Great Plains, USA. Biol. Environ.: P. Royal Irish Acad. 101B:47–64.

Grimm, E. C., J. J. Donovan, and K. J. Brown. 2011. A high-resolution record of climate variability and landscape response from Kettle Lake, northern Great Plains, North America. Quat. Sci. Rev. 30:2626–50.

Grumbine, R. E. 1994. What is ecosystem management. Conserv. Biol. 8:27–38.

Haas, H. J., C. E. Evans, and E. R. Miles. 1957. Nitrogen and carbon changes in soil as influenced by cropping and soil treatments. USDA Tech. Bull. 1164, Govern. Printing Office, Washington, DC.

Hadley, E. B. 1969, Physiological ecology of *Pinus ponderosa* in southwestern North Dakota. Am. Midl. Nat. 81:289–314.

Hadley, E. B. 1970. Net productivity and burning responses of native eastern North Dakota prairie communities. Am. Midl. Nat. 84:121–35.

Hadley, E. B., and R. P. Buccos. 1967. Plant community composition and net primary productivity within a native eastern North Dakota Prairie. Am. Midl. Nat. 77:116–27.

Handy-Marchello, B. 2005. Women of the Northern Plains: Gender and Settlement in the Homestead Frontier. Minnesota Hist. Soc. Press, St. Paul.

Handy-Marchello, B., and F. E. Swenson. 2018. Traces: Early Peoples of North Dakota. State Histor. Soc. North Dakota, Bismarck.

Hansen, P. L., and G. R. Hoffman. 1988. The vegetation of the Grand River/Cedar River, Sioux, and Ashland Districts of the Custer National Forest: A habitat type classification. USDA Forest Service GTR-RM-157.

Hansen, P. L., G. R. Hoffman, and A. J. Bjugstad. 1984. The vegetation of Theodore Roosevelt National Park, North Dakota. USDA Forest Service GTR-RM-113.

Hanson, H. C., and W. C. Whitman. 1938. Characteristics of major grassland types in western North Dakota. Ecol. Monogr. 8:57–114.

Hanson, J. D. 1976. A phytosociological study of the oak savanna in southeastern North Dakota. Thesis, North Dakota State Univ., Fargo.

Hanten, R. L., and R. P. Hanten. 2007. Commercial fisheries and the baitfish industry. In C. Berry, K. F. Higgins, D. W. Willis, et al., eds. History of Fisheries and Fishing in South Dakota, 295–318. South Dakota Dept. Game, Fish, and Parks Publ., Pierre.

Harris, D. and S. E. Aldous. 1946. Beaver management in the northern Black Hills of South Dakota. J. Wildl. Manage. 10:348–53.

Hartman, M. D., E. R. Merchant, W. J. Parton, et al. 2011. Impact of historical land use changes in the U.S. Great Plains, 1883–2003. Ecol. Appl. 21:1105–19.

Haskell, B. J., D. Engstrom, and S. C. Fritz. 1996. Late Quaternary paleohydrology in the North American Great Plains inferred from the geochemistry of endogenic carbonate and fossil ostracodes from Devils Lake, North Dakota, USA. Palaeogeogr., Palaeoclimatol., Palaeoecol. 124:179–193.

Haugen, D. E., M. Kangas, M. Crocker, et al. 2005. North Dakota's forests 2005. USDA Forest Service Resource Bull. NRS-31.

Hauk, J. K. 1969. Badlands: The natural history story of Badlands National Park. Badlands Natural History Assn., Bull. 2.

Helzer, C. 2010. The Ecology and Management of Prairies in the Central United States. Univ. Iowa Press, Iowa City.

Henderson, N., and J. Thorpe. 2010. Ecosystems and biodiversity. In D. Sauchyn, et al., eds. The New Normal: The Canadian Prairies in a Changing Climate, 80–116. Canad. Plains Res. Center Press, Univ. Regina, Saskatchewan.

Henderson, N., E. Barrow, B. Dolter, et al. 2010. Climate change impacts and management options for isolated northern Great Plains forests. In D. Sauchyn, et al., eds. The New Normal: The Canadian Prairies in a Changing Climate, 308–321. Canad. Plains Res. Center Press, Univ. Regina, Saskatchewan.

Henderson, N., T. Hogg, E. Barrow, et al. 2002. Climate change impacts on the island forests of the Great Plains and the implications for nature conservation policy. Prairie Adaptation Research Collaborative, Univ. Regina, Regina, Saskatchewan.

Hendrix, P. F. 2006. Biological invasions belowground: earthworms as invasive species. Springer, New York.

Henry, A., and D. Thompson. 1965. New light on the early history of the greater northwest: the manuscript journals of Alexander Henry and David Thompson 1799–1814, E. Coues, ed. Ross and Haines, Minneapolis.

Henry, H. A. I. 2019. Biogeochemical cycling in grasslands under climate change. In D. J. Gibson and J. A. Newman, eds. Grasslands and Climate Change, 115–30. Cambridge Univ. Press, Cambridge.

Henshue, N., C. Mordhorst, and L. Perkins. 2018. Invasive earthworms in a Northern Great Plains prairie fragment. Biol. Invasions 20:29–32.

Herrick, F. M. 1968. Audubon The Naturalist: A History of his Life and Time. Volume II. Dover Publications, New York.

Hesse, L. W., et al. 1989. Missouri River fisheries resources in relation to past, present, and future stresses. In D. P. Dodge, ed. Proc. Intl. Large River Symp., 352–371. Canad. Spec. Publ. Fish. Aquatic Sci. 106. Dept. Fisheries and Oceans, Ottawa.

Hibbard, E. A. 1972. Vertebrate ecology and zoogeography of the Missouri River Valley in North Dakota. Diss., North Dakota State Univ., Fargo.

Higgins, K. F. 1984. Lightning fires in North Dakota grasslands and in pine-savanna lands of South Dakota and Montana. J. Range Manage. 37:100–103.

Higgins, K. F. 1986. Interpretation and compendium of historical fire accounts in the northern Great Plains. U.S. Dept. Fish and Wildlife Service Res. Publ. 161.

Higgins, K. F., A. D. Kruse, and J. L. Piehl. 1989. Effects of fire in the northern Great Plains. U.S. Fish and Wildlife Service and Cooperative Extens. Serv., South Dakota State Univ. and U.S. Dept of Agriculture. Publ. EC 761.

Higgins, K. F., E. D. Stukel, J. M. Goulet, et al. 2000. Wild mammals of South Dakota. South Dakota Game, Fish, and Parks, Pierre.

Hill, E., and M. Dixon. 2011. Mapping and characterizing calcareous fens in eastern South Dakota. (unpublished report)

Hocking, C. M., E. L. Krantz, and G. J. Josten. 2010. South Dakota Statewide Assessment of Forest Resources. South Dakota Dept. Agric., Pierre.

Hoffman, G. R., and R. R. Alexander. 1987. Forest vegetation of the Black Hills National Forest of South Dakota and Wyoming: A habitat type classification. U.S. Forest Service Res. Paper RM-276.

Hoffman, J. L., and R. W. Graham. 1998. The Paleo-Indian cultures of the Great Plains. In W. R. Wood, ed. Archaeology on the Great Plains, pp. 87–139. Univ. Kansas Press.

Homer, C. G., J. A. Dewitz, S. Jin, et al. 2020. Conterminous United States land cover change patterns 2001–2016 from the 2016 National Land Cover Database: ISPRS J. Photogramm. Remote Sens. 162:184–99. https://doi.org/10.1016/j.isprsjprs.2020.02.019

Hoover, H. T. 2009. Native peoples. In H. F. Thompson, ed. A New South Dakota History (2nd ed.), 40–46. Center for Western Studies, Augustana Univ., Sioux Fall, South Dakota.

Hufkens, K., T. F. Keenan, L. B. Flanagan, et al. 2016. Productivity of North American grasslands is increased under future climate scenarios despite rising aridity. Nat. Clim. Change 6:710–14.

Hulett, G. K., R. T. Coupland, and R. L. Dix. 1966. The vegetation of sand dune areas within the grassland region of Saskatchewan. Canad. J. Bot. 44:1307–31.

Hunt, F. 1951. Cap Mossman: Last of the Great Cowmen. Hastings House, New York.

Imhoff, D. 2006. A plea for bees. In Imhoff and Baumgartner, eds. Farming and the Fate of Wild Nature, pp. 96–102. Watershed Media, Healdsburg, California.

Isaacs, R., J. Tuell, A. Fiedler, et al. 2009. Maximizing arthropod-mediated ecosystem services in agricultural landscapes: The role of native plants. Front. Ecol. Environ. 7:196–203.

Isbell, F., D. Craven, J. Connolly, et al. 2015. Biodiversity increases the resistance of ecosystem productivity to climate extremes. Nature (doi: 10.1038/nature15374).

Isenberg, A. C. 2000. The Destruction of the Bison. An Environmental History, 1750–1920. Cambridge University Press, Cambridge.

Jackson, D. L., and L. L. Jackson. 2002. The Farm as Natural Habitat: Reconnecting Food Systems with Ecosystems. Island Press, New York.

Jackson, L. L. 2006. The farmer as conservationist. In D. Imhoff and J. Baumgartner, eds. Farming and the Fate of Wild Nature, pp. 48–59. Watershed Media, Healdsburg, California.

Jackson, S. T. 2021. Transformational ecology and climate change. Science 373:1085–86.

James, J. D., and E. R. Herbert. 2018. The many benefits of wetlands conservation. Ducks Unlimited 82(6): 48–54.

Jenks, J. A. 2018. Mountain Lions of the Black Hills: History and Ecology. Johns Hopkins Univ. Press, Baltimore.

Jensen, K. C., K. Higgins, and S. Vaa. 2014. History of waterfowl management, research, and hunting in South Dakota. South Dakota Dept. Game, Fish and Parks, Pierre.

Johnsgard, P. A. 2001. The Nature of Nebraska. Univ. Nebraska Press, Lincoln.

Johnsgard, P. A. 2003. Lewis and Clark on the Great Plains. Univ. Nebraska Press, Lincoln.

Johnson, C. M., S. A. Ahler, H. Haas, et al. 2007. A Chronology of Middle Missouri Plains Village Sites. Smithsonian Inst. Scholarly Press, Washington, DC.

Johnson, J. R., and G. E. Larson. 2016. Grassland plants of South Dakota and the Northern Great Plains. South Dakota State Univ. Extension, Brookings.

Johnson, R. G., and S. A. Temple. 1990. Nest predation and brood parasitism of tallgrass prairie birds. J. Wildl. Manage. 54:106–11,

Johnson, R. J., and M. M. Beck. 1988. Influences of shelterbelts on wildlife management and biology. Agric. Ecosyst. Environ. 22/23:301–35.

Johnson, W. C. 1992. Dams and riparian forests: Case study from the Upper Missouri River. Rivers 3:229–42.

Johnson, W. C. 2000. Tree recruitment and survival in rivers: Influence of hydrological processes. Hydrol. Processes 14:3051–74.

Johnson, W. C. 2019. Ecosystem services provided by prairie wetlands in northern rangelands. Rangelands 41:44–48

Johnson, W. C. and M. A. Volke. 2015. The Mortenson Ranch story: Balancing environment and economics. iGrow Extension Publication. College of Agriculture and Biological Sciences, South Dakota State Univ. 44 pp.

Johnson, W. C., and K. A. Poiani. 2016. Climate change effects on prairie pothole wetlands: findings from a 25-year numerical modeling project. Wetlands (Suppl. 2): S273–S286.

Johnson, W. C., and M. Volke. 2015. The Mortenson Ranch Story: Balancing Environment and Economics. South Dakota State Univ. Extension, igrow.org publication, Brookings.

Johnson, W. C., B. V. Millett, T. Gilmanov, et al. 2005. Vulnerability of northern prairie wetlands to climate change. BioScience 55:863–72.

Johnson, W. C., B. Werner, G. R. Guntenspergen, et al. 2010. Prairie wetland complexes as landscape functional units in a changing climate. BioScience 60:128–40.

Johnson, W. C., M. A. Volke, M. L. Scott, et al. 2014. The dammed Missouri: prospects for recovering Lewis and Clark's river. Ecohydrol, Hydrobiol. 8:765–71.

Johnson, W. C., M. D. Dixon, M. L. Scott, et al. 2012. Forty years of vegetation change on the Missouri River floodplain. BioScience 62:123–35.

Johnson, W. C., R. L. Burgess, and W. R. Keammerer. 1976. Forest overstory vegetation and environment along the Missouri River in North Dakota. Ecol. Monogr. 46:59–84.

Johnson, W. C., S. E. Boettcher, K. A. Poiani, et al. 2004. Influence of weather extremes on the water levels of glaciated prairie wetlands. Wetlands 24:385–98.

Johnston, C. A. 2012. Beaver wetlands. In D. P. Batzer and A. Baldwin, eds. Wetland habitats of North America: Ecology and Conservation Concerns, 161–171. Univ. California Press, Berkeley.

Johnston, C. A. 2013. Wetland losses due to row crop expansion in the Dakota Prairie Pothole Region. Wetlands 33:175–82.

Johnston, C. A. 2014. Agricultural expansion: Land use shell game in the US Northern Plains. Landscape Ecol. 29:81–95.

Jones, M. B. 2019. Projected climate change and the global distribution of grasslands. In D. J. Gibson and J. A. Newman, eds. Grasslands and Climate Change, 67–81. Cambridge Univ. Press, Cambridge.

Junk, W. J., P. B. Bayley, and R. E. Sparks. 1989. The flood pulse concept in river-floodplain systems. In D. P. Dodge, ed. Proc. Intl. Large River Ecosystems, pp. 112–27. Canad. Spec. Public.

Fisheries and Aquatic Sciences 106. Dept. Fisheries and Oceans, Ottawa.

Kaeming, M. A., B. D. S. Graeb, C. W. Hoagstrom, et al. 2007. Patterns of fish diversity in mainstem Missouri River reservoir and associated delta in South Dakota and Nebraska, USA. River Res. Appl. 23: 786–91.

Kantrud, H. A., and R. E. Stewart. 1977. Use of natural basin wetlands by breeding waterfowl in North Dakota. J. Wildl. Manage. 41:243–53.

Kantrud, H. A., G. L. Krapu, G. A. Swanson. 1989. Prairie basin wetlands of the Dakotas: A community profile. Biological Report 85(7.28). U.S. Department of the Interior, Fish and Wildlife Service, Washington, DC.

Karl, T. R., J. M. Melillo, and T. C. Peterson, eds. 2009. Global Climate Change Impacts in the United States. Cambridge University Press, New York.

Karle, S. T., and D. Karle. 2017. Conserving the Dust Bowl: The new deal's prairie states forestry project. Louisiana State Univ. Press, Baton Rouge.

Kaufmann, M. R., and R. K. Watkins. 1990. Characteristics of high- and low-vigor lodgepole pine trees in old-growth stands. Tree Physiol. 7:239–46.

Kaye, B. M., and H. A. Schoch. 1993. Theodore Roosevelt National Park: The Story Behind the Scenery. KC Publications, Inc., Wickenburg, Arizona.

Kaye, M. W., C. A. Woodhouse, and S. T. Jackson. 2010. Persistence and expansion of ponderosa pine woodlands in the west-central Great Plains during the past two centuries. J. Biogeogr. 37:1668–83.

Kayes, L. J., and D. B. Tinker. 2012. Forest structure and regeneration following a mountain pine beetle epidemic in southern Wyoming. For. Ecol. Manage. 263:57–66.

Keating, W. H. 1824. Narrative of an expedition to the sources of St. Peter's River, Lake Winnepeek, Lake of the Woods, &c. &c. performed in the year 1823, by order of the Hon. J. C. Calhoun, Secretary of War under the command of S. H. Long, Major U.S.T.E.; compiled from the notes of Major Long, Messrs. Say, Keating, and Calhoun, by W. H. Keating. 2 vols. Philadelphia: H. C. Carey & I. Lea. (Severson and Sieg 2006, p. 35).

Keddy, P. A., L. H. Fraser, A. I. Solomeshch, et al. 2009. Wet and wonderful: the world's largest wetlands are conservation priorities. BioScience 59:39–51.

Keyser, T. L., L. B. Lentile, F. W. Smith, et al. 2008. Changes in forest structure after a large, mixed-severity wildfire in ponderosa pine forests of the Black Hills, South Dakota, USA. Forest Sci. 54:328–38.

Kharel, G., H. Zheng, and A. Kirilenko. 2016. Can land-use change mitigate long-term flood risks in the Prairie Pothole Region? The case of Devils Lake, North Dakota, USA. Region. Environ. Change 16:2443–56.

King, D. A., D. M. Bachelet, and A. J. Symstad. 2013. Climate change and fire effects on a prairie-woodland ecotone: Projecting species range shifts with a dynamic global vegetation model. Ecol. Evol. 3:5076–97.

Klebnikov, P. 2018. Big plans for a mighty land. Environmental Defense Fund, Solutions, vol. 49:16–17.

Knapp, A. K., C. Beier, D. D. Briske, et al. 2008. Consequences of more extreme precipitation regimes for terrestrial ecosystems. BioScience 58:811–21.

Knapp, A. K., J. M. Blair, J. M. Briggs, et al. 1999. The keystone role of bison in North American tallgrass prairie. BioScience 49:39–50.

Knight, D. H., G. P. Jones, W. A. Reiners, et al. 2014. Mountains and Plains: The Ecology of Wyoming Landscapes. Yale Univ. Press, New Haven.

Knopf, F. L., and F. B. Samson. 1997. Conservation of grassland vertebrates. In F. L. Knopf and F. B. Sampson, eds. Ecology and Conservation of Great Plains Vertebrates, 273–89. Springer-Verlag, New York.

Koel, T. M., and J. J. Peterka. 2003. Stream fish communities and environmental correlates in the Red River of the North, Minnesota and North Dakota. Environ. Biol. Fishes 67:137–55.

Koh, I., E. V. Lonsdorf, N. M. Williams, et al. 2016. Modeling the status, trends, and impacts of wild bee abundance in the United States. Proc. Natl. Acad. Sci. U.S.A. 113:140–45.

Kornfeld, M., and A. J. Osborn, eds., 2003. Islands on the Plains: Ecological, social, and ritual use of landscapes. Univ. Utah Press. Salt Lake City.

Kort, J., and R. Turnock. 1999. Carbon reservoir and biomass in Canadian prairie shelterbelts. Agrofor. Syst. 44:175–86.

Kotliar, N. B., Miller, B. J., Reading, R. P. et al., 2006. The prairie dog as a keystone species. In J. L. Hoogland, ed. Conservation of the black-tailed prairie dog, 53–64. Island Press, Washington, DC.

Küchler, A. W. 1964. Potential natural vegetation of the conterminous United States. Amer. Geogr. Soc. Special Publ. 36, New York.

Kurtz, C. 2013. A practical guide to prairie reconstruction (2nd ed.). Univ. Iowa Press, Iowa City.

Lark, T. J., J. M. Salmon, and H. K. Gibbs. 2015. Cropland expansion outpaces agricultural and biofuel policies in the United States. Environ. Res. Lett. 10:044003.

Larson, D. L. 2002. Native weeds and exotic plants: Relationships to disturbance and mixed-grass prairie. Plant Ecol. 169:317–33.

Larson, D. W. 2012. Runaway Devils Lake. Amer. Sci. 100:47–53.

Larson, F., and W. Whitman. 1942. A comparison of used and unused grassland mesas in the badlands of South Dakota. Ecology 23:438–45.

Larson, G. E., and J. R. Johnson. 1999. Plants of the Black Hills and Bear Lodge Mountains. South Dakota Agric. Exp. Sta. B732, Brookings.

Lass, W. E. 2008. Navigating the Missouri. Steam boating on nature's highway, 1819–1935. The Arthur H. Clark Company, Norman, Oklahoma.

Lauenroth, W. K., D. G. Milchunas, J. L. Dodd, et al. 1994. Effects of grazing on ecosystems of the Great Plains. In M. Vavra, W. A. Laycock, and R. D. Pieper, eds. Ecological Implications of Livestock Herbivory in the West, 69–100. Society for Range Management, Denver, CO.

Lauenroth, W. K., I. C. Burke, and M. P. Gutmann. 1999. The structure and function of ecosystems in the central North American grassland region. Great Plains Res. 9:223–59.

Lavelle, M. 2015. Last dance? Science 348:1300–05.

Lavelle, P. 2012. Soil as habitat. In D. H. Wall, ed., Soil Ecology and Ecosystem Services. Oxford Univ. Press, Oxford.

Lavin, S. J., F. M. Shelley, and J. C. Archer, eds. 2011. Atlas of the Great Plains. Univ. Nebraska Press, Lincoln.

Lavorel, S. 2019. Climate change effects on grassland ecosystem services. In D. J. Gibson and J. A. Newman, eds. Grasslands and Climate Change, 131–146. Cambridge Univ. Press, Cambridge.

Lawson, M. L. 2009. Dammed Indians Revisited. South Dakota State Hist. Soc. Press, Pierre.

Lehmer, D. J. 1971. Introduction to Middle Missouri Archaeology. Anthropological Papers 1, National Park Service, U.S. Dept. of Interior, Washington, DC

Lentile, L. B., F. W. Smith, and W. D. Shepperd. 2006. Influence of topography and forest structure on patterns of mixed severity fire in ponderosa pine forests of the South Dakota Black Hills, USA. Int. J. Wildland Fire 15:557–66.

Leopold, L. B. 1994. A View of the River. Harvard University Press, Cambridge.

Leoschke, M., and J. Pearson.1988. Fen: A special kind of wetland. Iowa Conservationist 47:16–19.

Ley, M. J. 2012. Riparian forest vegetation patterns and historic channel dynamics of the Big Sioux River, South Dakota. Thesis, Univ. South Dakota, Vermillion.

Licht, D. S. 1997. Ecology and Economics of the Great Plains. Univ. Nebraska Press, Lincoln.

Licht, D. S., J. J. Millspaugh, K. E. Kunkel, et al. 2010. Using small populations of wolves for ecosystem restoration and stewardship. BioScience 60:147–53.

Little, Jr., E. L. 1971. USDA Forest Service.

Liu, G., F. W. Schwartz, C. K. Wright, et al. 2016. Characterizing the climate-driven collapses and expansions of wetlands habits with a fully integrated surface-subsurface hydrologic model. Wetlands 36 (Suppl. 2): S287-S298.

Liu, M., and D. L. Swanson. 2014. Physiological evidence that anthropogenic woodlots can substitute for native woodlands as stopover habitat for migrant birds. Physiol. Biochem. Zool. 87:183–95.

Lockwood, J. A. 2004. Locust: The devastating rise and mysterious disappearance of the insect that shaped the American frontier. Basic Books, New York.

Lott, D. F. 2002. American Bison: A Natural History. Univ. California Press, Oakland.

Lovaas, A. L. 1976. Introduction of prescribed burning to Wind Cave National Park. Wildl. Soc. Bull. 4:69-73.

Ludlow, W. 1875. Report of a reconnaissance of the Black Hills of Dakota made in the summer of 1874. U.S. Army, Dept. of Engineers, U.S. Government Printing Office, Washington, DC.

MacCracken, J. G., D. W. Uresk, and R. M. Hansen. 1983. Plant community variability on a small area in southeastern Montana. Great Basin Nat. 43:660–68.

Madson, J. 1985. Where the Sky Began: Land of the Tallgrass Prairie. Sierra Club Books, San Francisco.

Main, A. R., J. V. Headley, K. M. Peru, et al. 2014. Widespread use and frequent detection of neonicotinoid insecticides in wetlands of Canada's Prairie Pothole Region. PLoS/One: doi.org10.1371.

Mangan, J. M., J. T. Overpeck, R. S. Webb, et al. 2004. Response of Nebraska sand hills natural vegetation to drought, fire, grazing, and plant functional type shifts as simulated by the century model. Clim. Change 63:49–90.

Mann, C. C. 2005. 1491: New Revelations of the Americas Before Columbus. Alfred A. Knopf, New York.

Mann, M. 2017. Anger and indifference on Lake Winnipeg. The Walrus (July/August),

Manning, R. 1995. Grassland: The History, Biology, Politics, and Promise of the American Prairie. Penguin Books, London.

Manning, R. 2009. Rewilding the West. Univ. California Press, Berkeley.

Manske, L. L. 1980. Habitat, phenology and growth of selected sandhills range plants. Diss., North Dakota State Univ., Fargo.

Manske, L. L., and W. T. Barker. 1988. Habitat usage by prairie grouse on the Sheyenne National Grassland. USDA Forest Service GTR-RM-159, pp. 8–20.

Marriott, H. J. 2012. Survey and mapping of Black Hills montane grasslands. Report. South Dakota Dept. Game, Fish, and Parks, Pierre.

Marriott, H. J., and D. Faber-Langendoen. 2000a. Riparian and wetland plant communities of the Black Hills. The Nature Conservancy and The Association for Biodiversity Information (http://conserveonline.org/library/)

Marriott, H. J., and D. Faber-Langendoen. 2000b. Black Hills community inventory (vol. 2): Plant community descriptions. Report to The Nature Conservancy and Association for Biodiversity Information. (http://conserveonline.org/library/)

Martin P. S. and H. E. Wright, Jr., eds. 1967. Pleistocene Extinctions: The search for a cause. Yale University Press, New Haven.

Maslin, M. 2014. Climate Change. A Very Short Introduction. Oxford University Press, Oxford.

Mason, J. A., J. B. Swinehart, P. R. Hanson, et al. 2011. Late Pleistocene dune activity in the central Great Plains, USA. Quat. Sci. Rev. 30:3858–3870.

McAndrews, J. H. 1966. Postglacial history of prairie, savanna, and forest in northwestern Minnesota. Mem. Torrey Bot. Club 22:1–72.

McIntosh, A. C. 1931. A botanical survey of the Black Hills of South Dakota. Black Hills Engr. 19:159–276.

McIntyre, N. E., G. Liu, J. Gorzo, et al. 2019. Simulating the effects of climate variability on waterbodies and wetland-dependent birds in the Prairie Pothole Region. Ecosphere 10: e02711.

McLean, K. I., D. M. Mushet, and C. A. Stockwell. 2016. From "duck factory" to "fish factory": Climate induced changes in vertebrate communities of prairie pothole wetlands and small lakes. Wetlands DOI 10.1007/s13157-016-0766-3.

McMillion, S. 2006. Wild work crew. In Imhoff and Baumgartner, eds. Farming and the Fate of Wild Nature, pp. 80–83. Watershed Media, Healdsburg, California.

Meneguzzo, D. M., A. J. Lister, and C. Sullivan. 2018. Summary of Findings from the Great Plains Tree and Forest Invasives Initiative. USDA Forest Service GTR-NRS-177.

Mercado, J. E., R. W. Hofstetter, D. M. Reboletti, et al., 2014. Phoretic symbionts of the mountain pine beetle (*Dendroctonus ponderosae* Hopkins). Forest Sci. 60:512–26.

Miller, B. J., R. P. Reading, D. E. Biggins, et al. 2007. Prairie dogs: An ecological review and current biopolitics. J. Wildl. Manage. 71:2801–10.

Millett, B., and W. C. Johnson. 2009. Climate trends of the North American prairie pothole region 1906–2000. Clim. Change 93:243–67.

Mitsch, W. J., and J. G. Gosselink. 2000. Wetlands (3rd ed.) John Wiley & Sons, Inc., Hoboken, New Jersey.

Montgomery, D. R. 2012. Dirt: The Erosion of Civilizations. Univ. Calif. Press, Oakland.

Montgomery, D. R. 2017. Growing a Revolution: Bringing our Soil Back to Life. W. W. Norton & Company, New York.

Moore, P. M. K., and L. D. Flake. 1994. Forest characteristics in eastern and central South Dakota. Proc. South Dakota Acad. Sci. 73:163–74.

Morehouse, K., T. Johns, J. Kaye, and M. Kaye. 2008. Carbon and nitrogen cycling immediately following bark beetle outbreaks in southwestern ponderosa pine forests. For. Ecol. Manage. 255:2698–2708.

Morgan, J. A., D. R. LeCain, E. Pendall, et al. 2011. C4 grasses prosper as carbon dioxide eliminates desiccation in warmed semiarid grassland. Nature 476:202–5.

Moulton, G., ed. 1987. The Journals of the Lewis & Clark Expedition, vols. 3, 4 and 8. Univ. Nebraska Press, Lincoln.

Muhs, D. R., and P. B. Maat. 1993. The potential response of eolian sands to greenhouse warming and precipitation reduction on the Great Plains of the U.S.A. J. Arid Environ. 25:351–61.

Muhs, D. R., and S. A. Wolfe. 1999. Sand dunes of the northern Great Plains of Canada and the United States. Geol. Surv. Canada, Bull. 534, pages 183–97.

Muhs, D. R., T. W. Stafford, Jr., J. Been, et al. 1997. Holocene eolian activity in the Minot dune field, North Dakota. Canad. J. Earth Sci. 34:1442–59.

Mushet, D. M., M. B. Goldhaber, C. T. Mills, et al. 2015. Chemical and biotic characteristics of prairie lakes and large wetlands in south-central North Dakota—effects of a changing climate. U.S. Geological Survey Scientific Investigations Report 2015–5126.

Myers, R. 2009. Exploration and the fur trade. In H. F. Thompson, ed. A New South Dakota History (2nd ed.), 47–68. Center for Western Studies, Augustana Univ., Sioux Fall, South Dakota.

Nagel, L. M., B. J. Palik, M. A. Battaglia, et al. 2017. Adaptive silviculture for climate change. J. For. 115:1–12.

Naiman, R. J., H. Decamps, and M. E. McClain. 2005. Riparia: Ecology, Conservation, and Management of Streamside Communities. Elsevier Academic Press, New York.

National Agricultural Statistics Service. 2017. Statistics by state. U.S. Department of Agriculture, Washington, DC.

National Research Council. 1995. Wetlands. National Academy Press, Washington, DC.

National Research Council. 2002. The Missouri River Ecosystem: Exploring the Prospects for Recovery. Natl. Acad. Press, Washington, DC.

National Research Council. 2011. Missouri River Planning: Recognizing and Incorporating Sediment Management. Natl. Acad. Press. Washington, DC.

Naugle, D. E., K. F. Higgins, S. M. Nusser, et al. 1999. Scale-dependent habitat use in three species of prairie wetland birds. Landscape Ecol. 14:267–76.

Negrón, J. F., K. Allen, A. Ambourn, et al. 2017. Large-scale thinnings, ponderosa pine, and mountain pine beetle in the Black Hills, USA. Forest Sci. 63:529–36.

Nekola, J. C. 1994. The environment and vascular flora of northeastern Iowa fen communities. Rhodora 96:44–68.

Nelson, P. 1986. After the West was Won: Homesteaders and Town-builders in Western South Dakota, 1900–1917. Univ. Iowa Press, Iowa City.

Nelson, P. 1996. The Prairie Winnows out its Own: The West River Country of South Dakota in the Years of Depression and Dust. Univ. Iowa Press, Iowa City.

Nelson, P. W. 1964. The forests of the lower Sheyenne River Valley, North Dakota. Thesis, North Dakota State Univ., Fargo.

Newton, H., and W. P. Jenney. 1880. Report on the Geology and Resources of the Black Hills of Dakota, with Atlas, p. 322. U.S. Gov. Printing Office, Washington, DC.

Niehus, C. A., A. V. Vecchia, and R. F. Thompson. 1999. Lake-level frequency analysis for the Waubay Lakes chain, northeastern South Dakota. Water-Resources Investigations Report 99–4122.

Niemuth, N. D., K. K. Fleming, and R. E. Reynolds. 2014. Waterfowl conservation in the US Prairie Pothole Region: Confronting the complexities of climate change. PLOS ONE 9 (6) e100034.

Nordt, L., J. Von Fisher, L. Tieszen, et al. 2008. Coherent changes in relative C4 plant productivity and climate during the late Quaternary in the North American Great Plains. Quaternary Sci. Rev. 27:1600–11.

Norland, J., J. T. Larson, C. Dixon, et al. 2015. Outcomes of past grassland reconstructions in North Dakota and Northwestern Minnesota. Ecol. Restoration 33:408–17.

Norris, J. R., J. L. Betancourt, and S. T. Jackson. 2016. Late Holocene expansion of ponderosa pine (Pinus ponderosa) in the central Rocky Mountains. J. Biogeogr. 43:778–90.

O'Brien, D. 2017. Great Plains Bison. Univ. Nebraska Press, Lincoln.

O'Connor, J. E., J. J. Duda, and G. E. Grant. 2015. 1000 dams down and counting. Science 348:495–97.

Odum, E. P. 1969. The strategy of ecosystem development. Science 164:262–70.

Ojima, D., D. Steiner, J. McNeeley et al. eds. 2015. Great Plains Regional Technical Input Report. Island Press, Washington.

Olson, V. 1988. The steppe vegetation of the north unit of Badlands National Park. Thesis, Univ. South Dakota, Vermillion, DC.

Orr, H. K. 1975. Watershed management in the Black Hills: The status of our knowledge. USDA For. Ser. Res Paper RM-141.

Orth, J. 2007. The shelterbelt project: Cooperative conservation in 1930s America. Agr. Hist. 81:333–57.

Osterkamp, W. R., M. L. Scott, and G. T. Auble. 1998. Downstream effects of dams on channel geometry and bottomland vegetation: Regional patterns in the Great Plains. Wetlands 18:619–33.

Otto, C. R. V., C. L. Roth, B. L. Carlson, et al. 2016. Land-use change reduces habitat suitability for supporting managed honey-bee colonies in the Northern Great Plains. P. Natl. Acad. Sci. U.S.A. 113:10430–35.

Palmer, M. A., J. B. Zedler, and D. A. Falk. 2016. Foundations of Restoration Ecology. Island Press, Washington, DC.

Patel-Weynand, T., G. Bentrup, and M. Schoeneberger. 2017. Agroforestry: Enhancing resiliency in U.S. agricultural landscapes under changing conditions. U.S. Dept. Agric. Gen. Tech. Report WO-96a, Washington, DC.

Patterson, D. 2016. The Missouri River Journals of John James Audubon. Univ. Nebraska Press, Lincoln.

Pearsall, P. 2013. Farmer-philosopher Fred Kirschenmann on food and the warming future. YES! Magazine, Food and Warming Future, Feb. 22.

Pejchar, L., Y. Clough, J. Ekroos, K. A. Nicholas, O. Olsson, D. Ram, M. Tschumi, and H. G. Smith. 2018. Net effects of birds in agroecosystems. BioScience 68:896–904.

Perry, L. G., D. C. Anderson, L. V. Reynolds, et al. 2012. Vulnerability of riparian ecosystems to elevated CO2 and climate change in arid and semiarid western North America. Global Change Biol. 18:821–42.

Petrie, M. D., J. B. Bradford, R. M. Hubbard, et al. 2017. Climate change may restrict dryland forest regeneration in the 21st century. Ecology 98:1548–59.

Pfeifer, E. M., J. A. Hicke, and A. J. H. Meddens. 2011. Observations and modeling of aboveground tree carbon stocks and fluxes following a bark beetle outbreak in the western United States. Global Change Biol. 17:339–50.

Pielou, E. C. 1991. After the Ice Age. Univ. Chicago Press, Chicago.

Pierce II, R. A., D. T. Farrand, and W. B. Kurtz. 2001. Projecting the bird community response resulting from the adoption of shelterbelt agroforestry practices in Eastern Nebraska. Agrofor. Syst. 53:333–50.

Pierce, R. 2012. Northern Pike: Ecology, Conservation, and Management History. Univ. Minnesota Press.

Pietz, P. J., D. A. Buhl, J. A. Shaffer, et al. 2009. Influence of trees in the landscape on parasitism rates of grassland passerine nests in southeastern North Dakota. Condor 111:36–42.

Piva, R.J, B. F. Walters, D. D. Haugan, et al. 2013. South Dakota forests 2010. USDA For. Serv. Res. Bull. NRS-81.

Plumb, G. E., and R. Sucec. 2006. A bison conservation history in the U.S. National Parks. Journal of the West 45:22–28.

Pollan, M. 2006. The Omnivore's Dilemma: A Natural History of Four Meals. Penguin Press, New York.

Pool, D. B., and J. E. Austin, eds. 2006. Migratory Bird Management for the Northern Great Plains Joint Venture: Implementation Plan. Gen. Tech. Rep. TC-01. Bismarck, ND: Northern Great Plains Joint Venture. U.S. Fish and Wildlife Service.

Potter, L. D., and D. L. Green. 1964a. Ecology of ponderosa pine in western North Dakota. Ecology 45:10–23.

Potter, L. D., and D. L. Green. 1964b. Ecology of a northeastern outlying stand of *Pinus flexilis*. Ecology 45:866–68.

Potter, L. D., and D. R. Moir. 1961. Phytosociological study of burned deciduous woods, Turtle Mountains, North Dakota. Ecology 42:468–80.

Preisler, H. K., J. A. Hicke, A. A. Ager, et al. 2012. Climate and weather influences on spatial temporal patterns of mountain pine beetle populations in Washington and Oregon. Ecology 93:2421–34.

Prince, H. 1997. Wetlands of the American Midwest: A Historical Geography of Changing Attitudes. Univ. Chicago Press, Chicago.

Progulske, D. R 1974. Yellow Ore, Yellow Hair, Yellow Pine: A Photographic Study of a Century of Forest Ecology. South Dakota. Agric. Exp. Sta. Bull. 616:1–169.

Pyne, S. J. 2017. The Great Plains: A Fire Survey. Univ. Arizona Press, Tucson.

Quinnild, C. L., and H. E. Cosby. 1958. Relicts of climax vegetation on two mesas in western North Dakota. Ecology 39:29–33.

Raffa, K. F., B. H. Aukema, B. J. Bentz, et al. 2008. Cross-scale drivers of natural disturbances prone to anthropogenic amplification: Dynamics of biome-wide bark beetle eruptions. BioScience 58:501–517.

Ralston, R. C. 1960. The structure and ecology of the north slope juniper stands in the Little Missouri Badlands. Thesis, Univ. Utah, Salt Lake City.

Ralston, R. D. 1968. The grasslands of the Red River Valley. Diss., Univ. Saskatchewan, Saskatoon.

Raventon, E. 1994. Island in the Plains: A Black Hills Natural History. Johnson Books, Boulder, Colorado.

Raventon, E. 2003. Buffalo Country: A Northern Plains Narrative. Johnson Books, Boulder, Colorado.

Read, R. A. 1958. The Great Plains shelterbelt in 1954. Great Plains Agricultural Council Publ. No. 16.

Redman, R. E. 1972. Plant communities and soils of an eastern North Dakota prairie. Bull. Torrey Bot. Club 99:65–76.

Regniere, J., and B. Bentz. 2007. Modeling cold tolerance in the mountain pine beetle, *Dendroctonus ponderosae*. J. Insect Physiol. 53:559–72.

Rehfeldt, G. E., N. L. Crookston, M. V. Warwell, et al. 2006. Empirical analyses of plant-climate relationships for the western United States. Int. J. Plant Sci. 167:1123–50

Reiger, B. A., K. F. Higgins, J. A. Jenks, et al. 2006. Demographics of western South Dakota wetlands and basins. College Agric. Biol. Sci., South Dakota State Univ., Publ.

Reily, P. W., and W. C. Johnson. 1982. The effects of altered hydrologic regime on tree growth along the Missouri River in North Dakota. Can. J. Bot. 60:2410–23.

Reitsma, K. D., B. H. Dunn, U. Mishra, et al. 2015. Land-use change impact on soil sustainability in a climate and vegetation transition zone. Agronomy J. 107:2363–72.

Reitsma, K. D., S. A. Clay, B. H. Dunn, et al. 2016. Does the U.S. Cropland Data Layer provide an accurate benchmark for land-use change estimates? Agron. J. 108:266–72.

Renton, D. A., D. M. Mushet, and E. S. DeKeyser. 2015. Climate Change and Prairie Pothole Wetlands: Mitigating Water-level and Hydroperiod Effects through Upland Management. U.S. Geol. Surv. Sci. Invest. Rep. 2015-5004.

Risjord, N. K. 2012. Dakota: The Story of the Northern Plains. Univ. Nebraska Press, Lincoln.

Roberts, C. P., V. M. Donovan, C. L. Wonkka, et al. 2019. Fire legacies in eastern ponderosa pine forests. Ecol. Evol. 9:1869–79.

Robertson, D. S., M. C. McKenna, O. B. Toon, et al. 2004. Survival in the first hours of the Cenozoic. Geol. Soc. Amer. Bull. 116:760–68.

Robinson, W. B. 1966. History of North Dakota. Univ. Nebraska Press, Lincoln.

Rodewald, A. D., P. Arcese, J. Sarra, etal. 2020. Innovative Finance for Conservation: Roles for Ecologists and Practitioners. Issues in Ecology, Rep. No. 22, Ecol. Soc. Amer., Washington, DC.

Roe, F. G. 1972. The North American Buffalo. Univ. Toronto Press, Toronto, Ontario.

Romme, W. H., D. H. Knight, and J. B. Yavitt. 1986. Mountain pine beetle outbreaks in the Rocky Mountains: Regulators of primary productivity? Am. Nat. 127:484–94.

Rosenburg et al. 2005. Nelson and Churchill river basins. In A. C. Benke and C. E. Cushing, eds. Rivers of North America, pp. 853–903. Elsevier Academic Press.

Ross, A. 1856. The Red River Settlement: Its Rise, Progress, and Present State. Smith, Elder, and Co., London (Minneapolis: Ross and Haines Inc., 1957).

Royte, E. 2017. The same pesticides linked to bee decline might also threaten birds. Audubon Magazine (Spring).

Rumble, M. M., C. H. Sieg, D. W. Uresk, et al. 1998. Native woodlands and birds of South Dakota: Past and Present. U.S. Forest Service RMRS-RP-8.

Sagoff, M. 2002. On the value of natural ecosystems: The Catskills parable. Polit. Life Sci. 21:16–21.

Samson, F. B., F. L. Knopf, and W. R. Ostlie. 2004. Great Plains ecosystems: Past, present, and future. Wildl. Soc. Bull 32:6–15.

Sandom, C., S. Faurby, B. Sandel, et al. 2014. Global late Quaternary megafauna extinctions linked to humans, not climate change. Proc. R. Soc. London, Ser. B 281:20133254.

Sargent, C. S. 1897. The Silva of North America, vol. 11. Peter Smith, New York (1947).

Sauchyn, D., H. Diaz, and S. Kulshreshtha, eds. The New Normal: The Canadian prairies in a changing climate. Canad. Plains Res. Center Press, Univ. Regina, Saskatchewan.

Savage, C. 2004. Prairie: A Natural History. Greystone Books Ltd., Vancouver.

Savory, A. (with Jody Butterfield). 2016. Holistic Management: A Commonsense Revolution to Restore our Environment. Island Press, New York.

Schook, D. M. 2017. Seeing the river through the trees: Using cottonwood dendrochronology to reconstruct river dynamics in the Upper Missouri River Basin. Diss., Colorado State Univ., Fort Collins.

Schuh, W. M., M. H. Hove, and D. J. Farrell. 2017. Historical, current and prospective future surface water management in the Little Missouri River scenic river basin. Response of the request of Governor Burgum for assessment of basin water management needs and policies. Office of the North Dakota State Engineer. 32 pp.

Schulte, L. A., J. Niemi, M. J. Helmers, et al. 2017. Prairie strips improve biodiversity and the delivery of multiple ecosystem services from corn-soybean croplands. Proc. Natl. Acad. Sci. 114:11247–52.

Schumm, S. A. 1977. The Fluvial System. John Wiley & Sons, New York.

Schumm, S. A. 2005. River Variability and Complexity. Cambridge University Press, Cambridge.

Sedgwick, J. A., and F. L. Knopf. 1991. Prescribed grazing as a secondary impact in a western riparian floodplain. J. Range Manage. 44:369–73.

Severson, K. E., and C. H. Sieg. 2006. The Nature of Eastern North Dakota: Pre-1880 Historical Ecology. North Dakota Inst. Reg. Studies, Fargo.

Severson, K. E., and J. F. Thilenius. 1976. Classification of quaking aspen stands in the Black Hills and Bear Lodge Mountains. U.S. For. Ser. Res. Paper RM-166.

Shapley, M. D., W. C. Johnson, D. R. Engstrom, et al. 2005. Late-Holocene flooding and drought in the Northern Great Plains, USA, reconstructed from tree rings, lake sediments and ancient shorelines. The Holocene 15:29–41.

Shay, C. T. 1967. Vegetation history of the southern Lake Agassiz basin during the past 12,000 years. In E. W. J. Mayor-Oakes, ed. Life, Land, and Water: Proceedings of the 1966 conference on environmental studies of the Glacial Lake Agassiz region, 231–51. Univ. Manitoba, Winnipeg.

Shearer, J., and C. Berry. 2003. Fish community persistence in eastern North and South Dakota rivers. Great Plains Res. 13:139–59.

Shelby, A. 2003. Red River Rising: The anatomy of a flood and the survival of an American city. Borealis Books, Minnesota Hist. Soc. Press, St. Paul.

Shepperd, W. D., and M. A. Battaglia. 2002. Ecology, silviculture, and management of Black Hills ponderosa pine. USDA Forest Service RMRS-GTR-97.

Shinneman, D. J., and W. L. Baker. 1997. Nonequilibrium dynamics between catastrophic disturbances and old-growth forests in ponderosa pine landscapes of the Black Hills. Conserv. Biol. 11:1276–88.

Shjeflo, J. B. 1968. Evapotranspiration and the water budget of prairie potholes in North Dakota. U.S. Geol. Survey Prof. Paper 585-B.

Shuman, B. 2012. Recent Wyoming temperature trends, their drivers, and impacts in a 14,000-year context. Clim. Change 112:429–47.

Silcox, F. 1935. The Problem. Pages 1–2 in Possibilities of Shelterbelt Planting in the Plains Region. Lake States Forest Experiment Station, USDA Forest Service.

Singh, J. S., W. K. Lauenroth, R. K. Heitschmidt, et al. 1983. Structural and functional attributes of northern mixed prairie of North America. Bot. Rev. 49:117–49.

Skalak, K. J., A. J. Benthem, E. R. Schenk, et al., 2013. Large dams and alluvial rivers in the Anthropocene: The impacts of the Garrison and Oahe Dams on the Upper Missouri River. Anthropocene 2:51–64.

Smeins, F. E. 1967. The wetland vegetation of the Red River Valley and drift prairie regions of Minnesota, North Dakota, and Manitoba. Diss., Univ. Saskatchewan, Saskatoon.

Smith, A. G., J. H. Stoudt, and J. B. Gollop. 1964. Prairie potholes and marshes. In J. P. Linduska, ed. Waterfowl Tomorrow, 39–50. U.S. Fish and Wildlife Service, Washington, DC.

Smith, D., D. Williams, G. Houseal, et al. 2010. The Tallgrass Prairie Center guide to prairie restoration in the Upper Midwest. Univ. Iowa Press, Iowa City.

South Dakota Department of Environment and Natural Resources. 2018. The South Dakota Integrated Report for Surface Water Quality Assessment. Environmental Protection Agency, Washington, DC.

Spencer, C., M. Van Essen, E. Renner, et al. 2015. Prairie or woodland? Reconstructing past plant communities at Good Earth State Park via soil core and tree ring analysis. Proc. South Dakota Acad. Sci. 94:227–36.

Spencer, C., S. Matzner, J. Smalley, et al. 2009. Forest expansion and soil carbon exchanges in the Loess Hills of eastern South Dakota. Am. Midl. Nat. 161:273–85.

Spiering, D. J., and R. L. Knight. 2005. Snag density and use by cavity-nesting birds in managed stands of the Black Hills National Forest. For. Ecol. Manage. 214:40–52.

Stark, K. J., E. Iverson, M. R. Komp, et al. 2011. Jewel Cave National Monument: Natural Resource Condition Assessment. Natural Resources Report NPS/JECA/NRTR—2011/477.

Steen, V., S. K. Skagen, and B. R. Noon. 2016. Vulnerability of breeding waterbirds to climate change in the Prairie Pothole Region, USA. PLoS One doi.10.1371/journal.pone.0096747

Stein, B. A., and M. R. Shaw. 2013. Biodiversity conservation for a climate-altered future. In S. C. Moser and M. T. Boykoff, eds. Successful Adaptation to Climate Change, 50–66. Routledge, New York.

Stephens, S. E., J. A. Walker, D. R. Blunck, et al. 2008. Predicting risk of habitat conversion in native temperate prairie. Conser. Biol. 22:1320–30.

Stevens-Rumann, C. S., K. B. Hemp, P. E. Higuera, et al. 2018. Evidence for declining forest resilience to wildfires under climate change. Ecol. Letters 21:243–52.

Stewart, R. E., and H. A. Kantrud. 1971. Classification of natural ponds and lakes in the glaciated prairie region. U.S. Department of the Interior, Bureau of Sport Fisheries and Wildlife, Fish and Wildlife Service, Resource Publication 92, Washington, DC.

Stoeckeler, J. H., and R. A. Williams. 1949. Windbreaks and shelterbelts. U.S. Dept. Agric., Yearbook of Agriculture, 191–99.

Stohlgren, T. J., K. A. Bull, Y. Otsuki, et al. 1998. Riparian zones are havens for exotic plant species in central grasslands. Plant Ecol. 138:113–25.

Stoner, J. D., D. L. Lorenz, G. J. Wiche, et al. 1993. Red River of the North basin, Minnesota, North Dakota, and South Dakota. Water Resources Bull. 29:575–615.

Striped Face-Collins, M., and C. A. Johnston. 2007. Rangeland drought mitigation by beaver (*Castor canadensis*) impounded water. Proc. South Dakota Acad. Sci. 86:228.

Svedarsky, W. D., and P. E. Buckley. 1975. Some interactions of fire, prairie and aspen in northwest Minnesota. In M. K. Wali, ed. Prairie: A Multiple View, 115–121. Univ. North Dakota Press, Grand Forks.

Swanson, D. L., H. A. Carlisle, and E. T. Liknes. 2003. Abundance and richness of neotropical migrants during stopover at farmstead woodlots and associated habitats in southeastern South Dakota. Amer. Midl. Nat. 149:176–91.

Swanson, G. A., T. C. Winter, V. A. Adomaitis, et al. 1988. Chemical characteristics of prairie lakes in south-central North Dakota—their potential for influencing use by fish and wildlife. U.S. Dept. Interior, Fish and Wildlife Service, Fish and Wildlife Technical Report 18. Washington, DC

Symstad, A. J. and J. L. Jonas. 2011. Incorporating biodiversity into rangeland health: Plant species richness and diversity in Great Plains grasslands. Rangeland Ecol. Manage. 64:555- 72.

Symstad, A. J., W. E. Newton, and D. J. Swanson. 2014. Strategies for preventing invasive plant outbreaks after prescribed fire in ponderosa pine forest. For. Ecol. Manage. 324:81–88.

Tack, J. D., F. R. Quamen, K. Kelsey, et al. 2017. Doing more with less: Removing trees in a prairie system improves value of grasslands for obligate bird species. J. Environ. Sci. 198:163–69.

Tangen, B. A., and R. G. Finocchiaro. 2017. A case study examining the efficacy of drainage setbacks for limiting effects to wetlands in the Prairie Pothole Region, USA. J. Fish Wildl. Manage. 8:513–29.

Teller, J. T., and D. W. Leverington. 2004. Glacial Lake Agassiz: A 5000 yr history of change and its relationship to the $\delta^{18}O$ record of Greenland. Geol. S. Amer. Bull. 116:729–42.

Tester, J. R., S. M. Galatowitsch, R. A. Montgomery, and J. J. Moriarty. 2020. Minnesota's Natural Heritage, second edition. Univ. Minnesota Press, Minneapolis.

Thebault, A., P. Mariotte, C. J. Lortie, et al. 2014. Land management trumps the effect of climate change and elevated CO2 on grassland functioning. J. Ecol. 102:896–904.

Thilenius, J. F. 1970. An isolated occurrence of limber pine (*Pinus flexilis* James) in the Black Hills of South Dakota. Am. Midl. Nat. 84:411–17.

Thogmartin, W. E., L. López-Hoffman, J. Rohweder, et al. 2017. Restoring monarch butterfly habitat in the Midwestern US: All hands on deck. Environ. Res. Lett. 12:074005.

Thompson, S. J., T. W. Arnold, J. Fieberg, et al. 2016. Grassland birds demonstrate delayed response to large-scale tree removal in central North America. J. Appl. Ecol. 53:284–94.

Thurman, L. I., B. A. Stein, E. A. Beever, et al. 2020. Persist in place or shift in space? Evaluating the adaptive capacity of species to climate change. Front. Ecol. Environ. 18:520–28.

Thwaites, R. G. 1905. Original journals of the Lewis and Clark Expedition, vol. 7. Dodd, Mead, New York.

Tieszen, L. L., and M. W. Pfau. 1995. Isotopic evidence for the replacement of prairie by forest in the Loess Hills of eastern South Dakota. In D. C. Hartnett, ed., Proc. 14th North American Prairie Conference: Prairie biodiversity, 153–65.

Tilman, D., C. Balzer, J. Hill, et al. 2011. Global food demand and the sustainable intensification of agriculture. Proc. Natl. Acad. Sci. 108:20260–64.

Tilman, D., P. B. Reich, and J. M. H. Knops. 2006. Biodiversity and ecosystem stability in a decade-long grassland experiment. Nature 441:1 (doi:10.1038/nature04742)

Tiner, R. 2009. Status report for the National Wetlands Inventory program. U.S. Dept. of the Interior, Fish and Wildlife Service, Division of Habitat and Resource Conservation. Branch of Revenue and Mapping Support, Arlington, Virginia.

Todhunter, P. E., and E. A. Knish. 2014. Lake flooding and synoptic weather-type frequency at Devils Lake, North Dakota, USA, between 1965 and 2010. Clim. Res. 61:191–201.

Todhunter, P. F., and B. C. Rundquist. 2004. Terminal lake flooding and wetland expansion in Nelson County, North Dakota. Phys. Geogr. 25:68–85.

Tolstead, W. L. 1941a. Germination habits of certain sand-hill plants in Nebraska. Ecology 10:393–397.

Tolstead, W. L. 1941b. Plant communities and secondary succession in south-central South Dakota. Ecology 22:322–28.

Tolstead, W. L. 1942. Vegetation of the northern part of Cherry County, Nebraska. Ecol. Monogr. 12:255–92.

Tolstead, W. L. 1947. Woodlands in northwestern Nebraska. Ecology 28:180–88.

Toom, D. L. 1992. Climate and sedentism in the Middle Missouri subarea of the Plains. Diss., Univ. Nebraska, Lincoln.

Travers, S. E., G. M. Fauske, K. Fox, et al. 2011. The hidden benefits of pollinator diversity for the rangelands of the Great Plains: Western prairie fringed orchids as a case study. Rangelands 33:20–26.

Tschumi, M., M. Albrecht, M. H. Entling, et al. 2015. High effectiveness of tailored flower strips in reducing pests and crop plant damage. Proc. R. Soc. London, Ser. B 282:20151369.

Tulbure, M. G., C. A. Johnston, and D. L Auger. 2008. Rapid invasion of a Great Lakes Coastal wetland by non-native *Phragmites australis* and *Typha*. J. Great Lakes Res. 33: 269–279.

Tweton, D. J., and E. C. Albers, eds. 1996. The Way It Was: The North Dakota Frontier Experience. Book One: The Sodbusters. The Grass Roots Press, Fessenden, North Dakota.

Tyndall, J., and J. Colletti. 2007. Mitigating swine odor with strategically designed shelterbelt systems: A review. Agrofor. Syst. 69:45–65.

U.S. Army Corps of Engineers. 2010. Cottonwood management plan/draft programmatic environmental assessment: Proposed implementation along six priority segments of the Missouri River. Internal report.

U.S. Dept. of Interior. 2014. "Secretary Jewell announces $1.1 billion to state wildlife agencies from excise taxes on anglers, hunters, and boaters." Press Release, March 25.

U.S. Fish and Wildlife Service. 2018. Waterfowl Population Status 2018. U.S. Department of the Interior, Washington, DC.

U.S. Environmental Protection Agency. 2016. What climate change means for South and North Dakota. EPA 430-F-16-043.

Urbanhower, C. E., Jr. 1992. Abundance, vegetation, and environment of four patch types in a northern mixed prairie. Can. J. Bot. 70:277–84.

Uresk, D. W. 2015. Classification and monitoring plains cottonwood ecological type in the Northern Great Plains. Proc. South Dakota Acad. Sci. 94:201–11.

Uresk, D. W., and W. W. Painter. 1985. Cattle diets in a ponderosa pine forest in the northern Black Hills. J. Range Manage. 38:440–42.

Uresk, D. W., D. E. Mergen, and J. Javersak. 2010. Model for classifying and monitoring hackberry (*Celtis occidentalis* L.)-shrub ecological type in sand hills prairie ecosystem. Proc. South Dakota Acad. Sci. 89:105–19.

Uresk, D. W., K. E. Severson, and J. Javersak. 2015. Model for classification and monitoring green ash-ecological type in the Northern Great Plains. Proc. South Dakota Acad. Sci. 94:213–26.

Van der Valk, A., and C. B. Davis. 1978. The role of seed banks in the vegetation dynamics of prairie glacial marshes. Ecology 59:322–35.

Van der Valk, A., and D. Mushet. 2016. Interannual water-level fluctuations and the vegetation of prairie potholes: potential impacts of climate change. Wetlands 36 (suppl. 2):397–406.

Van der Valk, A., ed. 1989. Northern Prairie Wetlands. Iowa State Univ. Press, Ames.

Van Tramp, J. C. 1868. Prairie and Rocky Mountain Adventures. Segner and Condit, Columbus, Ohio.

Vanderhoof, M. K., and L. C. Alexander. 2016. The role of lake expansion in altering the wetland landscape of the Prairie Pothole Region, United States. Wetlands 36 (Suppl. 2):S309–321.

Voldseth, R. A., W. C. Johnson, T. Gilmanov, et al. 2007. Model estimates of land-use effects on water levels of northern prairie wetlands. Ecol. Appl. 17:527–40.

Volke, M. A. 2015. Vegetation of the White River delta on the Missouri River in South Dakota. Diss., South Dakota State Univ., Brookings.

Volke, M. A., M. L. Scott, W. C. Johnson, et al. 2015. The ecological significance of emerging deltas in regulated rivers. BioScience 65: 598–611.

Volke, M. A., W. C. Johnson, M. D. Dixon, et al. 2019. Emerging reservoir delta-backwaters: Biophysical dynamics and riparian biodiversity. Ecol. Monogr. 89: e01363.

Von Loh, J., D. Cogan, J. Butler, et al. 2000. USGS-NPS vegetation mapping program: Theodore Roosevelt National Park, North Dakota. U.S. Dept. of Interior, Denver, Colorado.

Wagner, G. 1997. Studies of the northern leopard frog (*Rana pipiens*) in Alberta. Alberta Environmental Protection, Wildlife Management Division, Wildlife Status Report no. 9.

Wall, D. H., ed. 2012. Soil Ecology and Ecosystem Services, Oxford Univ. Press, Oxford.

Wanek, W. J. 1967. The gallery forest vegetation of the Red River of the North. Diss., North Dakota State Univ., Fargo.

Wanek, W. J., and R. L. Burgess. 1965. Floristic composition of the sand prairies of southeastern North Dakota. Proc. North Dakota Acad. Sci. 19:26–40.

Watts, W. A., and R. C. Bright. 1968. Pollen, seed, and mollusk analysis of a sediment core from Pickerel Lake, northeastern South Dakota. Geol. Soc. Amer. Bull. 79:855–76.

Weaver, J. E. 1943. Resurvey of grasses, forbs, and underground plant parts at the end of the great drought. Ecol. Monogr. 13:63–117.

Weaver, J. E. 1954. North American Prairie. Johnsen Publishing, Lincoln, Nebraska.

Weaver, J. E. 1965. Native Vegetation of Nebraska. Univ. Nebraska Press, Lincoln.

Wedin, W. F., and S. L. Fales, eds. 2009. Grassland: Quietness and Strength for a New American Agriculture. Amer. Soc. Agron., Crop Sci. Soc. America, and Soil Sci. Soc. Amer., Madison, Wisconsin.

Weller, M. W. 1981. Freshwater Marshes: Ecology and Wildlife Management. Univ. Minnesota Press, Minneapolis.

Weller, M. W. 1999. Wetland Birds: Habitat Resources and Conservation Implications. Cambridge University Press, Cambridge.

Weller, M. W., and L. H. Fredrickson. 1974. Avian ecology of a managed glacial marsh. Living Bird 12:269–91.

Werner, B., J. Tracy, W. C. Johnson, et al. 2016. Modeling the effects of tile drain placement on the hydrologic function of farmed prairie wetlands. J. Am. Water Resour. Assoc. 52:1482–1492.

Whitman, W. C., and H. C. Hanson. 1939. Vegetation on scoria and clay buttes in western North Dakota. Ecology. 20:455–57.

Whitman, W. C., and M. K. Wali. 1975. Grasslands of North Dakota. In M. K. Wali, ed. Prairie: A Multiple View, 53–73. Univ. North Dakota Press, Grand Forks.

Wiche, G. J., A. V. Vecchia, L. Osborne, et al. 2000. Climatology, hydrology, and simulation of an emergency outlet, Devils Lake Basin, North Dakota. USGS Water Resources Investigations Report 00–4174.

Widga, C., S. N. Lengyel, J. Saunders, et al. 2017. Late Pleistocene proboscidean population dynamics in the North American mid-continent. Boreas DOI 10.1111/bor.12235.

Wienk, C. L., C. H. Sieg, and G. R. McPherson. 2004. Evaluating the role of cutting treatments, fire and seed banks in an experimental framework in ponderosa pine forests of the Black Hills, South Dakota. For. Ecol. Manage. 192: 375–93.

Wilkins, R. P., and W. H. Wilkins. 1977. North Dakota: A Bicentennial History. W. W. Norton & Co., New York.

Williams, J. W., and S. T. Jackson. 2007. Novel climates, no-analog communities, and ecological surprises. Front. Ecol. Environ. 9:475–82. (see also Williams et al., 2007, Proc. Natl. Acad. Sci. U.S.A. 104:5738–42)

Williams, M. A., and W. L. Baker. 2012. Spatially extensive reconstructions show variable-severity fire and heterogeneous structure in historical western United States dry forests. Global Ecol. Biogeogr. 21:1042–52. See also (2014) 23:831–35.

Willis, D. W., C. G. Scalet, and L. D. Flake. 2009. Introduction to Wildlife and Fisheries: An Integrated Approach (2nd ed.). W. H. Freeman and Company, New York.

Wilson, G. L. 1987. Buffalo Bird Woman's Garden: The Classic Account of Hidatsa American Indian Gardening Techniques. Minnesota Historical Soc. Press, St. Paul.

Wilson, R. E. 1970. Succession in stands of *Populus deltoides* along the Missouri River in southeastern South Dakota. Am. Midl. Nat. 83:330–42.

Winter, T. C., and D. O. Rosenberry. 1998. Hydrology of prairie pothole wetlands during drought and deluge: A 17-year study of the Cottonwood Lake wetland complex in North Dakota in the perspective of longer term measured and proxy hydrologic records. Clim. Change 40:180–209.

Winter, T. C., and M. R. Carr. 1980. Hydrologic setting of wetlands in the Cottonwood Lake area, Stutsman County, North Dakota. U.S. Geological Survey WRI 80–99.

Withey, P., and G. C. van Kooten. 2011. The effect of climate change on optimal wetlands and waterfowl management in western Canada. Ecol. Econ. 70:798–805.

Witte, S. S., and M. V. Gallagher, eds. 2012. The North American Journals of Prince Maximilian of Wied: September 1833-August 1834. Univ. Oklahoma Press, Norman.

Wood, W. R., ed. 1998. Archaeology on the Great Plains. Univ. Press Kansas, Lawrence.

Wood, W. R., W. J. Hunt, Jr., and R. H. Williams. 2011. Fort Clark and its Indian Neighbors: An Upper Missouri River Trading Post. Univ. Oklahoma Press, Norman.

Wooster, D. 1979. Dust Bowl: The Southern Plains in the 1930s. Oxford Univ. Press, New York.

Wright, C. K., and M. C. Wimberly. 2013. Recent land use change in the Western Corn Belt threatens grasslands and wetlands. P. Natl. Acad. Sci. U.S.A. 110:4134–139.

Wright, H. E., Jr. 1970. Vegetational history of the Great Plains. In W. Dort, Jr., and J. K. Jones, Jr., eds., Pleistocene and recent environments of the central Great Plains, 157–72. Univ. Press Kansas, Lawrence.

Wright, H. E., Jr. and A. W. Bailey. 1982. Fire Ecology. John Wiley and Sons, New York.

Wuebbles, D. J., D. W. Fahey, K. A. Hibbard, et al. 2017. Climate Science Special Report: Fourth National Climate Assessment. U.S. Global Change Research Program, Washington, DC.

Yansa, C. H., and A. C. Ashworth. 2005. Late Pleistocene paleoenvironments of the southern Lake Agassiz basin, USA. J. Quat. Sci. 20:255–67.

Young, R. T. 1924. The Life of Devils Lake, North Dakota. Publication of the North Dakota Biological Station.

Yurkonis, K. A. 2013. Can we reconstruct grasslands to better resist invasion? Ecol. Restoration 31:120–23.

Yurkonis, K. A., and W. Harris. 2019. Keeping up: Climate-driven evolutionary change, dispersal, and migration. In D. J. Gibson and J. A. Newman, eds. Grasslands and Climate Change, 218–33. Cambridge Univ. Press, New York.

Yurkonis, K. A., J. Dillon, D. A. McCranahan, et al. 2019. Seasonality of prescribed fire weather windows and predicted fire behavior in the northern Great Plains, USA. Fire Ecol. 15: https://doi.org/10.1186/s42408-019-0027-y.

Zedler, J. B. 2003. Wetlands at your service: reducing impacts of agriculture at the watershed scale. Front. Ecol. Environ. 1:65–72.

Zedler, J. B., and S. Kercher. 2005. Wetland resources: Status, trends, ecosystem services, and restorability. Annu. Rev. Environ. Resour. 30:39–74.

Zheng, H., D. Barta, and X. Zhang. 2014. Lessons learned from adaptation response to Devils Lake flooding in North Dakota, USA. Reg. Environ. Change 14:185–94.

Ziegler, J. P., C. M. Hoffman, P. J. Fornwalt, et al. 2017. Tree regeneration spatial patterns in ponderosa pine forests following stand-replacing fire: Influence of topography and neighbors. Forests 8:391–406.

Zilverberg, C. J., K. Heimerl, T. E. Schumacher, et al. 2018. Landscape dependent changes in soil properties due to long-term cultivation and subsequent conversion to native grass agriculture. Catena 160:282–97.

Zilverberg, C. J., W. C. Johnson, A. Boe, et al. 2014. Growing *Spartina pectinata* in previously farmed prairie wetlands for economic and ecological benefits. Wetlands 4:853–64.

Zilverberg, C. J., W. C. Johnson, D. W. Archer, et al. 2015. The EcoSun Prairie Farm: An experiment in bioenergy production, landscape restoration, and ecological sustainability. South Dakota State Univ. Extension, igrow.org, Brookings.

Zilverberg, C., W. C. Johnson, D. Archer, et al. 2014a. Profitable prairie restoration: The EcoSun Prairie Farm experiment. J. Soil Water Conserv. 69:22A–25A.

Zilverberg, C., W. C. Johnson, V. Owens, et al. 2014b. Biomass yield from planted mixtures and monocultures of native prairie vegetation across a heterogeneous farm landscape. Agric., Ecosyst. Environ. 186:148–59.

Zimmerman, K. M., J. A. Lockwood, and A. V. Latchininski. 2004. A spatial, Markovian model of rangeland grasshopper (Orthoptera: Acrididae) population dynamics: Do long-term benefits justify suppression of infestations. Environ. Entomol. 33:257–66.

Zon, R. 1935. What the study discloses. In Possibilities of Shelterbelt Planting in the Plains Region, 3–10. Lake States Forest Experiment Station, USDA Forest Service.

Index

Page locators refer to boxes, figures, and tables as well as the text. An "n" indicates the page has one or more (nn) pertinent end notes.